Endocrine Biomarkers

**Clinical Aspects and Laboratory Determination
of Biomarkers Series**

Series Editor: Amitava Dasgupta

Volume 1
Alcohol and Its Biomarkers: Clinical Aspects and Laboratory Determination

Volume 2
Biomarkers in Inborn Errors of Metabolism: Clinical Aspects and Laboratory
Determination

Volume 3
Endocrine Biomarkers: Clinical Aspects and Laboratory Determination

Endocrine Biomarkers
Clinicians and Clinical Chemists in Partnership

Edited by

Hossein Sadrzadeh, PhD

*Professor, Department of Pathology and Laboratory Medicine,
and Graduate Studies, University of Calgary and Section
Chief of Clinical Biochemistry, Calgary Laboratory Services,
Calgary, Alberta, Canada*

Gregory Kline, MD

*Clinical Professor, Department of Medicine,
Division of Endocrinology, University of Calgary,
Calgary, Alberta, Canada*

ELSEVIER

Elsevier
Radarweg 29, PO Box 211, 1000 AE Amsterdam, Netherlands
The Boulevard, Langford Lane, Kidlington, Oxford OX5 1GB, United Kingdom
50 Hampshire Street, 5th Floor, Cambridge, MA 02139, United States

Notices

Knowledge and best practice in this field are constantly changing. As new research and experience broaden our understanding, changes in research methods, professional practices, or medical treatment may become necessary.

Practitioners and researchers must always rely on their own experience and knowledge in evaluating and using any information, methods, compounds, or experiments described herein. In using such information or methods they should be mindful of their own safety and the safety of others, including parties for whom they have a professional responsibility.

To the fullest extent of the law, neither the Publisher nor the authors, contributors, or editors, assume any liability for any injury and/or damage to persons or property as a matter of products liability, negligence or otherwise, or from any use or operation of any methods, products, instructions, or ideas contained in the material herein.

British Library Cataloguing-in-Publication Data
A catalogue record for this book is available from the British Library

Library of Congress Cataloging-in-Publication Data
A catalog record for this book is available from the Library of Congress

ISBN: 978-0-12-803412-5

For Information on all Elsevier publications
visit our website at https://www.elsevier.com/books-and-journals

Working together
to grow libraries in
developing countries

www.elsevier.com • www.bookaid.org

Publisher: Mica Haley
Acquisition Editor: Tari Broderick
Editorial Project Manager: Timothy Bennett
Production Project Manager: Laura Jackson
Cover Designer: Matthew Limbert

Typeset by MPS Limited, Chennai, India

I would like to dedicate this book to my family, Maryam, Parisa, Sepehr, and Daniel, for their continuous support and encouragements to complete this work, especially when I was overwhelmed with other responsibilities. Also, I would like to dedicate this work to my colleague Dr. Kline for his outstanding contribution and support throughout this work. In addition, I would like to thank all the authors who devoted so much time and energy to writing about their experiences and their knowledge of the latest in the field. Finally, I dedicate this book to all the young endocrinologists and clinical chemists as our main goal for creating this book was to encourage a continuous communication between the clinicians and laboratorians for the best care of our patients.

Hossein Sadrzadeh, PhD

I would like to dedicate this work to my colleague Dr. Sadrzadeh who continually inspires me to seek excellence through collaboration and to my family who never stop encouraging me and who always bring me joy at the end of a long working day.

Gregory Kline, MD

Contents

CHAPTER 3   Disorders related to calcium metabolism................. 95

Christopher Symonds, MD and Joshua Buse, PhD

CHAPTER 4   Bone metabolism... **157**
Gregory Kline, MD, Dennis Orton, PhD
and Hossein Sadrzadeh, PhD

List of Contributors

Leland Baskin, MD
Associate Professor, Department of Pathology and Laboratory Medicine, University of Calgary and VP of Medical Operations Calgary Laboratory Services, Calgary, Alberta, Canada

Jessica Boyd, PhD
Assistant Professor, Department of Pathology and Laboratory Medicine, University of Calgary, and Clinical Biochemist Calgary Laboratory Services, Calgary, Alberta, Canada

Joshua Buse, PhD
Biochemistry Fellow, Department of Pathology and Laboratory Medicine, University of Calgary and Calgary Laboratory Services, Calgary, Alberta, Canada

Alex C. Chin, PhD
Assistant Professor, Department of Pathology and Laboratory Medicine, University of Calgary and, Clinical Biochemist, Calgary Laboratory Services, Calgary, Alberta, Canada

Bernard Corenblum, MD
Professor, Department of Medicine, Division of Endocrinology, University of Calgary, Calgary, Alberta, Canada

Lawrence de Koning, PhD
Associate Professor, Department of Pathology and Laboratory Medicine, University of Calgary, and Clinical Biochemist, Calgary Laboratory Services, Alberta, Canada

Ethan A. Flynn, MD
Associate Professor, Department of Pathology and Laboratory Medicine, University of Calgary, and Section Chief of General Pathology, Calgary Laboratory Services, Calgary, Alberta, Canada

Gregory Kline, MD
Clinical Professor, Department of Medicine, Division of Endocrinology, University of Calgary, Calgary, Alberta, Canada

Dennis Orton, PhD
Clinical Biochemist, Dr. CJ Coady Associates. Surrey Memorial Hospital, Surrey, British Columbia, Canada

Otto P. Rorstad, MD
Professor Emeritus, Department of Medicine, Division of Endocrinology, University of Calgary, Calgary, Alberta, Canada

Hossein Sadrzadeh, PhD
Professor, Department of Pathology and Laboratory Medicine, and Graduate Studies, University of Calgary and Section Chief of Clinical Biochemistry, Calgary Laboratory Services, Calgary, Alberta, Canada

Christopher Symonds, MD
Associate Professor, Department of Medicine, Division of Endocrinology, University of Calgary, Calgary, Alberta, Canada

Erik Venos, MD
Lecturer, Division of Endocrinology, Department of Medicine, Division of Endocrinology, University of Calgary, Calgary, Alberta, Canada

Preface

<hr>

WHO NEEDS THIS BOOK AND WHY

The power of therapeutics in modern western medicine lies in the ability to understand the precise genesis and mechanism of each and every disease. Now, with the progress of our understanding down to the molecular genetic level, there are more (and more effective) disease treatments than ever before in human history. The corollary to this hopeful news however, is that such advanced treatments are often only effective when appropriately applied to the right disease process. Thus the translation of clinical benefit to individual patients is increasingly beholden to accurate diagnosis. Without clear diagnosis the choice of a therapy is at best an educated guess and at worst a failure to realize the promise of modern medicine for our patients.

Diagnosis, in turn, has been made possible through major advances in genetics, radiology, pathology, and laboratory medicine, especially clinical chemistry. While each of these fields brings critical information to the bedside, it is clinical chemistry that carries the bulk of the diagnostic burden, at least as pertains the largest number of patients and with the highest frequency of use. This is a high calling for clinical chemists who thus shoulder a lot of the (often invisible) responsibility to ensure that a high-volume laboratory serves its population with accuracy, efficiency, and understanding.

Patients and doctors alike understand the importance of clinical chemistry for both routine and difficult diagnosis as well as the routine monitoring of many chronic conditions. In years past, many endocrinologic conditions were diagnosed by clinical features alone; however, most endocrinologists know that the spectrum of endocrine disease is much wider than previously thought; nuances are important and influence both diagnosis and therapy decisions. Clinical diagnosis alone is simply not able to deliver the kind of detailed disease understanding necessary for practice. The clinical chemistry laboratory thus has become, in many cases, the final arbiter of much endocrine diagnosis and therapy. With such an important clinical role, we hope that this book will improve the understanding, collaboration, and ultimately clinical care by offering useful information to healthcare providers on both sides of the laboratory bench.

For clinical chemists: Each section in this book has been written in part by an experienced academic endocrinologist whose role has been to ensure that the clinical context is supplied for each and every endocrine biomarker under discussion. This focus on endocrine disease applications of laboratory tests will help the clinical chemist to understand why the endocrinologist has ordered the test and will help to better explain the clinical situation that the endocrinologist is facing when he or she calls with a request to repeat or reanalyze a result. Despite their heavy reliance on laboratory results, endocrinologists are usually highly committed to clinical assessment; more often than not, discrepant results between the laboratory and clinical picture will result in the endocrinologist casting doubt upon the laboratory results (while the clinical chemist may seek to question the endocrinologist's clinical diagnosis). These clinical puzzles require collaboration between both parties and collaboration is best when each understands what the other can do. Secondly, the clinical endocrinology overview may help clinical chemists who are subtly asked to be endocrinologists when contacted by a community physician who wants assistance with test interpretation. Since such interpretation goes far beyond deviations from the reference ranges, some clinical background may help the clinical chemist to at least understand what is being asked, if not to also point the requester in the right direction.

For endocrinologists: A wise endocrinologist once said that "the patient is the bioassay"—in other words, look at the patient carefully if you want to figure out something about their endocrine system. As true as that may be, in modern day practice, virtually all clinical diagnoses and decisions need to be supported by some kind of quantifiable evidence. The clinical chemistry laboratory is often able to provide exactly the evidence needed but even clinical chemists are bound by the limitations of their instruments. Good clinical chemists will be able to alert the clinician to any potential problems arising from the analytical phase but avoidance of errors in the preanalytical phase (patient preparation and test ordering) and the postanalytical phase (test report and interpretation) is the shared responsibility of both clinical chemists and endocrinologists. Therefore it is not possible to be a high functioning endocrinologist while remaining ignorant of the many factors besides assay technique that go into a test result. Every section of this book is cowritten by an experienced clinical chemist who presents all the relevant analytical issues that would be important to a clinician. The presentation is deliberately nontechnical and this book is not a laboratory manual, nor it is a simple listing of various test confounders. Rather, it is written in such a way as to help the endocrinologist to understand exactly what is involved in complete testing process ranging from patient preparation for specimen collection to the highly sophisticated technologies used in modern laboratories to correct calculation of the results and explanatory comments reported with the results. Also, this book discusses circumstances in which endocrine tests may sometimes generate confusing or unexpected results and what needs to be done to address that. With this knowledge, endocrinologists can adjust their test selection and patient preparation as well as be aware of the kind of help they may be able to get through ongoing collaboration with a clinical chemist.

For primary care doctors: Endocrine disorders make up some of the most common diseases in medicine and yet most medical students learn that endocrinology is also the home of the "weird and rare" diseases that are critically important to diagnose when present. Therefore primary care health providers are the highest volume users of the endocrine laboratory. If patients are not going to be seeing an endocrinologist, this book will help the primary care providers to review their clinical endocrinology in an logical and easy format; the clinical sections are full of practical suggestions as to what may (and what may not) be easily diagnosed and how. Each section also contains a subsection that points out—in an educational fashion—the most commonly seen errors in the ordering or use of endocrinology tests. These are practice pearls that are unlikely to found elsewhere, all in one place. Their use will help the reader to become a wise chooser of both common and rare endocrine tests.

For clinical and chemistry trainees: For far too long, clinical chemists and clinicians have worked in separate worlds where clinical chemists do not know why tests are ordered and clinicians do not know what happens outside of an instrument printing test results. This highly unsatisfactory situation fails to harness the power of the laboratory for patient care and likely generates large, unnecessary expenditures through the performance of many tests which are either unnecessary or unhelpful. A little communication goes a long way to fixing this problem. We urge clinical and chemistry trainees to avoid this trap right from the beginning; this book will help you to pass your exams in endocrinology or laboratory medicine but more importantly, will help you to see the power of partnership that is yours to use once you understand the role and skills of the person on the other side of the lab bench.

Collaboration is the key: If you read the history of discovery of many of the endocrine disorders known today, you will notice that the bulk of classical endocrinology has come about through the shared work of an astute clinician who asked specific questions of a thoughtful clinical chemist. Such work was often slow and accomplished through the exchange of letters sent by mail. But nonetheless, neither person was able to elucidate and explain the endocrine syndrome alone; it was always a team approach. We will argue that even today, that kind of clinic–laboratory collaboration is vital if we are to unlock the potential of the laboratory to support new diagnoses and therapies for our patients. It is our hope that this book will encourage clinical chemists and endocrinologists to communicate on regular basis. At our institution, clinical biochemistry department has established regular quarterly meetings with other clinical departments (endocrinology, emergency medicine, clinical toxicology, poison center, etc.). At these meetings the clinicians discuss their issues with their clinical chemist colleagues and resolve most of their issues. Also, new technologies, tests, and other related topics of interest to both groups are discussed. These meetings have not only resolved many of the issues for both the clinicians and laboratorians but also resulted in many fruitful collaboration. This book is the first successful product of one of these partnerships.

Gregory Kline, MD and Hossein Sadrzadeh, PhD

Variables affecting endocrine tests results, errors prevention and mitigation

Hossein Sadrzadeh, PhD[1], Leland Baskin, MD[2] and Gregory Kline, MD[3]

[1]Professor, Department of Pathology and Laboratory Medicine, and Graduate Studies, University of Calgary and Section Chief of Clinical Biochemistry, Calgary Laboratory Services, Calgary, Alberta, Canada
[2]Associate Professor, Department of Pathology and Laboratory Medicine, University of Calgary and VP of Medical Operations Calgary Laboratory Services, Calgary, Alberta, Canada
[3]Clinical Professor, Department of Medicine, Division of Endocrinology, University of Calgary, Calgary, Alberta, Canada

CHAPTER OUTLINE

Endocrine Biomarkers. DOI: http://dx.doi.org/10.1016/B978-0-12-803412-5.00001-X

1.1 INTRODUCTION

To make no mistakes is not in the power of man; but from their errors and mistakes the wise and good learn wisdom for the future

Plutarch

Errors inevitably occur in life, from errors of transcription to errors in judgement, and human beings, like any other creature, are not immune from making or experiencing errors in all aspects of their everyday lives, including in their health care. Indeed the Institute of Medicine of the National Academies in its 1999 report entitled "To Err is Human: Building a Safer Health System" estimated that up to 98,000 patients per year in the United States die due to medical errors [1]. Although the number of people affected by medical errors has been questioned by Brennan [2], the important fact is that errors do occur and each year many patients are deleteriously affected. In the United States, it is estimated that 22.8 million individuals have experienced at least one medical error either personally or via a family member [1], with an annual cost of 17−29 billion dollars (USD) [3]. The United States Agency for Health Care Research and Quality estimates that medical errors are the eighth leading cause of death in the country, higher than cancer, AIDS, and motor vehicle accidents [1,4].

With more than 7 billion laboratory tests performed in the United States each year and a common belief that ∼70% of medical decisions are based on laboratory results, it can be expected that laboratory test results are a significant contributor to medical errors. This belief is supported by considering both the numerous responsibilities of the laboratory and its continued emphasis on reducing errors to improve patient safety and reduce adverse events. The importance of these endeavors is a result of the belief that 50% of errors were due to failure to do the requested tests, 32% were failure to act appropriately on the results, and in general, more than 50% were related to avoidable delays in diagnosis [5]. Therefore the recognition and minimization of preanalytical, analytical, and postanalytical variables that can lead to erroneous results demonstrates that the generation of "accurate results" in the clinical laboratory does not simply depend on having an expensive test run on a state-of-the-art instrument.

1.2 PREANALYTICAL VARIABLES

Preanalytical variables impact the laboratory test result prior to the analysis of the specimen, including physiological (e.g., biological rhythms, gender, age, pregnancy, fasting, and nonfasting) and nonphysiological (e.g., specimen collection process, tube material/type, stoppers, separating gel, preservatives in collection tubes, specimen processing, effect of drugs patient is taking, patient posture, tourniquets, and hemolysis effect). These variables must be considered and their impacts should be completely understood by both the laboratory and clinical team. These preanalytical variables, when applicable, are discussed for different tests in their corresponding chapters.

1.3 ANALYTICAL VARIABLES

Analytical factors are those affecting the results during the analysis of specimens and mostly depend on characteristic of a given method and instrument such as precision, accuracy, linearity, limit of detection, and its comparison to a reference method. All methods used for patient testing must be evaluated and verified according to the Clinical Laboratory Standard Institute (CLSI) guideline [6]. Discussion of analytical methods is beyond the scope of this chapter.

1.4 POSTANALYTICAL VARIABLES

Postanalytical errors are mostly clerical in nature and can be significant when results are calculated and reported manually. These include reporting results for patients with the same first and last name, miscalculation (specifically when specimen requires dilution and dilution factors must be used to calculate results), and reading or writing errors due to dyslexia (e.g., "76" instead of "67"). It is a good idea that all the results generated from any manual method be checked by senior technologists before reporting.

KEY POINT

Errors occur on daily basis in clinical laboratories. Most errors occur due to preanalytical factors that affect test results before specimens are tested. Clinicians should be familiar with these errors.

1.5 COMPLETE TEST PROCESS

To understand and prevent errors in laboratory testing, one should consider all stages of the testing process as indicated below:

1. Preanalytical
   a. Physician visit and test ordering
   b. Patient physiology
   c. Patient specimen (collection, transport, storage)

2. Analytical
    a. Sample preparation
    b. Instrumentation
3. Postanalytical
    a. Reporting of results
    b. Physician action

In this chapter, we will focus on an area of preanalytical variables that is most susceptible to error which is the specimen collection. Specifically, we will talk about sites of blood collection, tubes, and appropriate anticoagulants that are used for proper blood collection and the factors that can impact these processes.

There are many publications on preanalytical variables and their effects on test results. The important one is a set of books written by Donald Young describing all preanalytical variables, including the effects of drugs and diseases on test results [7].

1.6 PREANALYTICAL

The term "pre-analytical variables" refers to factors that affect patient results before the specimen is tested and contribute the most to laboratory error. As mentioned earlier, preanalytical variables can be of physiologic and nonphysiologic nature. Unfortunately, not much can be done to prevent the impact of physiological factors on the test results. Fortunately, nonphysiological factors that can occur at any point from the time of patient's visit to a doctor to the final stages of specimen collection are mostly preventable. Therefore a thorough understanding of these factors by both the clinicians and laboratorians is of extreme importance for the collection of high-quality specimens and generating highly accurate results. Most of these preanalytical variables, when applicable, are discussed for different tests in their corresponding chapters.

1.6.1 PHYSICIAN VISITS AND TEST ORDERING

A patient's visit to their doctor's office is the first step as the consultation and examination performed by the physician provides the basis for test selection. Preanalytical errors can occur at this point in the process if the physician's examination of the patient is flawed or if the appropriate test(s) are not selected by the physician. If the physician is unsure of the criteria for test selection or how to interpret the results, undertaking a consultation with the laboratory's medical staff can ensure the appropriateness of the ordered test(s).

1.6.1.1 Patient physiology

It is well established that most laboratory errors (48%−68%) are related to the preanalytical phase of testing [8]. Patient preparation is an additional preanalytical consideration that is highly susceptible to the introduction of errors. Fortunately, patient adherence to preparation instruction and the implementation of specific reference ranges can reduce the impact of some physiological variables on test results. The laboratory, however, has limited ability to prevent physiological variables such as biological rhythms and nutritional status of the patient. As mentioned earlier, it is important that both the medical and laboratory staff be familiar with factors that can impact patient preparation and specimen collection.

1.6.1.2 Patient specimen

Specimen collection is an important phase of the preanalytical process that is highly susceptible to preanalytical errors. Fortunately, good laboratory practice can limit or eliminate most nonphysiological variables that can affect test results, significantly improving the overall accuracy of the results because quality results are achieved from quality specimens. The different stages of preanalytical phase will be described in the following sections.

1.6.1.3 Specimen collection area

The environments that patients encounter during the specimen collection phase of the preanalytical process should be quiet and relaxing. Maintaining a relaxed patient demeanor can be accomplished by ensuring that the waiting area is equipped with proper lighting that does not strain the eyes and decor that promotes relaxation through the inclusion of pictures of nature, plants, or a fish tank. In addition, ensuring that there are enough chairs for each patient to sit for 5−10 min prior to their blood collection allows for their transition into a more rested state. If there are not enough chairs and patients are standing while waiting to give a specimen, blood will move to their lower extremities, causing plasma and small molecules to leave the circulating blood, reducing plasma volume, and erroneously elevating proteins concentrations in blood. The impact of elevated protein concentrations further affects the blood concentrations of protein-bound analytes such as ionized calcium; free hormones and free drugs will be elevated eventually affecting the results of those analytes [9].

Most patients experience some degree of stress prior to their blood collections, with patients scheduled for surgery and/or medical procedures potentially being nervous about their health or outcome of their surgery/procedure. Catecholamines are one such analyte that can be influenced by stress, with patients who are afraid of needles having elevated levels of some catecholamines measured in the blood specimens. Similarly, patient posture during the blood collection can have significant impact on some tests. For example, plasma metanephrine levels can be significantly modified depending on whether the patient is in the supine or sitting position during collection; further discussion can be found in Chapter 5, Adrenal Disorders, with the recommendation that specimen collection occurs in patients after they have rested in the supine position for 30 min [10]. Therefore collection staff and phlebotomists need to be aware of preanalytical variables that are influenced during collection, which includes their role in ensuring that patients are relaxed and calm both prior to and during specimen collection.

1.6.1.4 Patient test entry

Blood, urine, oral fluids, and solid tissues biopsies (e.g., needle biopsy) are the types of specimens used in endocrine assessment, with blood and urine specimens being ideal for the measurement of biomarkers that correspond to different physiological and pathophysiological processes. The collection of these specimens in the preanalytical phase of the testing process is specifically susceptible to errors, including the misidentification of the patient, incorrect test ordering, and/or mislabeling of specimen containers. Although it seems reasonable to assume that errors such as these should not occur in a modern laboratory, that assumption is not correct and everyday laboratory staff are required to not only identify errors but also correct them.

The ability of both physicians and members of the patient's medical team (resident, fellow, nurse, etc.) to order laboratory tests could allow for the potential introduction of errors. For example, when the test order is verbally given to a second person to process, an error in the transcription of the test order can prevent the correct test from being ordered, including BMP (basic metabolic panel, an abbreviation for a set of chemistry tests mostly used in the United States) being mistaken with BNP (B-type Natriuretic peptide) or hs-CRP (high sensitivity C-reactive Protein) for regular CRP. Thus it is important that the individual ordering the test writes the name of the test clearly for the person who is going to process the order. Similarly a patient's name can be a source of error due to the similarity between patients first and last names. For instance, it is not unusual to have two patients with the same first or last name and the same dates of birth. The chances of making error in ordering one patient's tests for the other and/or reporting one patient's results for another patient with the same names and date of birth are very high. Thus it is important for the laboratory staff to check at least two identifiers prior to specimen collection and order entry, including first and last name, healthcare identification number, and social security/identification number.

Although correct patient identification is a source of preanalytical errors, a device called the Personal Data Assistant (PDA) was recently introduced into clinical laboratories to significantly reduce some of the preanalytical errors related to specimen collection [11]. These portable devices are equipped with barcode scanners which are used by the laboratory staff to scan their identifications in order to access the PDA software and get the list of patients who need phlebotomy. Following correct identification of the patient, using at least two unique identifiers (e.g., patient name and social security number or healthcare number), phlebotomist scans the patient's wrist band and can see the patient's name, the list of the tests, the number and the types of the test tubes, and the proper order that the test tubes must be collected in. After collecting each tube the phlebotomist can scan the tube and print a label for each tube at the patient's bedside. This can not only reduce mislabeling, but can also provide the exact time of specimen collection for each sample. At the end of the sample collection, PDA can be downloaded to a laboratory information system, capturing all the information relative to the collected specimens and can be used as a tracking mechanism for transportation of the specimens to the laboratory. Therefore by ensuring correct identification of the patient, proper specimen collection (correct tubes with correct order of draw), labeling the tubes at the patient's bedside, and acting as tracking mechanism, this new technology (PDA) can significantly reduce preanalytical errors. This system is relatively new and currently only in use in a few hospitals in the United States; however, as soon as more information is published on its significant positive impact on reducing preanalytical error, more hospitals will implement this great system. Obviously, not all hospitals can have enough funding resource to acquire the PDA system and must rely on following a well-established procedure to prevent most preanalytical errors.

1.6.2 SITES OF BLOOD COLLECTION

The phlebotomist's awareness of preanalytical variables and the manner by which they may introduce an error allow for their collection of optimal specimens. As such the phlebotomist should try to reassure and calm the patient through their friendly and professional demeanor, outlining the procedure for the patient and providing warnings during any portion of the procedure that may cause discomfort. The phlebotomist demonstrates their knowledge of the correct collection procedure for each blood specimen type, which includes the correct preparation of the site of collection with either isopropanol or benzalkonium chloride followed by 30−60 s of drying time. Although blood is typically collected as a venous specimen, it can also be collected from other sites such as arteries, skin punctures, catheters, and intravenous (IV) lines.

1.6.2.1 Venous blood collection

The collection of blood specimens typically occurs at venous sites on the body, with the median cubital vein in the antecubital fossa being the typical collection site because this vein is relatively large and easily accessible [12]. Other veins, cephalic and basilic, as well as veins on the dorsal surface on the hand, wrist and ankle, can also be used for venous blood collection. It is important to avoid collecting blood from scarred skin and veins, sites distal to IV lines, bruised areas, and arms ipsilateral to a mastectomy site [12].

Application of the tourniquet 3−4 in. above the site of collection applies pressure to the vein, impeding its flow of blood back to the heart and increasing the peripheral vein's definition making them easier to locate and pierce during blood collection. Improper application of a tourniquet, however, can introduce numerous preanalytical errors. For example, creating a pressure >76 mmHg with a tourniquet induces anaerobic metabolism, increasing lactate and ammonia concentrations while decreasing pH [13]; most laboratories do not apply a tourniquet for the collection of blood for lactate testing. Application of a tourniquet for longer than 1 min can cause hemoconcentration with destruction of tissue and release of intracellular components such as potassium, enzymes, proteins, and protein-bound substances. In addition, application of tourniquet for 3 min can result in the increase of proteins (4.9%), lipid (4.7%), cholesterol (5.1%), iron (6.7%), bilirubin (8.4%), and aspartate aminotransferase (9.3%) [14]. Similarly, repeated fist clenching, increased muscle contraction, or high stress can increase the concentrations of many analytes such as potassium, cortisol, glucose, free fatty acids, and muscle enzymes [13]. Therefore it is critical to maintain the venous occlusion during the blood collection to <1 min [13].

1.6.2.2 Skin puncture blood collection

Skin puncture has been the method of choice for collecting blood from infants (and children <2 years of age), point of care testing (POCT) as well as patients with thrombotic tendencies, severe burn, and obesity [15]. A skin puncture releases blood from capillaries; however, due to high arteriolar pressure (which is greater than venules and capillaries), this specimen is primarily arterial blood that has been slightly diluted with interstitial and intracellular fluids. For proper collection the puncture site should be cleaned with isopropanol and left to dry for 30−60 s prior to the skin being punctured with a lancet. Although iodine is sometimes used to clean the site prior to collection, it should be avoided for skin puncture as it falsely increases uric acid, potassium, and phosphorus [15]. Following skin puncture the first drop should be wiped and the following free flowing drops should be collected. It is important to not squeeze or milk the region around the puncture site in order to increase blood flow because it can dilute the blood specimen with intracellular or intestinal fluid. Blood collected from a skin puncture is appropriate for the measurement of most analytes including pH and CO_2, except for blood culture, erythrocyte sedimentation rate, and coagulation studies [16].

1.6.2.3 Arterial blood collection

Collection of a blood specimen from an arterial site is more complicated and requires the phlebotomist have special training to properly perform the technique; otherwise a physician or a nurse should collect the blood specimen from this site. Arterial blood is used only for blood gas analysis including pO_2, pCO_2, pH, and sometimes lactate. The site of arterial blood collection, in their order of preference, is radial, brachial, and femoral arteries [16]. For neonates, umbilical artery catheter is the best site to collect the blood [16]; however, these sites need to be free of inflammation, infection, or edema.

1.6.2.4 Blood collection from intravenous lines and catheters

Owing to the potential for contamination from IV fluids, it is best to avoid collecting blood from areas near an IV line. However, for clinically ill patients or those who require several blood collection, specimens may be collected through central venous lines or arterial catheters if proper precautions are employed. The proper collection technique that avoids contamination and dilution of specimens with IV fluid requires that the catheter valve must be closed for at least 3 min before blood collection [13]. It is recommended that the first 6−10 mL (or equal to the volume of the catheter) of collected blood be discarded to avoid contamination [13]. It should also be noted that specimens collected from a catheter or IV lines are not appropriate for blood culture.

1.6.3 BLOOD SPECIMEN CLASSIFICATIONS

The average blood volume in males is about 5.5 L and in females is about 5.0 L [17], with blood representing about 8% of the total human body weight. Its primary functions are to deliver oxygen/nutrients to tissues while distributing carbon dioxide/waste products to their sites of disposal. In addition the capacity of blood to rapidly distribute hormones, proteins, and other messengers throughout the body is critical in communication as well as providing cellular and humoral immunity against invading organisms and foreign material.

Blood can be divided into two fractions: cellular and liquid. The cellular fraction consists of red blood cells (erythrocytes), white blood cells (leukocytes), and platelets (thrombocytes). The liquid fraction of blood is called plasma and it consists of approximately 93% water and 7% proteins, electrolytes, and small organic molecules [17]. Blood specimens can be easily collected and are ideal to measure biomarkers that correspond to different physiological and pathophysiological processes. The procedure for blood specimen collection and processing is dependent upon the test(s) ordered by the physician and the requirements of the instrumentation that analyzes the specimen.

1.6.3.1 Whole blood

Whole blood is an ideal specimen because it can be analyzed immediately following collection without the need for separation. This is of great importance in the emergency department or when dealing with seriously ill patients where a short turn-around-time (TAT) can be the difference between life and death. An ideal instrument for any laboratory is the one that can use whole blood for the majority of the analytes on its test menu and there is a major effort from diagnostic industry to make more analyzers that can use whole blood as specimen. Currently, whole blood can be used only for a limited number of tests, including blood gas analysis, POCT devices, and dried blood specimen (DBS). POCT devices are those that are used either by the patients (e.g., glucometers, pregnancy testing, and coagulometer) or by healthcare providers at the patient bedside. Removing testing from the laboratory, however, increases its susceptibility to errors. The most common errors associated with POCT are lack of compliance by the users to follow the manufacturer's or laboratory's instructions for operating the POCT device or inappropriate use of quality control materials before testing patient specimens.

1.6.3.2 Dried blood spots

A blood specimen collected by skin puncture is ideal for preparing DBS and has been traditionally used in newborn screening for inborn errors of metabolism. In addition to collecting specimens from very small patients (e.g., pediatrics), DBS has the potentials to be the future specimen of choice for large clinical trials and patients who are difficult to draw specimens by venipuncture from, which includes patients who are obese, afraid of needles, critically ill patients, or who live far away from the laboratory. The main advantages of DBS are the ability to collect the specimen with the assistance of a phlebotomist and simplified transportation because they can be mailed to the laboratory. These benefits have led to an increasing interest in using DBS as specimen of choice for other tests, including as drugs of abuse, drug monitoring, and hormones. Once the patient knows how to collect the specimen appropriately, the only issue is mislabeling of the specimen or clerical error in reporting the results.

Although several advantages exist for the use of DBS and many laboratories are trying to develop tests that utilize DBS, the specimen composition differs in several ways from serum, plasma, and whole blood. As the source of blood for DBS is a combination of capillary, arteriolar, and venous blood, it may contain some interstitial and cellular fluids; the introduction of interstitial and intercellular fluid increases if the finger is vigorously massaged or milked during blood collection. For example the concentration of hydrophilic or water-soluble molecules such as electrolytes, glucose, or drugs is lower in capillary whole blood than whole blood collected by venipuncture [12]. Indeed, International Federation of Clinical Chemistry (IFCC) proposed a conversion factor of 1.11 for obtaining plasma-equivalent glucose molarity on all glucose meters [18,19]. However, the IFCC also suggested that the conversion factor ignores the wide variations in plasma water content and hematocrit (% of red blood cells in blood) that is usually seen in some patient subpopulation that can result in 10%−15% error in glucose measurements [20]. Therefore interpreting results in whole blood must be done with great caution and in relation to the overall hemodynamic status of the patient.

Dried specimens are very convenient specimens to collect, process, transport, and store. Indeed, using DBS can eliminate most preanalytical variables that occur during specimen collection and delivery. The patient can use a lancet to pierce their skin, deposit their own blood onto a laboratory-provided collection card and place the card in an envelope, and mail it to the laboratory. This provides an opportunity for the patient to collect their blood at the most appropriate time to minimize diurnal variation. DBS can be stored in the laboratory until analysis, thus eliminating the need for refrigeration. Therefore it is clear that DBS will a specimen of choice for most modern laboratories in the near future because of their capacity to significantly reduce costs, including eliminating the need for phlebotomists, expensive blood collection tubes, and transfer and storage costs. Currently the most widely used DSB is in newborn screening for inborn error of metabolism. Without a doubt, dried specimens (blood, urine, etc.) will become the specimens of choice for many tests such as hormones, tumor markers, therapeutic drug monitoring, and drugs of abuse in the near future. For general routine chemistry, hematology and coagulation tests using dried blood spots will be more challenging.

1.6.3.3 Serum

Serum is obtained from a blood specimen that is collected in tubes without anticoagulant, with hay-colored cell-free liquid remaining after clotting has finished. The main difference between serum and plasma is that plasma contains fibrinogen and serum does not because fibrinogen in serum is converted to fibrin during clotting. During clotting, some substances are used (e.g., fibrinogen, platelets, and glucose) and some other substances are released from the cells due to physical pressure from fibrin strings that covered the cells. These substituents include potassium, phosphate, lactate dehydrogenase, lactate, and ammonia 21 and require a plasma sample be collected to confirm a suspicious result [21]. For example a potassium result from a serum sample that is questionably high can be confirmed by drawing a new blood specimen in a tube containing the appropriate anticoagulant (e.g., lithium heparin). It is also important to ensure that fibrinogen is not present in measurable amounts within a serum sample as it can interfere with the interpretation of serum protein electrophoresis due to it being mistaken for a monoclonal protein. The presence of fibrinogen in a serum can be due to the following reasons: (1) wrong specimen (blood is collected in a tube with an anticoagulant), (2) patient is on anticoagulant medication (e.g., Coumadin or heparin), and (3) patient has a clotting problem. The latter can be extremely important, as patient is in danger of bleeding, if not treated.

One of the advantages of using serum is the lack of interference caused by the anticoagulants and other additives that are added to plasma tubes. These additives can impact the analyte of interest, changing its physiochemical properties, and affecting its stability and protein binding. Also, when the preservatives are added to the tubes in liquid form, they can result in dilution of the analyte of interest and falsely reduce its concentration. Thus it is better to add the additives in a dry form. Although serum samples are the preferred specimens for most clinical chemistry tests, they have the disadvantage of requiring additional time for the blood to completely clot, up to 30 min. Following clotting, specimens are centrifuged for about 5−10 min, following which the serum supernatant is ready for analysis. Therefore it can take about 60 min from sample collection for a serum specimen to be ready for analysis. An additional disadvantage of the serum samples is a false increase in the concentration of some analytes if the serum is not separated from the cells within 6 h of collection. When serum and cells are allowed to remain in contact with one another, analytes can be released from the cells following clotting. These include ammonia ($\sim$38%), potassium ($\sim$6%), lactate ($\sim$22%), and inorganic phosphate ($\sim$11%) [16].

To reduce the time between specimen collection and analysis a Rapid Serum Tube (RST) containing "clot activator" has been introduced to promote complete clotting in 5 min. These tubes can significantly improve the TAT for tests done in serum, allowing for serum to now be an acceptable specimen for many urgent (STAT, from the Latin Statum) STAT tests. Recent study has shown that the RST tubes can provide highly reproducible results with minimal false positive results compared with lithium heparin tubes [22].

1.6.3.4 Plasma

Plasma samples are obtained from blood that is collected in tubes containing an anticoagulant agent such as heparin, EDTA, and citrate. One of the main advantages of using plasma is that it can be separated from the cellular fraction quickly, requiring only centrifugation, which allows for its use in urgent situations (STAT testing). Blood is collected in a tube that contains an anticoagulant, with the use of the appropriate anticoagulant being critical for ensuring the integrity of results. Proper mixing is important for collecting quality specimens as inadequate mixing of blood with the anticoagulant can result in clot formation as well as possible interference with the measurement of analytes. In addition, inadequate clotting can also give rise to "microclots" that can get caught or block the narrow capillary tubes found within instrumentation, leading to an instrument malfunction and significantly delaying the testing process.

1.6.4 BLOOD COLLECTION TUBES

Collection of high-quality specimen requires appropriate type of collection tube, with or without appropriate preservatives. In the past 10 years there has been significant advances in developing new tubes that not only preserve the integrity of the blood but also significantly reduce the time for processing samples and improving TAT. The new generation of blood collection tubes has different preservatives or small solid polymer inside tubes that has greatly improved plasma or serum preparation. It is imperative that clinical chemists and endocrinologists be familiar with these tubes, their contents, advantages or disadvantages to reduce the effect of preanalytical variables and prepare a high-quality specimen. The following sections discuss these issues in more detail.

1.6.4.1 Order of draw

Another important factor to consider when collecting more than one blood specimens using different tube types is the order in which individual specimen is collected. Proper collection order prevents the contamination of specimens with anticoagulants or bacteria (in case of bacterial culture tubes) and some laboratories collect a white/clear top tube with no additive first to ensure a clean draw; this tube is discarded and not used.

In general the order of specimen collection should be as follows:

1. tubes or bottles with sterile media for blood culture
2. tubes with no additive (e.g., for trace elements)

3. sodium citrate (coagulation analysis)
4. serum tubes with or without gel or clot activator
5. plasma tubes with heparin
6. plasma tubes with EDTA
7. plasma tubes with acid citrate dextrose
8. plasma tubes with sodium fluoride and potassium oxalate

Most the current blood collection tubes with or without additive or serum separator gel with their applications, mixing instructions, and correct order of draw are described in Table 1.1. Table 1.1 is prepared by Calgary Laboratory Services (CLS) in Calgary, Alberta, Canada and used for specimen collection. As mentioned previously, tubes with additives must be thoroughly mixed and the phlebotomist should follow the manufacturer's recommendations for proper volume and mixing. Incomplete mixing not only results in "microclot" formation that can cause instrument malfunction, but it can also cause falsely elevated results specially for coagulation tests (prothrombin time, INR, and partial thromboplastin time) [23]. It is important that each laboratory establishes its own policy based on CLSI [6] recommendations and ensures that the ratio of anticoagulant to blood is appropriate and the draw volume is within 10% of the stated volume by the tube manufacturer.

It is important to be familiar with both the materials that comprise the different tube types (glass, plastic, polymers, etc.) and additives (anticoagulants, separator gels, preservatives, etc.) that are used in each tube as well as their ability to influence test results. Table 1.2 shows most commonly used blood tubes in the laboratory.

1.6.4.2 Anticoagulants

Although the majority of the chemistry tests are done using serum samples, the advantages of plasma samples mentioned earlier have allowed for their continued use in the laboratory. However, for most hematology tests, in which cellular components of blood are analyzed, an appropriate anticoagulant is added to the tube to ensure the blood cells are kept in good condition until analysis. One important preanalytical variable to consider during the collection of specimens in a tube with an anticoagulant is the proper ratio of blood to anticoagulant. For example the measurement of coagulation factor requires that the blood specimen be collected in tubes with sodium citrate (tubes with light blue stoppers) and it is important that tubes be filled completely (up to the line) otherwise the results will not be accurate [23]. Therefore when the results (i.e., prothrombin time or partial thromboplastin time, platelet aggregation assessment) do not match other findings in the patient, the clinician should consider preanalytical effects and repeat the test ensuring that the specimen is collected appropriately.

Anticoagulants such as heparin (tubes with green stoppers) are used in plasma collecting tubes for Stat tests because plasma tubes can be centrifuged immediately following collection. Ensuring result integrity requires that the anticoagulants be completely mixed after specimen collection, with 5−8 complete tube

Table 1.1 Order of Draw for Various Specimen Collection

Order of Draw	Color of Stopper	Invert	Additive	Comments and Common Tests
1	Clear	Not required	No additive	Tube used ONLY as a discard tube
2	Blood culture bottle (aerobic, anaerobic and pediatric bottles have different colors of label and top)	Invert gently to mix	Bacterial growth medium and activated charcoal	When a culture is ordered along with any other blood work, the blood cultures MUST be drawn first
3	Yellow with clear label	8–10 times	Sodium polyanethol sulfonate (SPS)	Tube used for mycobacteria (AFB) blood culture
4	Royal Blue (with red Band on label)	Not required	Glass tube with no additive	Tube used for serum trace element tests.
5	Red GLASS	Not required	Glass tube with no additive	Tube used for serum tests, which CANNOT be collected in serum separator tubes (SSTs). NOTE: red PLASTIC tubes are preferable for lab tests
6	Light blue	3–4 times	3.2% buffered sodium citrate anticoagulant	Tube used mainly for PT (INR), PTT, and other coagulation studies
7	Black GLASS	3–4 times	3.2% sodium citrate anticoagulant	Tube used for ESR ONLY
8	Red	5 times	Clot activator, and no anticoagulant	Tube used or serum tests, which CANNOT be collected in SST

#	Tube color	Inversions	Additive	Description
9	Gold	5 times	Gel separator and clot activator	Usually referred to as "SST". After centrifugation the gel forms a barrier between the clot and the serum
10	Dark green	8–10 times	Sodium heparin anticoagulant	Tube used mainly for amino acids and cytogenetics tests
11	Light green (mint)	8–10 times	Lithium heparin anticoagulant and gel separator	Usually referred to as "PST" (plasma separator tube). After centrifugation the gel forms a barrier between the blood cells and the plasma. Tube used mainly for chemistry tests
12	Royal blue (with blue band on label)	8–10 times	K_2EDTA anticoagulant	Tube used for plasma and red cell trace element tests.
13	Royal blue (with lavender band on label)	8–10 times	Na_2EDTA anticoagulant	Tube used for whole blood trace element tests.
14	Lavender	8–10 times	K_2EDTA anticoagulant	Tube used mainly for complete blood count (CBC), pretransfusion testing, hemoglobin A1C, and antirejection drugs.
15	Yellow with yellow banded white label	8–10 times	Acid citrate dextrose solution "A"	Tube used for tissue typing and some flow cytometry testing
16	Gray	8–10 times	Sodium fluoride and potassium oxalate anticoagulant	Tube used for lactate

Table 1.2 Common Blood Collection Tubes Used in Clinical Laboratory

Tube Type	Specimen	Stopper Color	Content
Plastic	Serum	Red	No additive
Plastic (SST)	Serum	Gold	Clot activator plus gel separator
Plastic	Serum	Red/black	Clot activator
Plastic	Serum	Orange	Thrombin plus gel separator. These new tubes have fast clotting (5 min)
Glass	Serum	Red	No additive
Plastic	Plasma	Green	Heparin with sodium, lithium, or ammonium salt
Plastic (PST)	Plasma	Light mint green	Lithium heparin plus gel separator
Plastic	Plasma	Lavender	Potassium EDTA (K_2EDTA or K_3EDTA)
Plastic	Plasma	Light blue	Sodium citrate
Plastic	Plasma	Dark green	Sodium heparin
Plastic	Plasma	Royal blue (with lavender or blue band)	Sodium EDTA or potassium EDTA
Plastic	Plasma	Yellow (pale)	Acid citrate dextrose "A" solution
Glass	Plasma	Dark green	Sodium heparin
Glass	Plasma	Black	Sodium citrate
Glass	Plasma	Lavender	Potassium EDTA (K_2EDTA or K_3EDTA)

inversions ensuring blood is completely mixed with the preservative (e.g., anticoagulant). In general laboratory and other staff who collect specimens must follow the manufacturer's recommendation for specimen collection as closely as possible.

One of the most common preanalytical errors that are seen in laboratories on regular basis is the effect of anticoagulants on different tests. For example, using plasma tubes with heparin salts (sodium, lithium, and ammonium) analyses such as sodium, lithium or ammonia can generate erroneously high results. Obviously the results will be erroneously high for these analytes due to the addition of those salts to the samples and most anticoagulants currently used are in dry form because liquid anticoagulants may dilute the sample and cause an erroneously low result. Although the latter seems to be obvious and preventable, it happens on a regular basis in the laboratory. It mostly happens when physicians request Stat additional tests ("add on") on a previously collected specimen and the technologist runs the specimen to report the results as quickly as possible. Most of the times, these errors are detected in the laboratory via a "delta check" which compares the patient's previous result with the current one. Any difference greater than a previously established limit will alert the technologist to investigate the cause and request a new specimen if necessary. However, if this is the patient's first test or a patient is in intensive care unit, the erroneous results. will be reported and that may impact the patient's management. Therefore it is imperative that clinicians and clinical chemists communicate with each other and discuss suspicious results to ensure the most accurate results are reported, and patients receive the most appropriate care.

1.6.4.3 Serum separator gel

Blood cells live in blood samples for several hours and during that time continue to consume nutrients (e.g., glucose) and release others (e.g., K, LD (lactate dehydrogenase), and AST(aspartate amino transferase)); therefore to prevent the above, cells must be separated from plasma/serum as soon as possible. These tubes are referred to as SSTs and contain a separator gel. The separator gels are made of thixotropic materials that during centrifugation are mixed with blood and because of its specific gravity (higher than serum and lower than cells) at the end of centrifugation forms a physiochemical barrier between cells and plasma/serum based on specific gravity. The cells that have the highest specific gravity are precipitating at the bottom, separator gel above the cells and serum or plasma with the lowest specific gravity at the top. In general, separation of cells from serum/plasma must take place as soon as possible, ideally within 90 min following blood collection, as delay in this process will result in the release of different analytes such as lactate dehydrogenase, potassium phosphate, and magnesium into the sample (plasma, serum as whole blood) [24]. Serum in SST is stable for several hours, unless the tube is centrifuged again. The second centrifugation can falsely increase concentrations of analytes such as potassium, magnesium, or lactate dehydrogenase [25]. The latter happens mostly when additional serum or plasma is for repeating a test or performing additional new tests.

Addition of separator gel has significantly improved the quality of serum samples by increasing the stability of most analytes and keeping their concentrations constant as well as eliminating more than 90% of preanalytical issues related to blood cell metabolisms [24]. However, some substances, especially drugs, can bind to the separator gels resulting in falsely low serum drug levels that can impact the treatment and harm the patient. Indeed, there is a report showing that hydrophobic drugs such as phenobarbital, phenytoin, lidocaine, and quinidine are absorbed by the gel with significant decrease (59%−64%) in their serum concentrations [25]. Fortunately, reformulation of the separator gel (in SST II) has greatly reduced absorption of the drugs [26].

1.6.4.4 Tube materials

The collection tubes in the laboratory used to be made of glass several years ago as substrates and cells do not bind easily to glass. However, due to the risk of breakage, cutting, and exposure to blood, glass tubes were gradually replaced by plastic tubes [12]. Currently, all laboratories use plastic tubes for almost all their testing in order to comply with guideline from occupational health and safety standards. The plastic tubes are made of polyethylene terephthalate materials which can tolerate high centrifugation speeds without cracking or breaking, are lighter than glass tubes, and are inert to absorption of analytes. Although plastic tubes are safer to use, due to their relatively hydrophobic nature, their surface is not ideal for blood collection. In contrast, silica surface of glass is ideal for blood flow and complete clean separation of the blood clot from serum or plasma. Thus the interior surface of plastic tubes is sprayed with surfactant to make the tube surface properties similar to those of glass tubes (as described later). There are other compounds in the serum and/or plasma tubes such as anticoagulant, stopper, and stopper lubricants that are described later.

The surfactants that are used are silicone-based polymers. Surfactants can interfere with some immunoassays (IAs) (e.g., total T3) by displacing the capture or signal antibodies [27,28]. Fortunately, this interference is very much dependent on the IA method and how antibodies are attached to a solid surface. Not all IAs are affected in the same way by the surfactant. For instance the Abbott IA method for total triiodothyronine (T3) is not affected by surfactant [27,28]. Therefore questionable results should be investigated by the clinician and discussed with the laboratory. Following the routine trouble-shooting procedure for questionable results the analyte should be measured by another IA method that is not affected by the interfering compound (e.g., surfactant).

Exposure of blood to stoppers during and after blood collection may result in the introduction of inert plasticizers into the blood. In general these plasticizers do not interfere with any analytes in the blood. However, stoppers have lubricants such as glycerol or silicone, which make the capping and decapping easier for the technologists. This lubricant may interfere with some methods such as triglyceride that measures glycerol as the end product and falsely increase glycerol concentrations [29]. Also the standard red top stopper used for collecting serum samples may be contaminated with aluminum, zinc, magnesium, and heavy metals [29].

1.6.4.5 Hemolysis

Hemolysis is a preanalytical variable that is difficult to control, however, the collection of specimens by phlebotomists has the capacity to lower the occurrence of hemolyzed specimens because of their specialized training in the collection of specimens. Conversely, when specimens are collected by nonlaboratory staff, the rate of hemolysis goes up. One of the common causes of hemolysis in the past was to collect the blood by a syringe (especially in patients who are difficult to draw) and then insert the needle into a vacuum tube for fast transfer of blood into the tube. The forced flow of blood through a needle results in erythrocytes hemolysis, with small gauge needles producing a large amount of hemolysis. Although this practice is stopped in most laboratories, there are still occasions when a physician or a nurse uses this approach to collect blood from their patients. Obviously, not all blood specimens can be collected by phlebotomists and in many occasions nurses, residents, and medical students have to perform phlebotomy. Therefore it is important to educate the nurse educators about proper techniques for blood collection. Also, laboratory should always collect data about hemolysis, create and investigate the source of hemolysis, and share the results with the nurse educators at each hospital and provide feedback on how to improve blood collections.

Another common cause of hemolysis is the "resident pocket syndrome." In this case, blood is collected by a resident (or another staff) and placed in a resident's pocket and the resident forgets to send or deliver the specimen to the laboratory and the specimens are stored in a relatively warm environment (resident's pocket) sometimes for several hours before it is tested. Obviously, that specimen is affected by numerous preanalytical variables and results generated from it will not be accurate. To avoid this problem, all collected specimens in any unit should be kept together in a designated area for the next delivery to the laboratory. Hemolysis, especially when it increases the serum hemoglobin concentration to 20 g/dL (200 g/L), can erroneously increase the concentration of potassium, alanine aminotransferase, aspartate aminotransferase, lactate dehydrogenase, aldolase, acid phosphatase, isocitrate dehydrogenase, phosphate, and hemoglobin [13,30]. Fortunately the current instruments can estimate the amount of hemolysis in specimens and reject those that cannot be used for certain analysis.

1.6.5 SPECIMEN PROCESSING, TRANSFER, AND STORAGE

All specimens collected outside the laboratory for chemistry analysis should be centrifuged to separate serum or plasma from the cells within 90 min after collection at the collection sites (or preferably in the laboratory, if possible). After separation of serum or plasma from the cells, specimens should be stored at 4°C and transferred to the laboratory as soon as possible (within 3−5 h after collection). Obviously care must be exercised for proper labeling and processing of the specimens. All specimens collected at locations far from the laboratory must be collected with proper preservative and handled properly. For example, for glucose analysis all specimens must be collected in a gray top tube that contains potassium oxalate and sodium fluoride to prevent glycolysis. Otherwise the glucose concentration will be significantly low at the time of analysis. Also, all specimens must be kept in a cooler with temperatures of about 4°C during transfer to the laboratory. Specimens should be analyzed within 6 h following collection, if possible, to minimize the effects of storage.

1.7 ANALYTICAL

The analytical factors are those affecting the results during the analysis of specimens and mostly depend on the characteristic of a given method and instrument. The discussion on analytical method is beyond the scope of this chapter. The methods and instruments in the laboratory must be appropriate and function properly for testing patient specimens. To evaluate a method, laboratories perform several different assessments based on the CLSI guideline for most automated analyzers that are used in a clinical chemistry laboratory [6].

The minimum assessment of a method includes the following:

1. Precision or reproducibility indicating the method can produce reproducible results all the times.
2. Accuracy that indicates the method can produce accurate results.
3. Linearity shows the concentration range the method can accurately and reproducibly measure an analyte.
4. Limit of detection indicates the lowest concentration that a method can measure.
5. Method comparison indicates how the tested method compares to a reference method.
6. Interference shows the effect of common interfering substances such as lipids, bilirubin, and hemoglobin on the method. Most manufacturers perform these studies before releasing the instrument and methods. For any specific drug analysis, it is important to study the effect of drugs with similar structure on the method.

As indicated above the methods used for patient testing must be evaluated and verified according to the CLSI guidelines.

The technical phase of testing also includes proper daily runs of control materials, at least once per shift, to ensure the instrument is in excellent condition prior for testing patient specimens.

In addition, each clinical laboratory should participate in an external proficiency testing program by an accredited agency such as College of American Pathologists (CAP). In Canada, The College of Physicians and Surgeons at each province provides the accreditation to Canadian Clinical Laboratories. To become accredited, laboratories receive unknown specimens from the proficiency testing provider 2−4 times per year (depending on the agency) that are tested, in the same manner as patients specimens are tested, and the results are reported back to the agency which will be compared with results reported by the laboratory's peer group (i.e., laboratories that use the exact same technologies). Based on established criteria, if the results are acceptable, the laboratory can test and report patient results. The accredited laboratories are also inspected every 2−4 years by the accrediting agency's inspectors to ensure compliance. Multiple failures can result in revoking of accreditation and the closure of the laboratory.

As indicated earlier, detailed discussion of technologies used in the laboratory is out of the scope of this book. However, as most of the endocrine markers are currently measured by different IA technologies, a brief discussion in this area is provided below to inform the clinicians about some of the important issues related to this technology.

1.7.1 IMPORTANT ISSUES IN IMMUNOASSAYS

IAs have been used to measure endocrine markers for many years. Although with the implementation of liquid chromatography tandem mass spectrometry (LC−MS/MS) in clinical laboratories, more modern laboratories use the latter technology to measure some of the hormones, the majority of clinical laboratories still rely on IAs for the daily measurement of endocrine markers. Therefore a brief introduction in IA techniques will be helpful to our clinician readers.

1.7.1.1 Immunoassay techniques

The extremely low concentration of endocrine markers requires measurement by highly sensitive techniques. For this reason, immunochemical techniques have been the most commonly used methods in endocrine laboratories. Initially developed in the 1950s [31], these require an antibody, usually an IgG immunoglobulin, raised to recognize unique regions of the target molecule known as epitopes. Complex molecules may contain multiple epitopes so that several different antibodies may bind to them. This presents both opportunities and risks in the development of assays using polyclonal or monoclonal antibodies. For immunochemical methods to perform optimally, balance between the concentration of antibodies and antigens (the markers that we want to measure) is critical. Although some methods require excess antigen whereas others require excess antibody, in all cases they should have similar concentrations (i.e., the concentration of the marker should be close to the concentration of the antibody used to detect the marker) [31].

Several different types and configurations of antibodies, labels, and antigens have been developed. Those techniques that use labeled target antigens are specifically called IAs. Depending on the nature of the labels, they are termed radioimmunoassay (RIA), enzyme immunoassay (EIA), fluorescence immunoassay (FIA), and chemiluminescence IA. Typically, RIA has been the first technique to be used due to its high sensitivity (10^{-15} M). FIA and EIA have usually followed, providing adequate sensitivity (10^{-10} M) without the need for radioactive materials. Finally, chemiluminescent labels have replaced enzyme and fluorescent labels due to their significantly improved sensitivity (10^{-13} M). Several refinements of these methods have been developed to either improve the sensitivity or eliminate steps in the process. Some of these adaptations include enzyme-linked immunosorbent assay (ELISA), enzyme-multiplied immunoassay technique (EMIT), cloned enzyme donor immunoassay, microparticle enzyme immunoassay, and fluorescence polarization immunoassay. In contrast, techniques using labeled antibodies are called "immunometric assays" and consist of immunoenzymometric, immunofluorometric, immunoradiometric, and immunochemiluminometric assays [32].

IAs are generally distinguished by the competitiveness of the antibody reaction format and need for separation of free and bound label prior to measurement.

Commonly used for low-molecular weight molecules such as drugs and hormones, competitive (equilibrium) techniques use a limited amount of reagent, usually labeled target antigen that competes with the antigen of interest in the patient's sample for the limited number of antibody binding sites. The quantity of measurable labeled antigen is inversely proportional to the amount of antigen in the patient's plasma. Sequential techniques are so called because the patient's plasma is initially mixed with the antibody and allowed to equilibrate before the labeled antigen is added. This may improve analytical sensitivity significantly.

In contrast, noncompetitive methods employ excess reagents, usually labeled antibodies, which do not compete for binding. The classic noncompetitive technique is the two-site sandwich immunometric assay that uses excess capture and detection antibodies directed toward different epitopes. This is particularly good for high molecular weight targets.

The quantity of the target antigen that is present in the patient's blood is estimated by measuring the amount of a label that is bound. All radioactive methods and most two-site immunometric methods require that free and bound labeled antigens be separated following adsorption, precipitation, or other techniques prior to measurement. These assays are termed heterogeneous. Conformational changes occurring upon antigen—antibody binding change the label's activity in most enzymatic, fluorescent and chemiluminescent methods, precluding the need for separation. These techniques are called homogeneous [31].

1.7.1.2 Immunoassay interferences

IAs are beset by various interferences including cross reactivity, heterogeneous antigens, high-dose hook effect, signal interference, matrix effects, and endogenous and exogenous antibodies [31—33]. For this reason among others, it is critical that physicians always question test results from IAs that do not correlate with clinical findings.

Cross reactivity refers to direct competition between the molecule of interest and other molecules for antibody binding sites due to structural similarity. This most commonly affects competitive IAs. The interfering molecule measures as target molecule. It typically binds rapidly but dissociates more quickly than the target due to weaker affinity and this may vary with relative concentrations of the target and interfering molecules. At least three techniques are typically used to overcome cross reactivity: (1) separating the target from the interfering molecules using chromatography, extraction, and ultrafiltration, (2) adjusting reaction conditions such as time and temperature or adding blocking substances, and (3) using more specific antibodies directed against different epitopes and two-site immunometric techniques.

Heterogeneous antigens refer to multiple similar forms or partially degraded forms of the target molecule in which epitopes may be blocked or lost. This leads to falsely low results and is most commonly seen in assays for proteins and peptide hormones.

High-dose hook effect is a reduction in measured signal that occurs in the presence of very high concentrations in a two-site immunometric assay, most of which are heterogeneous. In this case, binding of the excess target molecules may saturate the capture and detection antibodies disrupting the system. The labeled complex that is not captured will be removed, falsely lowering the measured value. Addition of a wash step before application of the labeled antibody will help to correct this. Also, diluting the patient sample will reduce the concentration of the marker in the sample to a concentration such that the antigen and antibody form a strong and stable complex that will not be washed away.

Signal interference is the alteration, blockage, or enhancement of the signal or its detection. Lipemia, hemolysis, and icterus interfere due to color or turbidity. Other examples include the presence of fluorescent compounds in fluorescence assays, macroalkaline phosphatase in alkaline phosphatase-based assays, and salicylate metabolites in EMIT assays.

Matrix effects are nonspecific biases due to the environment in which the antigen—antibody complex forms. Examples include variations in protein concentration, pH, and ionic strength.

Endogenous or exogenous antibodies include various antibodies and autoantibodies, of which the most commonly occurring are heterophile and human antianimal (HAAA) antibodies (e.g., human antimouse (HAMA) or human antirabbit) [34]. They are closely related and occasionally overlapping phenomena that interfere with these assays in a similar fashion and thus are often confused. Heterophile antibodies are poorly defined, polyspecific, low affinity antibodies that develop in response to no identifiable immunogen while HAAAs are monospecific, and high affinity antibodies that develop after exposure to mammals often by way of therapeutic or diagnostic treatments, including transfusion and vaccination but also through animal handling. The classic examples of heterophile antibodies include erythrocyte agglutinins associated with mononucleosis and rheumatoid factor (usually an IgM autoantibody directed against the Fc portion of an individual's IgG). Heterophile antibodies are reported to have a prevalence of 10%—40% in in-patients. These most commonly affect two-site immunometric assays by cross-linking capture and detection antibodies causing falsely high results or blocking binding sites sterically or specifically causing falsely low results. They also may interfere with competitive IAs by blocking the antigen binding site (paratope) of the antibody.

Several approaches have been developed to identify and eliminate some of these interferences [31]. These include adding blocking agents to the assay, taking a detailed medical history to identify exposure, applying parallelism, using comparative methods, and using specific heterophile antibody assays.

Blocking agents include nonimmune globulins as well as antihuman immunoglobulin antibodies. These are intended to remove heterophile and HAAA interferences. In most commercial assays some form of blocking agent is included due to the high prevalence of these interferences. Parallelism refers to serial dilution of a sample with a solution that is appropriate for the method of interest (e.g., zero calibrator, a solution that has a relatively similar matrix to that of the patient sample but is free of the analytes of interest). This is compared to a similar serial dilution of calibrator material or normal patient population. A nonparallel curve indicates the presence of an interfering agent. Most interference are method-specific because manufacturers use proprietary antibodies. In these cases, comparative methods from different manufacturers will respond differently demonstrating the presence of interference. Finally the addition of other proteins, polyethylene glycol, sulfhydryl agents, and detergents or heating the sample will reduce or eliminate the interference by precipitating the interfering substance or impairing its activity.

1.7.1.3 Biotin interferences in the clinical laboratory

Recently, there has been numerous publications, news release, and editorial comments on the impact of biotin on IA results. The biotin effect on IAs is probably one of the largest and most important interferences seen in modern laboratory medicine. The following paragraphs briefly describe biotin and its impact on IAs.

Biotin (also known as vitamin H, vitamin B7, and coenzyme R) is a water-soluble vitamin that acts as an enzyme cofactor in fatty acid biosynthesis, the citric acid cycle, and the metabolism of odd-numbered fatty acids and branched-chain amino acids. It has additional roles in gluconeogenesis and gene expression [35]. In the diet, biotin is bound to proteins found in liver, kidney, pancreas, eggs, yeast, and milk. Cereal grains, fruits, most vegetables, and meat are poor sources of biotin. In Western populations, dietary biotin intake is estimated to be $35-70\,\mu g/day$, a level that agrees with the recommended dietary allowance. Most multivitamin pills contain about $30\,\mu g$ of biotin.

Deficiency in biotin is uncommon in Western populations but can be seen with prolonged consumption of raw egg whites, in total parenteral nutrition without biotin supplementation, and in patients with a genetic deficiency in biotinidase, an enzyme that plays a key role in the metabolism of biotin [35]. Symptoms of biotin deficiency include anorexia, nausea, vomiting, glossitis, pallor, depression, dry scaly dermatitis, and hair loss [36]. High-dose supplementation of this vitamin (doses $>1\,mg/day$) plays a role in the therapy of several diseases including biotinidase deficiency, mitochondrial metabolic disorders, and multiple sclerosis [37,38]. Furthermore, doses up to $10\,mg$ a day are frequently encountered in nutritional supplements and are being taken to improve hair, skin, and nail health.

Many IAs employed by clinical laboratories use the interaction of biotin with the protein streptavidin as an immobilizing system to capture analytes. Biotin's small size enables it to be readily incorporated into a hormone or antibody, and streptavidin binds biotin with very high affinity. While the amount of usual dietary biotin intake is not expected to be high enough to affect IAs based on the biotin–streptavidin interaction, high levels of biotin in patient specimens can significantly interfere with these IAs. This interference manifests itself as either falsely low (sandwich IAs) or falsely high (competitive IAs) results depending on the design of the IA. It has been shown that biotin doses >1 mg per day can produce sufficient levels to cause this interference.

Biotin interference poses a particularly dangerous problem when (1) patients are in an emergency situation and do not realize they are taking high doses of biotin to inform the clinician, or (2) when the physician treating the patient does not know that their patient is taking high doses of this vitamin. A literature search of biotin interference reveals an increasing number of published cases, the majority of which describe the problem of biotin interference in thyroid function tests [39,40−44]. High-dose biotin can produce a dangerous combination of positive and negative interference among the thyroid tests (FT4, FT3, TSH, and TRAb) and paint a picture of Grave's disease in patients who have no clinical symptoms. Without good clinical observations by the physician, this could lead to unnecessary procedures and treatments. The interference of high-dose biotin on the thyroid tests is particularly troubling for patients with multiple sclerosis as large doses of this vitamin are emerging as a new treatment [37,38,42,43,45]. The interference of biotin on parathyroid hormone, follicle-stimulating hormone, luteinizing hormone, sex hormone−binding globulin, estradiol, progesterone, testosterone, cortisol, folate, vitamin B12, and ferritin has also been reported [43,45,46]. The list of affected IAs varies for each clinical diagnostic platform, and for a given test the manufacturer-supplied product information must be consulted to determine if biotin is interference. For some platforms the list is quite extensive and includes the aforementioned tests as well as those for cardiac function, beta-hCG, and cancer biomarkers [43,47]. In many of these IAs, biotin was an established interference but at doses thought to be rarely encountered in the general population. While the manufacturers are aware of the increasing use of high-dose biotin and its potential effects on patient care, they have not to date suggested any concrete solutions to this problem beyond having patients abstain from this vitamin (Different diagnostic companies, personal communication, 2016). Solutions proposed by local clinical laboratories include dilution of the specimen with a validated assay diluent, running of specimens on a different platform known to be unaffected by biotin, and using streptavidin-agarose beads to remove the biotin before the sample is run on the affected analyzer [48]. The interference of high-dose biotin on IAs in the clinical laboratory is an emerging issue and many questions have not yet been addressed. What is the prevalence of high-dose biotin among a given patient population? What are the pharmacokinetics of high-dose biotin? What is the effectiveness of the use of streptavidin-agarose beads to remove the biotin interference? If treatment of high biotin specimens with streptavidin-agarose beads is effective, how feasible this approach is for large laboratories that perform thousands of tests daily? As of now, the most practical approach for clinical laboratories to tackle this issue is to educate their customers; patients and healthcare providers. Currently the only safe approach is to ask patients on high-dose biotin to abstain from taking biotin for at least 48 h before specimen collection.

1.7.1.4 Alternative testing methods

Although the interference mentioned in the above sections has a deleterious impact on IAs, there are other methods that can be used to measure different molecules in the patient sample. These methods have improved sensitivity and specificity in comparison to IAs and are excellent alternatives for measuring endocrine markers. LC with tandem MS is a highly accurate method that has recently been used in most modern clinical labs. Chromatography refers to the separation of molecular species in a solution based on their physical and chemical properties as they pass through a column. The sample is carried in the "mobile phase" by either a carrier gas (for volatile analytes or those volatile upon chemical derivatization) or a liquid. Gas chromatography refers to a method that uses a gas as the mobile phase whereas LC uses a liquid carrier. Separation of compounds occurs based on their chemical structures and their relative affinity for the stationary phase (the materials in the column) with respect to the mobile phase. The amount of time that takes for a molecule to go through the column is called retention time. Prior to injection of the sample into the chromatograph, heterogeneous specimens such as biological fluids may require significant preanalytical effort to remove substances that can either interfere with analysis or damage the sensitive instrumentation. This can include precipitation techniques, solid or liquid phase extraction and dilution.

Following chromatographic separation and elution from the column the components are identified and their concentrations are measured by one or more of several techniques that use light absorbance or electrical, chemical or thermal properties. The optimal method depends on the various chemical and physical characteristics of the specimen and target molecules, sample size and volume, time constraint since these are essentially all analyzed in series and availability of instrumentation.

The current gold standard for detection is MS. In MS, molecules are fragmented and ionized by various means, generating a population of ions whose quantities depend upon the relative stability of the bonds within the parent molecule. The ions are accelerated and focused as they pass through magnetic fields to the detector which counts the number of ions and the mass/charge (M/Z) ratio of each as the particles strike it. The population distribution of ions is known as the molecule's mass spectrum, which is peculiar to each molecule and becomes its chemical "fingerprint." The mass spectrum of each molecule is compared to a library within the MS computer system to identify it. The relative size of the peak is compared to an internal standard to quantify its concentration. By programming the MS to collect only ions with certain charge (positive or negative), molecules may be identified specifically and measured reliably in very low concentrations.

Currently the "state of the art" of these methods consists of LC followed by serial mass spectrometers, known as LC–MS/MS. Among the advantages of LC–MS/MS include fewer constraints on the target molecule, high analytical specificity, precise quantitation, few supplies, and none of the biological reagents used by IAs. A by-product is much lower cost per assay. As mentioned earlier, although some interference may be seen, none involves the biological properties described with respect to IAs. Moreover the interferences are based on physical and chemical properties and are likely to be identified with clinical information. Obviously the major shortcoming of LC–MS/MS is the initial high cost and complexity of the system, requiring highly skilled technologists to operate and maintain the system. With these thoughts in mind, we recommend using LC–MS/MS for identification and measurement of most drugs, hormones, and tumor markers in human serum.

1.8 CLINICAL APPLICATIONS OF VALID LABORATORY RESULTS

The section above contains a full discussion of factors that must be considered and controlled by both the clinician and the laboratorian in order to ensure that the biochemical test ordered is valid for interpretation. Any deviation from these principles may lead to erroneously reported test results and erroneous clinical decisions thereafter. However, there are a number of factors that are almost exclusively in the domain of clinical interpretation, for which the ordering healthcare provider must take responsibility. The practice of ordering randomly selected laboratory tests and assuming validity and comprehensive diagnosis as a result is to be strongly discouraged. In modern medicine, this interpretive responsibility must often be explained to the patient as well; many patients seek and obtain copies of their laboratory tests. It is natural for patients to "look" up the meaning of their test results, with general interpretive comments widely available on a number of laboratory websites. Lacking understanding of the preanalytical, analytical, and postanalytical interpretive factors, patients will rely upon physicians for this information and thus physicians themselves must be knowledgeable about lab medicine.

1.8.1 CONSIDERATION OF CLINICAL APPLICATION OF RESULTS SHOULD PRECEDE ORDERING THE TEST

Although this point seems obvious, clinical practice shows that this is one of the most commonly violated laboratory testing principles. Tests ordered when the diagnosis is already patently obvious (e.g., thyroid-stimulating immunoglobulin in a man with hyperthyroidism, goiter, and exophthalmos) greatly increase the total cost of care without contributing new information. Tests for diseases of which there is absolutely no clinical evidence cannot contribute to diagnosis but can spawn an extraordinary amount of additional testing which may become necessary in order to "prove" a suspected false positive result. While a certain amount of test ordering is pragmatically necessary in a defensive clinical environment, clinicians are nonetheless encouraged to consider aforehand the potential quandary that may arise when a test (especially for a rare disease) returns as unexpectedly "abnormal." That is, ordering a laboratory test by a healthcare provider who is not fully familiar with the test or the condition in question can result in potential harm to the patient. A general rule of thumb is that if you would be greatly surprised and confused to receive a "positive" result on a given test, it is probable that the test should not be ordered in the first place.

1.8.2 CLINICIANS SHOULD SEEK ADVICE FROM SPECIALISTS PRIOR TO ORDERING TESTS WITH WHICH THEY ARE UNFAMILIAR

Every clinician develops their own clinical "comfort" zone in which they may be considered "expert." However, the modern complexity of patients often demands diagnostic and therapeutic measures that may fall outside one's expertise. Ordering laboratory tests of unfamiliar type is a potentially dangerous move; not only may a false diagnosis be made, but a serious diagnosis may be missed. This is particularly common in endocrinology given the high number of unusual endocrine syndromes that might only be seen once in a career by a primary care provider. In most areas of the developed world, endocrinologist consultation, even by phone or email, is readily available and may be the fastest way to order the appropriate sequence of tests, starting with the correct test first. A general rule of thumb is that one should consider speaking with an endocrinologist prior to ordering any given endocrine test for the first time.

1.8.3 TEST SENSITIVITY

Readers are referred to textbooks on medical statistics or "evidence based medicine" for a detailed discussion of how test sensitivity and specificity are calculated from a population dataset. Very few clinicians have the need (or time) to regularly do such calculations from published literature; however, the concept of sensitivity is critically important for use of a medical lab. Simply put the sensitivity of a test is the proportion of times that the test will register a positive/abnormal result in a population of patients who truly have the disease in question. For example a test with 70% sensitivity for disease X will have very limited use for the clinician since only 70% of people with disease X will return a positive/abnormal result. That leaves 30% of truly diseased patients being "missed" by the test; the clinician who suspects disease X will therefore have to do another test to continue the investigation. Similarly, it would be a mistake for the clinician, who suspects disease X, to fail to pursue additional testing when the test sensitivity is so poor. On the other hand, if a test has 99% sensitivity, that means only 1% of truly X-diseased patients will get a negative/normal result. In many cases, that is a sufficiently good test to rule out disease X. A negative/normal result does not tell the clinician what the disease is, but it lets the clinician know with increased probability, that it is not disease X.

There are still two practical problems that remain with a highly sensitive test. The first is that virtually no test is 100% sensitive for any single disease (in fact, in real life, a very high quality, sensitive test might only have a sensitivity of 95%). Thus the interpretation of even very sensitive tests must be informed by the clinician's "pre-test probability" which may either be based upon published clinical prediction rules that combine clinical risk factors, or, more commonly, clinical intuition, informed by experience. In these instances, communication with an endocrinologist or clinical chemist is highly recommended.

Examples:

- For a disease that is considered highly unlikely based upon the clinical presentation ("low pre-test probability"), a normal/negative result in a highly sensitive lab test almost always then excludes the disease from further consideration.
- For a disease considered somewhat unlikely based upon the clinical presentation ("moderate pre-test probability"), a normal/negative result in a poorly sensitive lab test makes it difficult to know whether the disease is still a consideration or not (an example of a scenario in which this lab test probably should not have been the first choice).
- For a disease considered highly likely in which a highly sensitive test returns positive/abnormal, the diagnosis is confirmed.
- For a disease considered highly likely in which a poorly sensitive test returns negative/normal, the diagnosis is not excluded and the clinician will usually need to do more testing (another example in which the chosen test may not have been ideal for the clinical scenario). This is a prime example of how a serious disease could be missed if input from the clinical chemist or specialist is not sought prior to test interpretation (or ordering).

The second problem is that most primary healthcare providers cannot be expected to know the disease-specific test sensitivity of every disease and test. Consultation with a specialist may actually save time and money by ensuring that the best test for the clinical scenario is ordered first. The best and perhaps most important rule of thumb for all laboratory endocrinology is this: If you really think the patient has a certain disease but the lab tests do not seem to match, never discard your clinical judgement before consulting with an endocrinologist.

1.8.4 TEST SPECIFICITY AND SENSITIVITY

The *specificity* of a laboratory test is quite different from its *sensitivity*. The clinician should understand the correct definition of specificity and sensitivity and the relationship between the two.

Specificity of a test defines its power to rule IN a diagnosis.

- That is a positive/abnormal result on a test with high specificity confirms the diagnosis.

Sensitivity, on the other hand, defines a test power to rule out a diagnosis.

- That is a negative/normal result on a test with high sensitivity excludes the diagnosis.

Together, these are two incredibly powerful concepts in diagnosis. Knowledge of each test's sensitivity and specificity allows a clinician to select one or more tests to logically include/exclude different possibilities in a differential diagnosis.

Once the clinician understands the concepts of sensitivity/specificity, it is next required to understand the general relationship between both properties. Ideally the "perfect" test will have 100% sensitivity and 100% specificity at the same time. However, this situation almost never exists in medical practice and the reason is that for most laboratory tests, the sensitivity and specificity have an inverse relationship. That is to say a highly sensitive test (i.e., a test that "captures" 100% of all affected patients) will usually have a poor specificity (i.e., positive/abnormal results might also be seen in other conditions as well).

Since many endocrine diseases exist across a spectrum (see Section 3.5.1), the reporting laboratory often needs to decide what kind or level of results will be flagged as abnormal and thus alerting the clinician to act. The process of choosing this "cut-off" is often done using a statistical maneuver called "Receiver-Operator Characteristic (ROC) Curve." The calculation of an "optimized" ROC point is beyond the scope of this text but it is important for readers to understand that this statistical procedure is designed to find the best compromise between sensitivity and specificity. While this is a generally good idea, the sophisticated laboratory user will remember that by using the ROC-derived cut-point, they are usually sacrificing some amount of both sensitivity and specificity.

Thus if a clinician wanted to rule out a serious disease with 100% certainly, they may choose to contact the clinical chemist to enquire as to the cut-off with 100% sensitivity, something that is readily available from the raw population data for any test.

Example:
Cushing's disease is a very serious and sometimes fatal disorder of excess cortisol production. One screening test for Cushing's disease is the overnight administration of dexamethasone followed by the measurement of serum cortisol the next morning. Based upon ROC curve analysis a medical lab might choose to report a sex-suppressed serum cortisol of " <75 nmol/L" (2.7 µg/dL) as the cut-off for normality. However, the sophisticated practitioner knows that "75" is a compromise value and at that level the sensitivity might only be 90%. Therefore faced with a patient for whom there is high clinical suspicion of Cushing's disease, a result of "69 nmol/L" (2.5 µg/dL) still leaves at least a 10% chance that the patient actually does have the disease. For a serious condition, that is far too high a possibility to just abandon the work up at that point.

After calling the endocrinologist a phone conference is held with the lab at which point the ROC curve (or a similar published ROC curve) is examined to find the level at which the dex-suppressed serum cortisol has near 100% sensitivity which turns out to be "50 nmol/L" (1.8 µg/dL). Therefore the patient's result of 69 nmol/L (2.5 µg/dL) still falls in the range at which Cushing's disease may be seen—the corollary being that Cushing's disease is not ruled out and further tests should be done.

Why did the clinician not choose the cut-off with 100% specificity (i.e., to rule in Cushing's)? It sounds at first like that would actually make more sense but remember the inverse relationship between sensitivity and specificity—a test cut-off with 100% specificity will have a lower sensitivity, meaning that if the test is positive/abnormal, that is great (disease confirmed) but the poor sensitivity means that a significant number of truly diseased people simply will not pass that confirmatory threshold and may thus get missed if testing stops there.

Revisit the example. If the ROC curve cut-off for 100% specificity was determined, it might be something like " > 200 nmol/L" (7.2 µg/dL). In other words, if your dexamethasone-suppressed cortisol was >200 nmol/L (7.2 µg/dL), you could be very confident in the Cushing's diagnosis. But, if that actual result was 69 nmol/L (2.5 µg/dL), it still falls in the range that might yet prove to be Cushing's but just cannot be confirmed yet by this test. One of the most powerful tools in medicine is the ability to eliminate diagnostic possibilities through the serial applications of sensitive tests; this is why first-choice tests for any given disease are ideally chosen for high sensitivity.

1.8.5 POSITIVE AND NEGATIVE PREDICTIVE VALUES

The positive and negative predictive values (PPV and NPV) are often confused with sensitivity and specificity and while they are related, they represent a slightly different way of interpreting a test result.

The PPV and NPV may be thought of as ways to express the "net effect of combining the sensitivity and specificity" of any given diagnostic cut-off, as performed in a specified population.

There are two parts to that sentence, let us examine each one separately.

The PPV may be explained as follows: using the numeric specific cut-off "A" to diagnose the disease in question, the PPV is the probability (0%−100%) that an abnormal result truly indicates the presence of the disease.

The NPV is the same concept, in reverse: using specific cut-off "A" to diagnose the disease in question, the NPV is the probability (0%−100%) that a normal result truly indicates the patient to be free of said disease.

A high positive predictive value then means that if the test is abnormal, the patient almost certainly has the disease. A high NPV means that if the test is normal, the patient almost certainly does not have the disease. But how is this different than sensitivity and specificity? The answer lies in the second part of the definition—both PPV and NPV are calculated specifically in a population of interest (unlike sensitivity and specificity).

If a certain disease is very common in a population and the test has a decent sensitivity and specificity, the calculated PPV will be high, indicating that for this common disease, an abnormal result is almost always indicative of disease. A good clinical example would be TSH testing for common hypothyroidism. Primary hypothyroidism is extremely common and high TSH measures are both sensitive and specific; therefore in this thyroid disease-prone population, an abnormally high TSH almost always indicates true disease; that is, it has a high PPV. The same thing is true for the TSH NPV—a normal TSH pretty much excludes primary hypothyroidism in a population even where hypothyroidism is common.

On the other hand, if a certain disease is quite rare, say 1 in 1,000,000 then a test may have a very good sensitivity and specificity and still have a very poor PPV. An example is the 24 h urine free cortisol (UFC) for the diagnosis of Cushing's disease. Even if the test has decently high sensitivity and specificity (determined in small samples of patients with Cushing's disease), when applied to a general population where Cushing's disease is rare, those few percentage ($<100\%$) in specificity still means that if the test was ordered in 10,000 patients, out of 100 abnormal UFC results, perhaps only 1 (or none!) of the patients would actually end up having true Cushing's disease.

The whole purpose of outlining this concept is to get to the following extremely important statement: When the disease under consideration is rare in a population, the vast majority of positive/abnormal test results will be false positives, even using a very good test.

The corollary messages for the ordering physician are as follows:

1. Do not order tests for rare diseases unless you are reasonably certain the disease is present (or at least you are willing to spend a lot of time investigating the subsequent false positive result).
2. If you do order the test and it is abnormal, do not assume you have diagnosed a rare disease; the odds are overwhelming that you have got a false positive and this is why rare diseases are best diagnosed with multiple complementary tests under specialist supervision. At the very least, return to Section 3.4 and consider the specificity (ruling "in") of the result you have obtained. If your result is far beyond a highly specific cut-off, then congratulations, you probably have picked up a true and rare disease.

1.8.5.1 Spectrum bias

Before leaving the principles of sensitivity, specificity, PPV, and NPV, it is particularly important for endocrinology's sake, to alert the reader to a common fallacy about some of the endocrine tests' diagnostic capabilities as a result of what is called "spectrum bias" [49].

A diagnostic test is considered useful if it can separate "diseased" patients from "non-diseased" patients. Every time a new test is introduced, researchers will compare the new test against either an older test or, even better, a recognized gold standard (which might be something like confirmed tumor pathology). Through this comparison the researchers generate an array of tests, results from both diseased and nondiseased subjects and are thus able to be calculate the ROC curve. Using the ROC curve we can show the best "cut-off" for achieving optimal specificity and sensitivity. Additionally the sensitivities and specificities may be known for each different "cut-off" used to define disease.

However, there are two scenarios where, if the new test is compared with an inferior test, the results can be misleading. That is, although the calculated specificity and sensitivity may look good, since the reference test that the new test was compared to is not a "gold standard," this type of research may be very misleading. The first is in the scenario where a new test is compared with an old test and no gold standard is applied. Although this may generate some apparently good sensitivity and specificity results, it must be remembered that the lack of gold standard comparison means that the new test really cannot be any better diagnostically than the old test (which was likely imperfect as well). There may be other advantages to the new test such as lower cost, easier acquisition, and faster turnaround, but it must be remembered that any sensitivity/specificity figures generated by such research are likely an overestimate since it ignores the inherent error in the comparison test. This kind of problem is usually very easy to spot and will be considered by the medical laboratory director before instituting it.

The second error (spectrum bias) is far more subtle since it may be otherwise very well performed research using a gold standard reference. The key lies in the recognition that endocrine diseases (and other diseases too) may have a spectrum of severity, acuity and may sometimes lack what are commonly called "classic" clinical features.

Therefore if a researcher chooses to evaluate a new test's ability to diagnose a disease X, they may choose to perform the test in 100 diseased people (confirmed by whatever is a gold standard) and 100 healthy volunteers. This type of comparison may well show the test to be highly sensitive and specific. However, for diseases that have a spectrum of presentation, it is an over-simplification of clinical practice to use healthy volunteers as the comparison group. If the comparison group has none of features of disease X, then arguably, those volunteers are not representative of the type of patient in whom such testing will be done. If the new test is evaluated in a population of patients with disease X and patients who look like they have disease X but do not, the test characteristics may show significantly more overlap of results between the two populations and thus much poorer sensitivity and specificity.

Example:

A new cortisol test is being researched for its ability to diagnose Cushing's disease. The test is performed in 34 patients who look like they have Cushing's disease and in fact, do—as confirmed when the pituitary is surgically explored. For comparison the researchers choose 47 healthy medical students and 59 patients who have some Cushing's features like obesity, diabetes, and hypertension. The medical students are young, healthy, and generally fit. It is obvious they do not have Cushing's disease! The researchers are pretty sure the other 59 obese control patients do not have Cushing's, but in truth, none of the control patients have had surgical pituitary exploration (the gold standard) and so it stands to reason that some of those control patients might actually have real Cushing's disease (and yet their results are analyzed as being in the control group). When the new test results are compared to the healthy medical students, there is a very clear separation and the test appears to have near perfect sensitivity and specificity. However, when compared with the "looks-like-Cushing's" group, it turns out there is a lot of overlap and the sensitivity and specificity is considerably poorer. The "true" test performance characteristics are those in the second group—because no one will be ordering the test in healthy people.

The whole point of this discussion is to point out the critical importance of experienced clinical assessment when it comes to either rare diseases or diseases which overlap other similar appearing but different health conditions. If you strongly believe that a diagnosis is present, or if the diagnosis in question is potentially fatal, do not let a surprisingly normal test necessarily end your investigation. Similarly, if you screen for a disease in a patient who does not really have much or any clinical features of the disease, remember that positive findings in a lab test may be false positive, even if the test is otherwise believed to be a useful test and performed according to the strictest lab procedures.

1.8.6 SCREENING AND CONFIRMATORY TESTS

As alluded to in subsequent specific endocrine sections, the "first" golden rule of endocrinology is that if you think a hormone is lacking, prove it by a stimulation test; if you think there is too much of a hormone, try to suppress it. The second golden rule recognizes that the clinician wants a sensitive test as a "screening" first step—sensitive enough to ensure that no diagnosis gets missed. However, if the result is positive it should be confirmed by a superior test to confirm the diagnosis, usually chosen for its specificity. If all goes according to the plan, this is a fairly straightforward test sequence that can confidently pick up most endocrine disorders.

The troublesome cases are the ones where the screening test is positive but the confirmatory test is negative, that is, a diagnosis is possible but unconfirmed. This situation may call for a repeat of the confirmatory test or, more commonly, the endocrinologist will choose a different confirmatory test to see if the results will be clearer when approached from a different angle. It is this two-step sequence that requires such careful collaboration with the chemistry lab. Many confirmatory tests are dynamic (stimulation or suppression tests) and require multiple sampling, often after administration of a stimulating/suppressing drug. Because the final diagnosis rests upon a positive confirmatory test, it is crucial that specimens are collected, labeled and reported correctly.

1.8.7 ACCURACY AND PRECISION APPLIED TO CLINICAL DECISION MAKING AROUND FIXED BIOCHEMICAL TARGETS OR THRESHOLDS

Accuracy and precision are two very different terms in biological measurement but are very relevant to endocrine practice. A test may have one or the other or neither or both. The *accuracy* of a test reflects its ability to be close to the actual biological result (i.e., knowing that there is a measurement error in all biological measures, an accurate test has very little deviation from the "true" value).

Precision describes the extent to which repeated measures differ from each other in the absence of true biological change.

For example, if a patient's "true cortisol" was 168 nmol/L, an accurate assay would generate values in the range of 160−176 nmol/L and if measured several times, the average might be 169 nmol/L. Over the course of several measures however, if the test is precise, the numbers will differ little from each other, as in this case.

As a second example, if the patient's true cortisol level was 168 nmol/L and multiple assays generated results of 120−130 nmol/L, we would say the assay is precise, but not very accurate.

Lastly, consider that same patient for whom the assay measures vary from 120 to 190 nmol/L. Averaged together, we might come up with a mean of 166 nmol/L giving reasonable overall accuracy but with very poor precision among the different individual results (and obviously dubious accuracy for some of the individual results as well).

Clearly both the lab and clinician are interested in having lab tests that are accurate but the point of this section is to illustrate the importance of reproducibility (precision). Many endocrine diagnoses are made (rightly or wrongly) on the basis of a single laboratory result. And, many endocrine diseases are followed longitudinally according to serial change in the biomarkers. Therefore reproducibility is a key factor for the endocrinologist.

For example:

- If a patient is diagnosed with hyperaldosteronism having an aldosterone−renin ratio of >50 pmol/L/ng/mL/h, a patient result of 45 pmol/L/ng/mL/h is very hard to interpret unless the test precision is known to be less than the "distance" between the measured result and the diagnostic threshold. If the reproducibility in this test is ± 20% then that "45" could really be as high as 54 pmol/L/ng/mL/h if repeated. At "45" the patient is deemed "normal", but at 54, he would be diagnosed with a disease.
- Key point: when using a firm threshold for any diagnostic test, beware of results that return close to that threshold, on either side. You may need to either repeat the test or talk to the clinical chemist to find out about the reproducibility of the test—if somewhat poor or uncertain, a repeat test is needed.

A second example using serial change:

- A patient with heart disease is receiving statin therapy to lower her cholesterol. The endocrinologist wants to achieve an LDL cholesterol lower than 2.0 mmol/L (per guidelines). The patient's baseline LDL cholesterol is 2.3 mmol/L and after going on a statin drug the next level is 2.1 mmol/L. Aiming to follow the guidelines perfectly the endocrinologist considers whether the statin dose should be increased.
- This example is even more complicated because it requires knowledge of the precision of the LDL result (which is actually calculated from other lipid measures) in addition to the actual change in LDL with the drug, in addition to the variation in actual change with serial measurements. If the LDL was measured with 100% accuracy and 100% precision every time, then any change would simply reflect the action of the drug. Even with such a perfect test the endocrinologist could measure the "real" LDL change daily for a week and may find results ranging from 1.8 to 2.3 mmol/L. When this biological variation is added to the assay reproducibility, then the range of measures that might be seen might include all numbers from 1.6 to 2.6 mmol/L. Notice that this range of results includes numbers that would be considered "at target" as well as numbers that seem to be even higher than before therapy!
- The point of the exercise is to help the clinician understand that test precision is exceedingly important in serial measures that are found to be around (just above or just below) a fixed target. Changes to therapy should likely only be considered after repeat clinical assessment and possibly repeat measurement; we would generally want to avoid over-or-under dosing a drug unnecessarily.

The sophisticated laboratory user thus becomes familiar with the term "Least Significant Change (LSC)" which denotes the expected range of values that may be seen with repeat measures of an unchanging biological variable. In statistical terms the LSC is the 95% confidence interval around any given result and that confidence interval can be determined simply by multiplying the assay coefficient of variation by 2.77. While this exercise is unlikely to be performed by most clinicians, the principle is what is most important; remember that lab results are an "estimate" of the true value. Hopefully, they are accurate and reproducible but often they may be less so than desired. Consult with the clinical chemist before making a major medical decision based on a borderline result.

1.9 THE ENDOCRINOLOGIST—CHEMIST RELATIONSHIP

Hopefully, this chapter has demonstrated the incredible scientific, procedural, and statistical nuances that underlie even the most routine clinical chemistry tests.

Modern medical practice in endocrinology may be on a collision course of ever-more precise and subdivided disease entities, often picked up at the preclinical stage—all running head long into the major reliance of clinicians upon the results written on a laboratory report often following a single measurement. The power of modern medicine lies in its ability to diagnose/categorize diseases down to the most subtle details relevant for treatment. But with this responsibility (and expectation by the patient) the whole process will only be possible in many cases by a collaborative approach between clinical chemist and healthcare practitioner. Understanding what results "can be believed" versus those that "should be questioned" is the interpretive job of the chemist after hearing from the clinician about the clinical scenario and differential diagnosis, reinforced with clinical experience on both sides.

We hope that this book begins a new era in the cross collaboration of local clinical chemists with health practitioners in the endocrine (and other) fields. It is this collaboration that unleashes the power of modern medicine and helps drive the best possible outcome for the patient.

In the previous sections, we discussed most of the variables associated with specimen collections. Physiologic and other nonphysiologic variables will be discussed for each test in the corresponding chapter.

In this chapter, we briefly and generally discussed the important preanalytical variables that affect test results prior, during and after specimen collection. There are several books written on this important topic and due to limited space, we tried to introduce the importance of preanalytical variables to the readers and focus on proper blood collection. All chapters discuss the importance of preanalytical variables specific to disorders and tests described in the chapter. One important approach to reduce and hopefully eliminate the effect of preanalytical factors on test results is to have an open communication between clinicians and clinical chemists and pathologists. Clinicians should contact their colleagues about any test results that do not correlate with clinical findings and "do not make sense." With an open, cordial, and mutually respectful communication, clinicians and clinical chemists/pathologists can work together to provide the most accurate results for best management of the patient.

REFERENCES

[1] Kohn LT, Corrigan JM, editors. To err is human: building a safe health system. Committee on Quality of Health Care in America. Institute of Medicine. Washington, DC: National Academy Press; 2000.

[2] Brennan TA. The Institute of Medicine report on medical errors—could it do harm? N Engl J Med 2000;342:1123−5.

[3] Davis K, Shoenbaum SC, Collins KS, Tenney K, Hughes DL, Audet AMJ. Room for improvement: patient report on the quality of their health care. Publication No. 534. New York: The Commonwealth Fund; 2002.

[4] AHRQ. Medical errors: the scope of the problem. Fact sheet. Rockville, MD: Agency for Healthcare Research and Quality; 2000. Publication No. AHRQ 00-P037, http://www.ahrq.gov/qual/errback.htm.

[5] Leape LL, Brennan TA, Laird NM. The nature of adverse events in hospitalized patients: results of the Harvard Medical Practice Study II. N Engl J Med 1991;324:377−84.

[6] Clinical Laboratory Standards Institute. Collection, transport and processing of blood specimens for testing plasma-based coagulation assays and molecular hemostasis assays. Approved guideline−fifth edition. CLSI Document H21-A5. Wayne, PA; 2008.

[7] Young DS. Effects of preanalytical variables on clinical laboratory tests. 2nd ed. Washington DC: AACC Press; 1997. p. 1281.

[8] Wallin O. Preanalytical errors in hospitals. Umeå University Medical Dissertations, New series No 1177, ISSN 0346-6612; ISBN 978-91-7264-562-2; 2008, p. 1−64.

[9] Haverstick DM, Groszbach AR. Specimen collection, processing nd other preanalytical variables. In: Burtis CA, Bruns DE, editors. Tietz fundamentals of clinical chemistry and molecular diagnostics. 7th ed. Amsterdam: Elsevier; 2015. p. 72−89.

[10] Lenders JW, Duh QY, Eisenhofer G, Gimenez-Roqueplo AP, Grebe SK, Murad MH, et al. Pheochromocytoma and paraganglioma: an endocrine society clinical practice guideline. J Clin Endocrinol Metabol 2014 Jun;99(6):1915−42.

[11] Nichols J. Preanalytical variation. In: Clarke W, editor. Contemporary practice in clinical chemistry. Washington, DC: AACC Press; 2016. p. 1−11.

[12] Baskin L, Dias V, Chin A, Abdullah A, Naugler C. Effect of patient preparation, specimen collection, anticoagulants and preservatives on laboratory test results. In: Dasgupta A, Sepulveda JL, editors. Accurate results in the clinical laboratory, a guide to error detection and correction. Amsterdam: Elsevier; 2013. p. 19−34.

[13] Young DS, Bernes EW, Haverstick DM. Specimen collection and processing. In: Burtis CA, Ashwood ER, Bruns DE, editors. Tietz textbook of clinical chemistry and molecular diagnostics. 4th ed. St. Louis, MO: Saunders; 2006. Chapter 2.

[14] Statland BE, Bokelund H, Winkel P. Factors contributing to intraindividual variation of serum constituents: 4. Effects of posture and tourniquet application on variation of serum constituents in healthy subjects. Clin Chem 1974;20:1513−19.

[15] Garza D, Becan-McBride K. Phlebotomy handbook: blood collection essentials. 7th ed. Upper Saddle River, NJ: Pearson; 2005.

[16] McCall RE, Tankersley CM. Phlebotomy essentials. 3rd ed. Baltimore: Lippincott Williams & Wilkins; 2005.

[17] Sherwood L. The blood. In: Sherwood L, editor. Human physiology: from cells to systems. Belmont, CA: Thomson/Brooks/Cole; 2004.

[18] D'orazio P, Burnett RW, Fogh-Andersen N, Jacobs E, Kuwa K, Kulpmann WR, et al. Approved IFCC recommendation on reporting results for blood glucose: International Federation of Clinical Chemistry and Laboratory Medicine Scientific Division, Working Group on Selective Electrodes and Point-of-Care Testing (IFCC-SD-WG-SEPOCT). Clin Chem Lab Med 2006;44:1486−90.

[19] Fogh-Andersen N, D'orazio P. Proposal for standardizing direct-reading biosensors for blood glucose. Clin Chem 1998;44:655−9.

[20] Lyon ME, Lyon AW. Patient acuity exacerbates discrepancy between whole blood and plasma methods through error in molality to molarity conversion: "Mind the gap!". Clin Biochem 2011;44:412−17.

[21] Guder WG, Narayanan S, Wisser H, Zawta B. Plasma or serum? Differences to be considered. Diagnostic samples: from the patient to the laboratory. 4th ed. Weinheim: Wiley-VCH; 2009.

[22] Strathmann FG, Ka MM, Rainey PM, Baird GS. Use of the BD vacutainer rapid serum tube reduces false-positive results for selected Beckman coulter Unicel DxI immunoassays. Am J Clin Pathol 2011;136:325−9.

[23] Chuang J, Sadler MA, Witt DM. Impact of evacuated collection tube fill volume and mixing on routine coagulation testing using 2.5-ml (pediatric) tubes. Chest 2004;126:1262−6.

[24] Boyanton Jr BL, Blick KE. Stability studies of twenty-four analytes in human plasma and serum. Clin Chem 2002;48:2242−7.

[25] Dasgupta A, Dean R, Saldana S, Kinnaman G, et al. Absorption of therapeutic drugs by barrier gels in serum separator blood collection tubes. Volume- and time-dependent reduction in total and free drug concentrations. Am J Clin Pathol 1994;101:456−61.

[26] Schouwers S, Brandt I, Willemse J, Van Regenmortel N, et al. Influence of separator gel in Sarstedt S-Monovette® serum tubes on various therapeutic drugs, hormones, and proteins. Clin Chim Acta 2012;413:100−4.

[27] Bowen RA, Chan Y, Cohen J, Rehak NN, et al. Effect of blood collection tubes on total triiodothyronine and other laboratory assays. Clin Chem 2005;51:424−33.

[28] Rodushkin I, Odman F. Assessment of the contamination from devices used for sampling and storage of whole blood and serum for element analysis. J Trace Elem Med Biol 2001;14:40−5.

[29] Drake SK, Bowen RA, Remaley AT, Hortin GL. Potential interferences from blood collection tubes in mass spectrometric analyses of serum polypeptides. Clin Chem 2004;50:2398−401.

[30] Pai SH, Cyr-Manthey M. Effects of hemolysis on Chemistry Tests. Lab Med 1991;22:408−10.

[31] Diamandis EP, Christopoulos TK, editors. Immunoassay. San Diego, CA: Academic Press; 1996.

[32] Kricka LJ, Park JY. Principles of immunochemical techniques. In: Burtis CA, Ashwood ER, Bruns DE, editors. Tietz: textbook of clinical chemistry and molecular diagnostics. 5th ed. St Louis MO: Elseviers Saunders; 2012. p. 379−99.

[33] Immunoassays. In: Clarke W, Sokol LJ, Rai AJ, Clarke W, editors. Contemporary practice in clinical chemistry. 3rd ed. Washington, DC: AACC Press; 2016.

[34] Boscato LM, Stwart MC. Heterophilic antibodies: a problem for all immunoassays. Clin Chem 1988;34:27−33.

[35] Burtis CA, Ashwood ER, Bruns DE, editors. Tietz textbook of clinical chemistry and molecular diagnostics. 5th ed. St. Louis: Elsevier Saunders; 2012.

[36] Combs GF. Biotin: the vitamins: fundamental aspects in nutrition and health. San Diego, California: Academic Press, 1992, p. 350−65.

[37] Sedel F, Papeix C, Bellanger A, Touitou V, Lebrun-Frenay C, Galanaud D, et al. High doses of biotin in chronic progressive multiple sclerosis: a pilot study. Mult Scler Relat Disord 2015;4:159−69.

[38] Tourbah A, Lebrun-Frenay C, Edan G, Clanet M, Papeix C, Vukusic S, et al. MD1003 (high-dose biotin) for the treatment of progressive multiple sclerosis: a randomized, double-blind, placebo-controlled study. Mult Scler 2016;22:1719−31.

[39] Henry JG, Sobki S, Arafat N. Interference by biotin therapy on measurement of TSH and FT4 by enzyme immunoassay on Boehringer Mannheim ES700 analyzer. Annal Clin Biochem 1996;33:162−3.

[40] Kwok JS, Chan IH, Chan ML. Biotin interference on TSH and free thyroid hormone measurement. Pathology 2012;44:278−80.

[41] Wijeratne NG, Doery JC, Lu ZX. Positive and negative interference in immunoassays following biotin ingestion: a pharmacokinetic study. Pathology 2012;44:674−5.

[42] Barbesino G. Misdiagnosis of Graves' disease with apparent severe hyperthyroidism in a patient taking biotin megadoses. Thyroid 2016;26:860−3.

[43] Elston MS, Sehgal S, Du Toit S, Yarndley T, Conaglen JV. Factitious Graves' disease due to biotin immunoassay interference—a case and review of the literature. J Clin Endocrinol Metabol 2016;101:3251−5.

[44] Kummer S, Hermsen D, Distelmaier F. Biotin treatment mimicking Graves' disease. N Engl J Med 2016;375:704−6.

[45] Batista MC, Ferreira CE, Fauhaber AC, Hidal JT, Lottemberg SA, Mangueira CL. Biotin interference in immunoassays mimicking subclinical Graves' disease and hyperestrogenism: a case series. Clin Chem Lab Med 2016;55(6):e99−e103.

[46] Waghray A, Milas M, Nyalakonda K, Siperstein AE. Falsely low parathyroid hormone secondary to biotin interference: a case series. Endocrinol Pract 2013;19: 451−5.

[47] Piketty ML, Polak M, Flechtner I, Gonzales-Briceño L, Souberbielle JC. False biochemical diagnosis of hyperthyroidism in streptavidin-biotin-based immunoassays: the problem of biotin intake and related interferences. Clin Chem Lab Med 2016;55 (6):780−8.

[48] Katzman BM, Rosemark C, Hendrix BK, Block DR, Baumann NA. Investigation of biotin interference in common thyroid function test using the Roche Elecsys® immunoassay system. In: Abstract presented at the AACC Annual Meeting; 2016.

[49] Ayala AR, Ilias I, Nieman LK. The spectrum effect in the evaluation of Cushing syndrome. Curr Opin Endocrinol Diabet Obes 2003 Aug 1;10(4):272−6.

Thyroid disorders

2

Gregory Kline, MD[1] and Hossein Sadrzadeh, PhD[2]

[1]*Clinical Professor, Department of Medicine, Division of Endocrinology, University of Calgary, Calgary, Alberta, Canada*

[2]*Professor, Department of Pathology and Laboratory Medicine, and Graduate Studies, University of Calgary and Section Chief of Clinical Biochemistry, Calgary Laboratory Services, Calgary, Alberta, Canada*

CHAPTER OUTLINE

Endocrine Biomarkers. DOI: http://dx.doi.org/10.1016/B978-0-12-803412-5.00002-1

2.1 OVERVIEW

Second only to diabetes, thyroid diseases are the most common group of endocrine disorders seen in clinical practice. Broadly speaking, clinical thyroid diseases may be either disorders of thyroid function (hyper- or hypo-function) or disorders of thyroid structure (nodules, cancer, or goiter). Diseases of the latter category are predominantly managed through collaboration of endocrinology, radiology, and anatomical pathology/cytology although the chemistry laboratory still plays a role in thyroid cancer tumor markers (analyzing the thyroid markers in fine needle biopsy specimens). Therefore the dominant clinical chemistry role in thyroid diseases has to do with the functional disorders of thyroid hormone production. At the extremes of severity, thyroid functional disorders may be confidently recognized by the clinician; however, the vast majority of such patients present with symptoms or signs that may be vague or undifferentiated and thus the clinician relies heavily upon accurate biochemical testing for all facets of diagnosis and management.

2.1.1 CLINICAL EPIDEMIOLOGY OF THYROID FUNCTIONAL DISORDERS

- Congenital hypothyroidism affects approximately 1 in 4000 newborns [1].
- Overt hypothyroidism (thyroid failure) affects between 1 and 10 women per 1000, and 1 and 4 men per 1000 in developed nations [2].
- Depending on age and biochemical definition, subtle (biochemical) evidence of thyroid hypo-function—also known as "subclinical hypothyroidism"—may be present in anywhere from 4% to 20% of adults [3].

- Overt hyperthyroidism (thyroid excess) affects up to 3−5 women per 1000 and 1−2 men per 1000, in predominantly developed nations [2].
- Subtle biochemical evidence of thyroid hormone excess—also known as "subclinical hyperthyroidism"—is less common than subclinical hypothyroidism, estimated at <1−2% of the population [3].
- Although still controversial in some definitional aspects, thyroid dysfunction during pregnancy may be seen in up to 10% of women [4].
- In iodine-deficient parts of the world such as some areas in Africa or Latin America, functional thyroid diseases may be several times more common than what is reported earlier [5].

KEY POINTS

Functional thyroid diseases are extremely common. Patients and healthcare providers will request thyroid function tests very frequently and the test results often form the entire basis of any diagnosis. Even subtle abnormalities in thyroid function are now considered as potential "disease states," necessitating both accurate biochemical testing and understanding of differing population definitions of "thyroid disease."

2.1.2 BASIC THYROID PHYSIOLOGY

The thyroid gland is located in the front of the neck along the trachea and composed of two lobes that are connected by a thin tissue called isthmus that gives the gland the appearance of a butterfly. In adult humans the thyroid gland weighs almost 15−20 g and it is approximately 10 cm long and 2−2.5 cm thick.

The thyroid gland contains two types of cells: the most abundant being follicular cells that are responsible for the production of thyroid hormones (T4, T3) and parafollicular or C-cells that produce calcitonin. For a complete description of thyroid hormone synthesis the reader is referred to a recent review [6].

The production, release, action, and self-monitoring (homeostasis) of thyroid hormones in the body is a complicated, integrated system requiring the coordinated actions of the brain, thyroid gland in the anterior neck, carrier proteins in the circulation, and tissue-based enzymes. Working together, this system is exquisitely regulated and able to respond to even minor variations in dietary iodine supply, physiologic needs (pregnancy), ageing and disease processes.

Simply stated, thyroid hormone may be thought of as being similar to the "gas pedal" in an automobile. A varying amount is needed on a moment-to-moment basis in order to drive many of the cellular functions related to metabolism of fuels/metabolic rates, cardiac function, skeletal and smooth muscle function, and play a role in brain/mental function among other things as well.

For a functional understanding of thyroid physiology, it may be summarized as follows:

- Thyroid hormone generally circulates in two forms: denoted "T4" (tetra-iodothyronine) and "T3" (tri-iodothyronine), the numbers 3 and 4 related to the number of attached iodine residues to the thyroid hormone molecule.
- T4 is the major circulating form of thyroid hormone (approximately 90%).

- Both T4 and T3 in circulation are bound to proteins, mostly thyroid-binding-globulin (TBG). When bound to TBG, the thyroid hormones are essentially not biologically active.
- T4 and T3 that are not protein bound are called "freeT4" and "freeT3" (0.03% and 0.3% of total thyroid hormones, respectively). These forms of thyroid hormone are available to exert a biological effect upon target tissues and it is these two forms of the hormone that are "sensed" by the body for the purpose of regulating further thyroid hormone synthesis and secretion.
- Upon arrival at a target tissue (such as muscle cell), local tissue enzymes convert "freeT4" into "T3" within the cell. This intracellular T3 then interacts with receptors at the cell nucleus to regulate the production of proteins and functions that are thyroid responsive.
- The level of thyroid hormones is sensed by neurons in the brain hypothalamus which determines whether there is enough, too little, or too much and then responds accordingly.
- The hypothalamus produces a hormone called thyrotropin-releasing hormone (TRH) when more thyroid hormone is needed; this acts upon the anterior pituitary to cause it to release thyroid-stimulating hormone (TSH).
- TSH then acts upon the thyroid gland to increase the uptake of dietary iodine for increased synthesis of new thyroid hormone.
- The overall amount of thyroid hormone produced is very carefully controlled by the hypothalamus which either increases or decreases TRH depending upon the perceived need.
- This overall process is often called a "feedback loop."
- Knowledge of the basic physiology and feedback processes form the cornerstone of interpreting measurements TSH, freeT4, and freeT3 in clinical practice.

KEY POINTS

T4 and T3 are the major thyroid hormones. The brain (hypothalamus) controls all thyroid gland actions. If there is too much T4/T3, the brain responds by decreasing the stimulatory hormones TRH and TSH. If there is too little T4/T3 for the body, the brain increases TRH and TSH to try and boost thyroid hormone production.

2.1.3 OVERVIEW PERSPECTIVE: THYROID DISORDERS—DISEASE STATE OR STATISTICAL OUTLIER?

In many populations the reference intervals ("normal range") for a biochemical test is defined according to the range of numbers seen between the bottom 2.5% and upper 97.5% of the normal population. However, it becomes very problematic to define this "normal" reference range if the "disease" in question is so very common among the population that it actually ends up influencing what appears to be "normal" biochemistry. This problem is particularly evident in thyroid medicine.

Therefore the challenge for the clinician is to determine the answers to the following questions:

- Do all abnormal thyroid function tests imply a thyroid "disease" state?
- Is there a reliable correlation between abnormal thyroid function tests and patient symptoms?
- Are all abnormal thyroid function tests indicative of an underlying process that has adverse health consequences if left untreated?
- Do treatments that restore thyroid function tests to normal provide health benefits and safety to all affected patients?

For some of the above questions, there is actually evidence to suggest that in some scenarios, clinical interventions are not always useful, safe, or necessary [7]. A number of continuously updated guideline papers are available that may help the clinician to have the best current evidence on these important points [8,9]. It is important for the clinical chemist to be aware of such guidelines in order to help the clinician in interpretation of thyroid test results. For example a reporting laboratory may choose to attach an interpretive comment to a report of thyroid function tests directing the ordering physician to the relevant resources needed to inform clinical actions.

KEY POINTS

The high prevalence of thyroid disease in the population can produce many "abnormal" chemistry test results. However, the abnormal results may not necessarily indicate the presence of thyroid disease requiring intervention. Ideally, clinical decisions should be made with the *combination* of biochemical results, patient assessment, and published evidence that supports any proposed intervention. Laboratory results should be interpreted based on solid knowledge of strength and weakness of the biochemical tests. Communications between clinicians and clinical chemists are of great importance in accurate diagnosis and management of patients with thyroid disorders.

2.2 HYPOTHYROIDISM: CLINICAL PICTURE

With the general ease of diagnosis, relatively straightforward treatment options and overall commonality in the population, a low-functioning thyroid is very often considered as a diagnostic possibility in a wide variety of clinical settings. The most common patient symptom is nonspecific fatigue or low energy although not everybody with fatigue has low thyroid hormone.

In addition to fatigue, other symptoms, including heart failure, coma, kidney failure, and hypothermia, may be seen in these patients. Death may occur in the most severe cases. Milder cases may also include depression, constipation, anemia, menstrual irregularities, electrolyte disorders, and high cholesterol. Although popularly thought of as a cause of obesity, the relationship between thyroid hormone economy and excess body weight is complex, with uncertainty surrounding whether mild hypothyroidism causes weight gain or vice versa [10].

2.2.1 **CAUSES**

There are numerous reasons as to why a person might lose thyroid function in addition to the obvious scenario of posttotal thyroidectomy (usually performed to treat thyroid nodules, cancer, or large goiter, enlargement of thyroid gland). In general, thyroid auto-immunity represents the vast majority of cases with hypothyroidism. Also, both low and high intake of dietary iodine may cause hypothyroidism along with many other rare disorders such as direct thyroid radiation and sometimes primary iron excess (hemochromatosis).

Autoimmune (hypo) thyroid disease is an often confusing concept since the presence of thyroid auto-immunity (positive serum auto-antibodies) may be present without any detectable changes in thyroid hormone levels. Thyroid biopsy specimens may also show evidence of lymphocytic infiltration and thus be called "autoimmune thyroiditis," without necessarily having any changes in T3, T4, or TSH. While the presence of tissue or auto-antibody evidence of "auto-immune thyroid disease" may predict the eventual development of endocrine thyroid dysfunction [11], the term "autoimmune hypothyroidism" should be reserved for cases showing both positive auto-antibodies and low-thyroid hormone production (usually detected as high serum TSH levels). The popular term "Hashimoto's Thyroiditis" is generally avoided by most current endocrinologists due to its lack of diagnostic clarity on the issue of actual thyroid dysfunction versus simple thyroid enlargement or antibody positivity.

In many other parts of the world, iodine deficiency remains the commonest cause of hypothyroidism. An updated list of global iodine sufficiency may be viewed at: http://www.who.int/vmnis/iodine/en.

The role of severe iodine deficiency must also be remembered when testing or treating people who are recently arrived from an iodine-deficient part of the globe [12].

Autoimmune thyroid diseases may be functionally categorized according to whether they induce a slowly progressive and usually permanent state of hypothyroidism versus a transient inflammatory condition that may produce temporary hyperthyroidism, hypothyroidism, or both (usually in that same sequence). The diagnostic terms commonly used are thus:

- *Primary autoimmune (Hashimoto's) hypothyroidism*—usually an insidious process characterized by high titer of antithyroid antibodies, lymphocytic infiltration of the thyroid glands with diffuse inflammatory reaction, goiter formation, and subsequent permanent loss of thyroid hormone production. Commonly affects women and may not always have a goiter, especially in the elderly.

- *Acute lymphocytic (silent) thyroiditis*—originally recognized as a common occurrence following pregnancy, now known to be a common cause of usually transient hypothyroidism. May be preceded by transient hyperthyroidism (due to destruction of many follicular cells at the same time releasing their hormone contents into circulation) and some cases may be left with permanent hypothyroidism especially those with high titer antithyroperoxidase antibodies [13]. This form may be a variant of the classic Hashimoto's thyroiditis but is typically a painless process which differentiates it from the other acute thyroiditis outlined later.
- *Acute (deQuervain's) thyroiditis*—thought to be an inflammatory condition precipitated by an upper respiratory tract viral infection and characterized by intense neck pain and fever. Like its painless cousin, this often presents initially as hyperthyroidism but may then pass through a transient hypothyroid phase of up to 10 weeks duration.

KEY POINTS

Depending on the global origins of the patient, the most common causes of primary hypothyroidism are either iodine deficiency or thyroid autoimmune–mediated injury. Autoimmune thyroid diseases may present with an initial hyperthyroid phase and then pass through a more prolonged hypothyroid phase that can be permanent or self-limited. Therefore clinicians should not assume that all biochemical hypothyroidism will be permanent or require lifelong thyroid hormone replacement.

2.2.2 SPECIAL SITUATIONS: PEDIATRICS, PREGNANCY, AND CRITICAL ILLNESS

2.2.2.1 Congenital hypothyroidism

While most hypothyroidism cases are relatively straightforward, there are some special situations that necessitate mentioning. Pediatric patients can certainly be affected by any of the above listed thyroid problems but congenital hypothyroidism must be added to this list. An embryonic defect in thyroid development or inherited abnormality in thyroid hormone production is the usual factor in neonatal hypothyroidism. Failure to detect and treat congenital hypothyroidism is linked to long term, severe intellectual disability and other adverse health outcomes that may persist throughout life. With the relative ease of diagnosis and treatment and potential devastating effects of missed diagnosis, population-wide neonatal screening programs are now in place in many countries and should ideally be implemented in every country around the world.

2.2.2.2 Gestational thyroid changes

Pregnancy represents a unique and challenging aspect of all types of functional thyroid disorders. Thyroid dysfunction is often detected for the first time in pregnancy or may develop on account of pregnancy itself. All of the usual causes of both hyper and hypothyroidism may present during pregnancy but the biochemical diagnosis (especially of the asymptomatic patient) may be substantially more difficult for the following reasons:

- The fetal thyroid does not develop until approximately 12 weeks gestation, and thus maternal thyroid hormone demands are increased during that time, in order to supply mother and fetus. In theory, this increased "demand" may expose previously compensated mild maternal thyroid dysfunction, resulting in the appearance of mild biochemical hypothyroidism (slightly high TSH) during early gestation in affected mothers.
- Normal pregnancy—induced rises in TBG and other thyroid hormone-binding protein production can significantly affect (total) T4 and T3 concentrations producing elevated total T4 and total T3 levels in circulation.
- During the first trimester in normal pregnancy the pregnancy hormone chorionic gonadotropin (HCG) rises rapidly. It has a structural similarity with normal TSH (shared alpha chain) and may thus stimulate both thyroid gland growth and hormone production, resulting in a decrease in actual TSH levels and mild elevations in T4 (both freeT4 and total T4). Uncommonly the rise in T4 production may even be sufficient to cause a mild "gestational transient thyrotoxicosis." During the second trimester, FT4 decreases by 20%−40% and TSH rises slightly in parallel with the decrease in HCG production. These changes in thyroid hormone economy have several implications for the clinical chemistry laboratory:
  - The TSH reference range in a normal pregnancy without thyroid disease is probably different from that of a nonpregnant population.
  - The TSH reference range in a normal pregnancy likely varies slightly according to the gestational trimester [14].
- FreeT4 measurements (previously thought to solve the problems of high TBG-T4 levels) likely still differ from a nonpregnant reference range and there is a controversy as to whether freeT4 or "adjusted" total T4 levels may become the preferred measurement during pregnancy [15,16].
- Ideally, each laboratory should establish its own, locally validated reference range for all thyroid measures across different trimesters of gestation. The reference population should be iodine sufficient [17].

Adding to the difficulty of the situation, it is now commonly held that the clinical implications of hypothyroidism in pregnancy are more acute, with fears of long lasting negative outcomes upon babies of hypothyroid mothers not treated early and fully during gestation. Although still controversial, there is some evidence that even mild degrees of thyroid dysfunction, untreated in pregnancy, may impair the child's future cognitive abilities [18]. This had lead to intense discussions among endocrinologists, obstetricians, and pediatricians regarding the role of routine screening of pregnant women to detect any and all thyroid dysfunction early in pregnancy, along with debate as to the role of thyroid hormone replacement according to "newer" pregnancy-specific definitions of thyroid dysfunction.

Nonetheless a randomized trial of screening for thyroid dysfunction in early pregnancy failed to demonstrate any benefit in neurocognitive function of offspring by 3 years of age [19] and so the value of routine TSH screening in early pregnancy is still unclear.

2.2.2.3 Critical illness

The vast majority of thyroid diseases are detected in the outpatient setting. However, hypothyroidism may also be first suspected during an inpatient stay; in some more severe cases, untreated hypothyroidism may be the cause of the problem that led to the hospital admission. Therefore thyroid function tests are commonly measured in hospital patients. However, T4, T3, or TSH may be affected due to an acute nonthyroidal illness, presenting a biochemical pattern that may at first seem to reflect hyperthyroidism (with isolated low TSH) or even low TSH-low freeT4 hypothyroidism. Simply stated, "Sick Euthyroid Syndrome" or "Non-Thyroidal Illness" is used to refer to conditions (acute illness) in which thyroid markers are abnormal in the absence of a thyroid disease. A full review of the pathophysiologic changes that underlie the changes in thyroid hormone measures is beyond the scope of this volume but several detailed reviews exist [20]. Careful clinical assessment is needed to differentiate true thyroid disease in a sick patient versus nonthyroidal illness.

KEY POINTS

Congenital hypothyroidism can have devastating effect on the fetus, if not diagnosed and treated thus neonatal screening is an important public health measure. Pregnancy-related changes in thyroid hormone measurements can complicate the diagnosis of hypothyroidism; screening and early intervention are being increasingly considered in this population but will require careful case definitions. If pregnant women undergo thyroid hormone testing, it should be interpreted using the laboratory's own established trimester-specific reference intervals. Critical illness can cause changes to thyroid function tests that are physiologic but may appear similar to either hypo or hyperthyroidism.

2.2.3 RARE SITUATIONS: PITUITARY DISEASE, MEDICATIONS, AND TUMORS

2.2.3.1 Pituitary disease

More than 99% of thyroid diseases involve a primary dysfunction of the actual thyroid gland. Thus the diagnostic approach to thyroid disease will usually be very uniform and largely directed by accurate measurement of the TSH level which will indicate whether there is too much or too little thyroid hormone. However, in the event of pituitary disease such as a tumor, infarct, head trauma or following cranial irradiation, the thyroid gland itself may be anatomically and functionally normal and yet fail to produce thyroid hormone if TSH is not being secreted by the pituitary. This condition is called "secondary hypothyroidism" to indicate that the problem is not the thyroid, but the control system (pituitary/hypothalamus). The clinical presentation may be similar to other forms of hypothyroidism; however, the hallmark high level of TSH (as seen in primary thyroid disease) is absent. This represents one of the few situations where TSH measurements can be clinically misleading, necessitating measurement of freeT4 to see the diagnosis as the freeT4 will be low despite a low or at least nonelevated TSH.

2.2.3.2 Medications

In the area of hypothyroidism the most important medication known to cause low thyroid function is amiodarone. The interaction of amiodarone and the thyroid is very complex and described in more detail in Section 2.2 but from a potential hypothyroid standpoint, it is the high iodine content of amiodarone that is thought to inhibit thyroid function [21]. Classical endocrinology teaches a mechanism known as the Wolff-Chikoff effect which basically describes the observation that a large iodine load is often sufficient to inhibit its own uptake by the thyroid gland and further inhibit the peripheral conversion of T4 to T3, resulting in a net decrease in thyroid hormone production and action. Other sources of high levels of iodine are known to produce a similar effect, such as may be seen in radiographic contrast agents or some complementary medicine products and nutritional additives [22]. While iodine-induced hypothyroidism may require thyroid hormone replacement, the thyroid dysfunction may be temporary, until the body's iodine pool has been depleted (through renal iodine excretion) enough to allow the thyroid function to return to normal.

Lithium is a second, common medication that may cause hypothyroidism through inhibition of thyroid hormone secretion. Tyrosine kinase inhibitors such as sunitinib (as used in some oncology treatments) are increasingly recognized as causing hypothyroidism by increasing thyroid hormone catabolism [23]. For those patients who are taking oral thyroid hormone replacement, there is a lengthy list of medications and supplements including iron and calcium supplements that may impair absorption causing biochemical hypothyroidism, see Table 2.1.

Heparin is another important drug that can increase free thyroid hormone levels. Heparin can induce lipoprotein lipase activity which can result in release of free fatty acids that can replace T4 and T3 from their binding proteins causing falsely elevated FT4 and T3. Even when administered subcutaneously, heparin (enoxaparin-2000) can significantly increase FT4 and FT3. Therefore it is suggested that specimens should be collected at least 10 h after the last dose of heparin (enoxaparin) injection and analyzed within 24 h following collection [24]. Other medications such as glucocorticoids, phenytoin, sulfonylureas, propranolol, diazepam, furosemide, cimetidine, and salicylates can affect FT4 and FT3 as shown in Table 2.1.

Table 2.1 Drug Effects on Oral Thyroxine Absorption and Measures of Thyroid Function

Mechanism	Common Drug Causes	Biochemical/Biological Effect
Inhibited absorption of orally administered thyroid hormones	Iron supplements, calcium supplements, cholestyramine, sucralfate, resonium calcium, sevelamer hydrochloride	High TSH, low freeT4
Inhibition of thyroid hormone synthesis or release from gland	Lithium	Increase in TSH, decrease in freeT4, freeT3
Inhibition of peripheral production of T3 by inhibiting T4 to T3 conversion	Iodine, amiodarone, propranolol, prednisone	Decreased freeT3, possible increased TSH
Inhibit T4 and T3 binding to transport proteins	Salicylates, sulfonylureas, phenytoin, furosemide	Increased freeT4 and freeT3
Increase hepatic metabolism of T4	Phenobarbital, rifampin, phenytoin, carbemazepine, sertraline	Increased thyroid hormone requirement for patients with hypothyroidism
Increase thyroid-binding globulin	Estrogen, tamoxifen, methadone, fluorouracil, mitotane	Increased thyroid hormone requirement for patients with hypothyroidism

2.2.3.3 "Consumptive" hypothyroidism

Mentioned only for completeness, there are rare reports of hepatic tumors, usually in the pediatric literature wherein an over-expression of the enzyme that degrades thyroid hormone causes a massively increased production demand upon the thyroid which is unable to be matched, resulting in hypothyroidism that can be very difficult to treat even with high-dose thyroid hormone replacement [25]. These patients present with high TSH and low freeT4, often resistant to normalization with thyroid hormone replacement.

KEY POINTS

Both low and high iodine can result in hypothyroidism. Large iodine loads from either prescription or alternative medications can temporarily impair thyroid function. Thyroid dysfunction due to a brain/pituitary injury may not be diagnosed by the usual biochemical approach to hypothyroidism investigation and should thus be considered in the rare case of clinical hypothyroidism with normal TSH levels. Several medications may affect freeT4 and freeT3 levels, requiring care in clinical interpretation.

2.2.4 DIAGNOSTIC APPROACH

2.2.4.1 Thyroid-stimulating hormone is always the first step

In adults, thyroid disease is so common that many primary care physicians routine screen for thyroid disorders in a variety of clinical presentations although population-wide screening is not yet endorsed. Since almost all hypothyroidism is due to a primary disorder of the thyroid gland, even the most subtle reductions in thyroid output may be detected by measurement of a serum TSH [8]. If the TSH is normal, primary hypothyroidism can be confidently excluded.

Guidelines on hypothyroidism diagnosis exist [9] but are largely empiric in nature, based upon the most common outpatient presentations of primary thyroid dysfunction. In our opinion, in the nonpregnant adult, if the TSH is markedly elevated (i.e., >10 mIU/L), this is usually indicative of severe thyroid dysfunction and may well correlate with a classical symptomatic presentation. Repeat TSH and/or measurement of a freeT4 may be considered but are unlikely to change the diagnosis or treatment approach when the first measurement is grossly abnormal. A clinical history and physical examination may help to discern whether the hypothyroidism may be part of a transient postthyroiditis phase and if such a diagnosis is considered based on suspicion of recent mild hyperthyroidism, repeat TSH testing after withdrawal of thyroid replacement 3−4 months later may help clarify the permanence of the thyroid failure.

In pregnant women suspected of hypothyroidism, we recommend that appropriate reference ranges be established for first, second, and third trimesters, using an appropriate patient population unique to each laboratory. This is another example in which a direct interaction between the clinical endocrinologists and clinical chemists is necessary for the best and evidence-based use of clinical laboratory and management of the patient.

There have been various opinions published regarding the role of adult population screening TSH measurements outside of pregnancy. Thus far there are no data to support routine measurements of TSH in healthy adults although some patient's subgroups such as those with other autoimmune diseases or diabetes may warrant thyroid disease screening [26].

2.2.4.2 Sometimes freeT4 is needed

Again, in the nonpregnant adult, if the TSH is modestly elevated (i.e., between 4 and 10 mIU/L), a repeat measurement of TSH with freeT4 may help clarify the overall severity of the hypothyroidism. This is particularly true in the elderly whose slightly elevated TSH rise may be rather unimpressive despite overtly low freeT4. If the freeT4 is overtly low, a formal diagnosis of hypothyroidism is made. However, if the freeT4 is still within the normal range, such patients are often classified as "subclinical" hypothyroidism (see further discussion later). The general idea is that these patients have very subtle defects in thyroid hormone synthesis, formerly called "compensated hypothyroidism" and debate continues as to whether such patients have clinical consequences or should routinely be treated with thyroid hormone [27,28].

For patients with a history of pituitary tumors, pituitary surgery, or cranial irradiation, pituitary/hypothalamic injury may lead to low thyroid hormone levels through insufficient TRH/TSH secretion. Although uncommon, it is possible that prior traumatic brain injury patients may also develop low thyroid function due to pituitary injury. In such cases the measured TSH may appear normal but may still be insufficient in amount or biologically inactive and thus cannot be relied upon for diagnosis. Therefore in cases where prior brain/pituitary injury is known or suspected, investigation of possible hypothyroidism must include measurement of freeT4—which will be low (with a normal or slightly low TSH) in true "secondary (pituitary)" hypothyroidism. Some endocrinologists have argued that TSH and freeT4 should be reflexively measured in order to avoid missing the rare case of secondary/pituitary-related thyroid failure but this approach is highly unlikely to be cost effective given the rarity of this diagnosis in contrast to the frequency with which TSH is measured in routine outpatient care.

2.2.4.3 Thyroid auto-antibodies

The role of thyroid auto-antibodies is supportive at best in diagnosis of hypothyroidism. Autoimmune thyroid disease is the most common cause of primary thyroid failure and most other causes may be excluded by history, medication list, and physical examination. Thyroid auto-antibodies are not inherently indicative of actual thyroid dysfunction and such antibodies have a relatively high background population prevalence [28]. A high titer may be occasionally useful if there is true doubt as to the autoimmune nature of the etiology but most endocrinologists do not measure such antibodies routinely. For patients with a transient thyroiditis picture, high titer thyroid antibodies may help to predict which patients will go on to have a permanent need for thyroid hormone replacement.

Thyroid auto-antibodies include:

- Thyroid peroxidase antibodies (TPO)—most closely associated with autoimmune biochemical hypothyroidism [29].
- Antithyroglobulin antibodies (anti-TGs)—may be present in autoimmune thyroid disease but not reliably associated with biochemical hypothyroidism [29]. These antibodies are more clinically relevant with respect to their interference with serum thyroglobulin (Tg) measures during thyroid cancer treatment (see Section 2.4).
- TSH receptor antibodies (anti-TSH R)—nonspecific antibodies that may act as blocking antibodies and block TSH receptors (causing hypothyroidism) or stimulatory antibodies that binds to TSH receptors and acts like TSH and (causing hyperthyroidism).
- Thyroid-stimulating immunoglobulin—as implied by the name, this antibody is part of the pathogenesis of hyperthyroidism (see appropriate section later).

As a general rule, thyroid imaging does not play a part in the routine investigation of thyroid failure. Ultrasonography and nuclear scintigraphy do not assist in the diagnosis of most adult hypothyroidism unless there are concomitant thyroid nodules or a question of hyperthyroidism (which is easily excluded by finding a high TSH level). As mentioned, primary thyroid failure is largely a clinical and biochemical diagnosis; ultrasonographic and nuclear studies of the thyroid do not confirm hypothyroidism or its etiology.

2.2.4.4 Pediatric considerations/newborn screening

It is now routinely recommended that newborns be screened for hypothyroidism between 2 and 4 days old and certainly before the end of the second week of life [30]. The most common screening test remains the serum TSH although some experts have suggested concomitant measurement of freeT4 to ensure that the rare case of congenital hypopituitarism causing low thyroid is not missed. An aggressive screening policy has ensured that no severe cases of congenital hypothyroidism are missed but has also demonstrated a marked increase in the detection of mild and/or transient hypothyroidism among neonates. There are a number of both congenital and acquired autoimmune conditions (such as Down syndrome, Turner syndrome, and type-1 diabetes mellitus), in which thyroid dysfunction (particularly hypothyroidism) is very common and for whom routine TSH testing is considered as part of regular medical follow up.

2.2.4.5 Pregnancy

As outlined earlier, a normal pregnancy induces changes in thyroid hormone levels that must be recognized for appropriate interpretation and diagnosis. It is now well established that TSH levels drop to levels lower than the nonpregnant reference range during the first trimester and then rise ever so slightly during trimesters 2 and 3. As such, there are now recommendations that TSH measures in pregnancy be interpreted according to a pregnancy-specific range established by the laboratory for their own patient population.

Owing to marked changes in thyroid-binding proteins, it is also now known that the levels of freeT4 may actually be in the lower half or even overtly "low" during pregnancy when standard references ranges are used [31]. This may be difficult to interpret, especially as TSH levels are also lower. However, it is also fairly well established that total T4 levels rise during normal pregnancy and may also require a validated, trimester-specific reference range to be developed. At the present time a widely accepted compromise is to use total T4 measures in pregnancy and interpret using reference limits that are $1.5 \times$ the normal nonpregnant range [15,31]. However, not all pregnancy-thyroid guidelines recommend this [32].

Although based on limited data, there is growing use of thyroid auto-antibody measurement during pregnancy or in women planning to become pregnant with a history of recurrent miscarriage, even in the absence of functional thyroid disease. It has been shown that a high titer of thyroperoxidase antibodies appears to predict an increased rate of miscarriage and preterm birth although a causative relationship is not yet established [33]. Nonetheless a single clinical trial endorses thyroid antibody measurement and even thyroid hormone replacement in otherwise euthyroid women with high titer who are or who wish to become pregnant [34].

KEY POINTS

If hypothyroidism is suspected, measure TSH and if it is markedly elevated (more than twice normal), the diagnosis is secured. However, if TSH is only slightly elevated consider a repeat measurement in 3—4 months' time with measurement of freeT4 to confirm the diagnosis. Thyroid antibodies usually play a minor role in confirming a coexistent autoimmune thyroid process although are increasingly viewed as a possible indication to offer thyroid hormone to pregnant women. Routine measurement of thyroid antibodies in pregnancy is not yet recommended.

2.2.5 DIAGNOSTIC DEFINITIONS

The standardization of TSH assays around the world would be relevant to all laboratories reporting thyroid indices such that a TSH measured in one lab should be comparable to a TSH measured in the same patient at a different lab. However, with global standardization comes a debate as to a globally applicable reference range.

Prior attempts to define a normal population TSH reference range may have been hampered by inclusion of subjects with subtle thyroid abnormalities such as slight goiter or positive thyroid antibodies. As such, it is now thought that older studies probably over-estimated the upper limit of the 95% confidence interval for TSH measures. The largest and most carefully performed population studies now suggest that the TSH upper limit probably lies between 3.0 and 3.7 mIU/L [35]. Although this might seem to be of minor overall importance in a healthcare system, there are enormous implications for diagnosis, treatment, and subsequent health resource utilization.

So, what is the best way to interpret a marginally elevated TSH? The answer is with caution.

As emphasized in recent guidelines statements [8], it is entirely unclear as to whether a marginally elevated TSH measurement necessitates a formal diagnosis or (usually) lifelong treatment of "hypothyroidism." Marginally elevated TSH measures have been shown to frequently spontaneously normalize with simple observation over time [36] and normal ageing itself is likely associated with a physiologic rise in TSH measures [37] that may actually predict longevity without treatment [38].

This subtlety in diagnosis may pose a problem for less experienced or excessively burdened clinicians who may see TSH measures as a dichotomous "disease/no disease" defining test. The result may be that many patients could inappropriately receive long-term thyroid hormone "replacement"—a situation known to result in over-replacement/hyperthyroidism in up to 10% of treated persons [39].

In accordance with evidence-based treatment guidelines the clinical chemistry laboratory may play an important role in both patient and physician education in this area. Whether through interpretive comments or careful reporting of reference ranges, the diagnostic use of thyroid measures may be improved. For TSH measures >10 mIU/L a diagnosis of "hypothyroidism" is usually appropriate and treatment offered without controversy. For TSH measures between the upper assay limit (2.5–4.5 mIU/L) and <10 mIU/L a freeT4 should be ordered and, if normal, the most appropriate diagnostic term will be "subclinical hypothyroidism." This particular term is encouraged as a means of alerting the ordering physician that standard "hypothyroidism" measures may not apply. Some laboratories may also choose to include comments suggesting repeat measures in a few months and/or consultation with local thyroid guidelines or endocrinologists to get the most up-to-date thinking as to whether and when such a finding should be treated.

Awareness of literature showing both age- and ethnicity-specific effects upon the TSH measures may add a level of increased sophistication to interpretation that may help with clinical decision making for such borderline cases [40].

2.2.5.1 Secondary hypothyroidism (pituitary disease)

As explained earlier, patients with pituitary disease may lose thyroid function via loss of TRH or TSH. TRH measurements are not performed in clinical medicine outside of research settings. TSH measures may be misleading in pituitary patients and thus the diagnosis of hypothyroidism depends upon demonstration of a low total or freeT4. Generally speaking a frankly low freeT4 level with normal TSH should raise the suspicion of central/secondary hypothyroidism although it is recognized this approach may miss more subtle cases.

2.2.5.2 Diagnosis in pregnancy

While awaiting global standardization of TSH and T4 interpretation in pregnant patients, current recommendations have been to use the following:

- first trimester TSH upper limit of normal 2.5 mIU/L
- second trimester TSH upper limit of normal 3.0 mIU/L
- third trimester TSH upper limit of normal 3.5 mIU/L
- total T4 of 150% the usual reference range in all trimesters
- freeT4 and freeT3 may be considered in place of normalized total T4 and total T3. Each laboratory should establish its own reference intervals based on its local population.

However, it is less certain as to whether all patients with values slightly exceeding these cut-offs should be treated or monitored. Further prospective studies will be needed to clarify the optimal overall approach to diagnosis in this special case. With the current lack of data proving value to intervention for such subtle biochemical abnormalities, some guidelines specifically recommend against TSH screening in pregnancy [32].

KEY POINTS

Hypothyroidism is reliably diagnosed when the TSH is >10 mIU/L, especially with a concomitant low freeT4. For TSH values between 4.5 and 10 mIU/L (or perhaps 2.5 and 10) with normal freeT4, the most appropriate diagnosis is "subclinical hypothyroidism." Pituitary disease patients require a low freeT4 for diagnosis and pregnant women may be better served using a pregnancy trimester-specific TSH and total T4 reference range.

2.2.6 THE ROLE OF THE CHEMISTRY LABORATORY IN TREATMENT/FOLLOW UP OF HYPOTHYROIDISM

Once a diagnosis of thyroid hormone deficiency has been made and treatment begun with thyroid replacement, the clinician will depend heavily upon the clinical chemistry laboratory to assure adequacy of therapy and provide long-term monitoring. Repeated measurements of both TSH and sometimes freeT4 will be the cornerstone tests. Precision (reproducibility) of the methods that are used to follow therapy is of great importance for successful treatment of patients with thyroid diseases. Therefore we recommend third generation assays for measuring TSH and a freeT4 and T3 methods with precision (%coefficient variation) of less than 5%.

For appropriate use of follow up testing, it is critical that the clinician understands the physiology of the hormones, the rates at which they change, and the extent to which medically administered thyroid hormone affects such measures. For example, TSH may change quickly or very slowly but usually takes weeks to achieve a new steady state after a dosing change. On the other hand, freeT4 has a much shorter half-life and may change within hours-to-days of starting thyroid hormone therapy.

General principles that must be kept in mind with respect to thyroid replacement:

- TSH may change rapidly in the setting of gross over-replacement (i.e., may quickly become suppressed/subnormal.
- TSH may rise rapidly to very high levels in hypothyroid patients who completely omit their replacement medicine for 1 or 2 weeks.
- Even following full thyroid replacement with patients' full adherence, it may take more than 8 weeks for a very high initial TSH to return to normal.
- Patients receiving chronic thyroid hormone replacement may take 6−8 weeks to achieve a stable TSH that reflects the overall effects of a minor dose change.

- In stable patients receiving chronic thyroid hormone replacement with previously normal TSH, it is possible to see a rise in TSH after missing as few as 2 days' doses in the 2 weeks preceding a TSH measurement.
- In a more urgent clinical situation where appropriate thyroid replacement dosing must be assured sooner than a 6-week wait period, it may be useful to measure freeT4 within days of starting or changing a thyroid replacement pill; normal or upper normal freeT4 should confirm the appropriateness of the current dose and the likely normalization of TSH.
- FreeT4 may also be useful for monitoring thyroid function during times of rapid change in thyroid status such as following radioiodine treatment of hyperthyroidism.
- Patients with pituitary disease receiving thyroid hormone must be monitored by freeT4 not TSH.

> **KEY POINTS**
>
> All patients on chronic thyroid hormone replacement therapy should be monitored by TSH every 6 weeks during any dose titration phases. Measurement of freeT4 may be useful in situations where TSH measures cannot be used, particularly in those with pituitary disease causing hypothyroidism.

2.2.7 OVERVIEW OF COMMON MISUSE OF LABORATORY TESTS IN HYPOTHYROIDISM DIAGNOSIS AND TREATMENT

2.2.7.1 Failure to consider watchful waiting in those with marginal thyroid-stimulating hormone elevations

As outlined earlier, marginal TSH elevations may be appropriate for age [38], do not usually correlate with clinical symptoms, and may revert to normal over time [36]. In the setting of a postthyroiditis, mild hypothyroidism is expected and often spontaneously resolves. Thus long-term thyroid hormone replacement should generally not be initiated on the basis of a single, slightly high TSH measure.

2.2.7.2 Failure to recognize when the thyroid-stimulating hormone and freeT4 results together do not make sense

The cornerstone of thyroid diagnostics rests in the understanding of normal hypothalamus—pituitary—thyroid hormone relationships and integrated feedback loop controls. Thus the TSH and freeT4 results are entirely predictable in virtually all common disorders of both high and low thyroid function. On occasion however, one may find a pair of TSH/freeT4 results that do not immediately make sense. The most common setting would be a low TSH and a low freeT4, which is frequently seen in any patient with serious medical illness but should also lead one to question possible pituitary disease—related thyroid failure. A less common pairing is the finding of high TSH and high freeT4, which may be seen in the very rare condition of congenital cellular thyroid hormone resistance or the even rarer TSH-secreting pituitary tumor. Both conditions are beyond the scope of this text but represent very important diagnoses in their own right thus should not be casually attributed to "laboratory error." Rather, if the TSH and freeT4 results seem physiologically discordant, endocrinologist consultation is strongly recommended.

2.2.7.3 Measurement of thyroid-stimulating hormone too frequently

Once hypothyroidism is diagnosed and treated, it usually takes several weeks for the TSH to normalize even with a perfect choice of thyroid replacement dose. Measurement of TSH sooner than 4−6 weeks as an absolute minimum can cause much confusion since it will often remain above normal which could lead the clinician to erroneously conclude that a dose increase is needed. The consequence is that an inappropriate dose increase will later be found to be excessive and causative of a suppressed TSH measurement. It is the author's practice not to measure TSH at intervals of <6 weeks following any dose change and no <8 weeks if the dose change has been quite small.

2.2.7.4 Serial measurement of thyroid antibodies

The presence of thyroid antibodies in a euthyroid, nonpregnant person is of very limited significance except perhaps in predicting future development of hypothyroidism. However, thyroid antibodies are common in the population and a change in titer has not been shown to be of any diagnostic or prognostic significance and therefore cannot be used to make any meaningful clinical decisions.

2.2.7.5 Confusion as to the role of thyroid antibodies in diagnosis

Outside of pregnancy or pregnancy planning, there is no proven rationale for offering any kind of treatment to patients with positive thyroid antibodies. Therefore in the patient presenting with possible symptoms of hypothyroidism, the only meaningful test is TSH. If the TSH is normal, a positive thyroid antibody does not provide any diagnostic options to explain the presenting symptoms.

2.2.7.6 Use of thyroid-stimulating hormone in patients with pituitary problems

As described earlier, TSH measures will be low, normal, or just slightly high even with severe hypothyroidism when the cause is due to pituitary injury. Therefore where a high clinical suspicion for hypothyroidism exists, a freeT4 should be measured even if the TSH is normal, in order to accurately make the diagnosis. Similarly, patients with pituitary failure or absence often have a low TSH and this is not an indication for them to have a dose reduction in their thyroid hormone replacement; only freeT4 can be used to gauge adequacy of thyroid replacement in such patients.

2.2.7.7 Measurement of freeT3 or total T3 in hypothyroid diagnosis or therapy monitoring

Readers will note that freeT3 and total T3 have not been mentioned thus far in the discussion of thyroid failure. This is intentional and emphasizes the primacy of TSH and freeT4 measures since freeT4/total T4 represents the vast proportion of circulating thyroid hormone. Patients with true hypothyroidism may have either low or normal freeT3/total T3 and thus its measurement contributes nothing to the diagnosis beyond what can be learned from TSH and freeT4 alone.

Thyroid hormone replacement therapy using compounds containing T3 is increasingly popular among many patients and some endocrinologists. However, the half-life of administered T3 preparations is very brief (often necessitating bid or tid dosing) and thus measurement of serum T3 is of little to no value as it will simply reflect the time since the last dose was taken. Current guidelines recommend TSH measures as the monitoring test of choice when T3-containing compounds are used [8] since it will reflect the overall exposure to thyroid hormone(s) over time.

2.2.7.8 Failure to recognize thyroid hormone over-replacement

Current evidence suggests that a large number of patients who receive thyroid hormone end up taking a dose that is supraphysiologic [39] and this is not without potential important adverse health consequences. Outside of thyroid cancer treatment, a suppressed TSH should not be ignored in a patient using chronic thyroid replacement.

2.2.7.9 Failure to recognize nonadherence to thyroid hormone replacement therapy

Periodic or annual TSH measurements in persons taking chronic thyroid replacement are considered standard care. It is very common to see a slight TSH elevation in such patients even after many years of stable dosing and normal TSH measures. In almost all such cases the "outlier" TSH represents some degree of missed medication dosing and careful patient questioning and examining of prescription refills is usually all that is required to establish the facts. In some cases, however, patients may not disclose their nonadherence which may be even further suspected if the freeT4 is upper normal (having just recently taken a dose) just with a slightly high TSH. In the author's opinion, in these cases, no dose changes should be made but rather, the TSH may be repeated in 8 weeks time after a discussion of how nonadherence affects biochemical assessment of therapy.

2.2.7.10 Failure to consider other medical reasons for an apparent need to increase thyroid replacement dosing

If nonadherence to therapy can be confidently excluded (sometimes requiring several days of direct observed therapy to be sure), the clinician should give consideration to concomitant medications impairing thyroid hormone absorption, the effect of significant weight change on dose requirements, possible pregnancy or malabsorption diseases like celiac sprue. Only after such diagnoses are considered and excluded should a previously stable thyroid replacement dose be changed. Following these general principles are usually effective at avoiding endless dose changes with the apparent TSH instability.

2.2.7.11 Failure to consider assay problems/heterophilic antibody interference in thyroid tests that do not match the clinical setting

Many ordering clinicians are not aware of the factors that can affect the specimen, methods, or other aspects of the thyroid testing producing erroneous results. Since the patient presentation may vary from asymptomatic to classical symptoms or atypical symptoms, it remains the clinician's responsibility to ultimately interpret the laboratory tests in light of the clinical scenario, which then may or not make sense. Unfortunately, many clinicians have trained in an environment where unexpected laboratory test results are simply dismissed as "laboratory error" without further thought or confirmation. However, when the patient actually has symptoms of importance, "laboratory error" is hardly an appropriate diagnosis. This underlines the critical importance of a clinical chemistry laboratory's relationship to the clinical community: knowing the clinical scenario, a chemist may help select a more appropriate test and knowing the clinical chemist, a clinician may know where to turn for help when the tests do not seem to fit what was expected.

2.2.8 LABORATORY CONSIDERATIONS TO ASSESS THYROID DISORDERS

As indicated before, a good understanding of normal physiology and pathophysiology of thyroid is a prerequisite for proper selection of the tests and accurate interpretation of the results. There is a sensitive balance between TSH secretion from pituitary and blood concentrations of thyroid hormones. In this reciprocal, sensitive and log-linear relationship between TSH and T4 and T3, a twofold change in freeT4 can result in 100-fold change in plasma TSH [41].

In previous sections, we briefly mentioned factors that can impact the laboratory tests and produce falsely elevated or reduced results. These factors can be divided into three major categories [42]

- preanalytical errors that affect the binding of T4 and T3 to different serum proteins
- genetic (proteins have less affinity to bind T4 and T3)
- medications (that can displace thyroid hormones from proteins or impact deiodination of T4 to T3. Please see Table 2.1 for details).
- diseases that can increase or decrease thyroid binding to TBG or albumin
- pregnancy that affects thyroid hormone levels due to multiple mechanisms.

Appropriate reference intervals for thyroid hormones should be used for thyroid patients at different ages. A major common mistake is to recommend a single reference interval for everyone or even all adults. From the first week of life until adulthood the hypothalamus—pituitary—thyroid axis continues to mature. During this period the concentrations of TSH and freeT4 are higher than those seen during the adulthood. This is especially important for cases such as congenital hypothyroidism that can be missed if appropriate age-specific reference interval is not used.

- Pregnancy is another condition that requires its own specific reference intervals based on gestational age. Each laboratory should establish its own reference intervals for first, second, and third trimester using its own patient population. Another important issue to be considered is the increase in concentrations of thyroid hormone-binding proteins during pregnancy. This leads to falsely elevated total T4 and total T3 *when using a conventional reference range*.

  As we indicated earlier, normalized reference intervals of total T4 (reference intervals for nonpregnant adult women 1.5×) may also be used by the laboratories . Ideally, trimester-specific reference ranges should be determined and validated for optimal precision.

- Specimens
  Appropriately, collected specimens are of great importance for accurate diagnosis of thyroid illness and following replacement therapy in patients. Although there is some diurnal variation in thyroid hormone synthesis and secretion, they do not significantly impact the laboratory results. However, a good laboratory practice indicates that all specimens be collected at approximately the same time of day (preferably between 8:00 and 10:00 am). Serum (blood collected in a tube with no anticoagulant) is the preferred sample. However, plasma (blood collected in a tube with EDTA heparin) is also acceptable. Specimens should be tested as soon as possible or can be stored in a refrigerator (2−8°C) for 24 h. If specimens cannot be processed within 24 h following collection, they may be frozen at −20°C for 1 month.

- Heparin artifact
  Intravenous or subcutaneous administration of heparin can significantly increase freeT4 in blood. This is based on lipoprotein lipase activity of heparin that can liberate free fatty acids from lipids which in turn displace thyroid hormones from thyroid-binding proteins leading to falsely elevated freeT4 and freeT3 in blood. In cases of subcutaneous heparin administration, specimens should be collected 24 h after the heparin administration [43].

- Technical variables
  Currently, almost all clinical chemistry laboratories around the world use immunoassays to measure thyroid hormones. These assays are affected by heterophile antibodies that can lead to erroneous results. Heterophile antibodies are produced in the body following exposure to animals (pet

handlers, sheep herdsmen, etc.), therapeutic agents (e.g., contrast dyes) that contain specific animal antigens or infection. These antibodies, also referred to as human antimouse antibody, interact with mouse antigens in test reagent and interfere with the assay. It is important to note that heterophile antibodies may result in a normal result in patient with true thyroid disease. Therefore discordant results should be questioned and clinician should ask the clinical chemist to investigate. When there is a suspicion about interference by heterophile antibodies, the specimen should be sent to another laboratory to be tested on another instrument that uses different source of antibodies (e.g., rabbit, sheep, goat, etc.) in their test. Also, specimens can be treated with antiheterophile antibodies to block them and retest.

• Endogenous thyroid hormone auto-antibodies
  Patients with autoimmune disease (thyroidal and nonthyroidal diseases) may have antibodies against T4 and T3 that can interfere with methods for free and total T4 and T3. These interferences are method specific and can be resolved by testing the specimens on another assay that may not be affected by those antibodies.

• TSH assays used in the clinical laboratory should be a third generation assay with a functional sensitivity below 0.01 U/L.

• Performance goals for most thyroid test:
  • For diagnostic applications the maximum bias and imprecision should be 15.7% and 11%, respectively.
  • For therapeutic applications, maximum bias and imprecision should be 2.6% and 5.2%, respectively.

• Dried blood specimens
  Dried blood spots can be used to assess thyroid function. The laboratory performing the test should routinely use dried specimens and has appropriate quality control and quality assurance practice to ensure the results are accurate and reproducible.
  The laboratory should also be accredited by a recognized and reputable accrediting body.
  In conclusion, good communication between clinical chemist/pathologist and endocrinologist is important for choosing the most appropriate tests and interpreting the results, especially when the results do not match the patient history. Also, thyroid tests should be patient specific based on the patient's-specific thyroid disease, stage of the disease, and coexisting disorders [44].

2.2.9 EXAMPLE CASE WITH DISCUSSION

A 35-year-old woman visits her primary care provider for a routine annual physical exam. She has no particular symptoms of note and takes no medications. She does report a family history of type-1 diabetes in her sister who also has been taking a thyroid replacement pill for the last 20 years. Her clinical examination is completely unremarkable. The care provider decides to order some "routine" blood tests as is their usual practice.

Question 1: Should the care provider order a TSH?

Answer: Although some guidelines have advocated for routine TSH measurements in all people starting at age 35, no one has yet demonstrated that doing so makes a difference in patient outcomes, especially when it is asymptomatic people being screened. Having said that, many primary care providers do consider ordering a TSH routinely with annual blood chemistries, perhaps more so among older patients. This practice has potentially large cost implications when applied population-wide and is probably deserving of more careful study. In the case at hand however, we think it would be reasonable for the care provider to consider checking a TSH given the family history of autoimmune disease and possible hypothyroidism in the sister. We say "possible" hypothyroidism because one may not know the certainty with which a remote diagnosis of thyroid hypofunction was made. It may be that long-term thyroid hormone replacement could have been given to the sister on the basis of a single, borderline elevated TSH in the past that may or may not have indicated permanent thyroid dysfunction. This emphasizes the importance of serial TSH measures over time in borderline cases before committing patients to lifelong thyroid hormone supplements. Once thyroid hormone replacement is started, it is virtually never questioned or stopped thereafter and so it is wise for the clinician to ensure that such therapy is truly indicated at the start.

Question 2: If they do order a TSH, should they also order a freeT4?

Answer: At this point of the case, there is nothing to indicate that the patient may have pituitary disease or past cranial irradiation and so asymptomatic, isolated pituitary disease causing hypothyroidism would be exceedingly unlikely. Therefore there is no reason to believe that a simple TSH measurement would be unreliable since it is the primary recommended first line test for all suspected thyroid disease in the absence of pituitary problems. Thus a freeT4 (or freeT3) would add nothing to patient care at this time.

Question 3: If they do order a TSH, should they also order antithyroid antibodies?

Answer: Speaking to this case, there is no role here for measurement of anti-TGs or TSH receptor antibodies. Anti-TPOs are associated with primary autoimmune hypothyroidism but the positive antibody titer by itself does not provide any functional information for the clinician beyond what is learned from measuring TSH alone. However, because of this patient's family history, we suppose one could consider measuring the anti-TPO titer if wanting see whether a thyroid autoimmune process was present and if so, perhaps counsel the patient that she may have a risk of developing hypothyroidism in the future. In our opinions, we think the patient could just as easily be counseled about her possible risk of future hypothyroidism by virtue of the sister's history alone (since autoimmune hypothyroidism is common in women especially with a family history), without requiring anti-TPO measurement. If our patient had a history of recurrent miscarriages (unexplained), there is a growing literature suggesting that anti-TPO measurement might be considered on account of its link to miscarriage and potential improved pregnancy outcomes with thyroid hormone even prior to the development of biochemical hypothyroidism. In summary though, we would not order TPO antibodies at this stage of the case.

The care provider orders a TSH alone. The result returns at 5.5 mIU/L.

Question 4: What should the provider do next in term of diagnosis or follow up?

Answer: According to the latest studies of normal thyroid function in populations of persons truly free of any thyroid abnormalities, the upper limit of normal on the TSH assay should be around 4.1−4.5 mIU/L with some advocating an upper limit of 3.0 mIU/L [35]. Either way, this patient's TSH level is supranormal. However, she would not be expected to have any symptoms related to this finding—in years past, this would even have been called "within normal." This TSH value alone does not diagnose permanent hypothyroidism. It is possible she could be in the recovery phase of a prior illness or mild thyroiditis. It is even possible that a repeat TSH measured right away could fall back within the normal range. Thus a repeat TSH now might well be <3.0 mIU/L in which case it would be quite inappropriate to diagnose a thyroid disease. It would be premature (in the absence of actively attempting pregnancy) to initiate thyroid hormone replacement at this stage, so the best next step is to repeat the TSH in 3−6 months' time. FreeT4 (or T3) measurement will not be useful at this stage (it will be normal). Anti-TPO, if measured and positive, might alert the clinician that this may well be the start of an insidious autoimmune thyroid failure but whether antibodies are positive or not will not really change the recommendation to simply repeat the TSH in a few months. The one exception might be the consideration of thyroid hormone administration to antibody positive women with a history of recurrent miscarriage although this practice needs further study and validation.

The care provider repeats the TSH 6 months later and it is now 6.9 mIU/L. The patient is prescribed levothyroxine 0.1 mg daily.

Question 5: What type of laboratory tests should be performed during the long-term follow up?

Answer: The repeatedly abnormally high TSH 6 months later is reasonable evidence that there is something wrong with this patient's thyroid function.

It would still be unlikely that this degree of thyroid dysfunction would cause reportable symptoms although if any "unexplained" patient symptoms are present, it can be challenging to convince all parties that such symptoms are not linked to the abnormal biochemistry. In many such cases, patients or their care providers will opt to start a low dose thyroid hormone replacement which may or may not have any discernable effect on patient symptoms [45]. Some endocrinologists are indeed in favor of treating "subclinical hypothyroidism" in young patients given a reported link to slightly higher cholesterol levels and possibly earlier onset ischemic heart disease [46]. Having said that, there are still no prospective randomized studies proving that this practice yields a clinical outcome benefit and the few completed studies specifically show no benefit to patient-level outcomes such as fewer heart attacks or decreased mortality. As a consequence, thyroid hormone treatment of subclinical hypothyroidism is considered reasonable but not mandatory. In the present case, I would recommend that the prescriber should remeasure TSH in 6—8 weeks time to prove that it has normalized (generally defined as a TSH <4.0 mIU/L). It is equally important to ensure that the TSH has not now become suppressed which would be indicative of thyroid hormone over-replacement. Once the TSH is normal, some clinicians recommend repeating it one more time 3 months later to ensure stability but the general long-term plan should be an annual TSH measurement as a routine with "as needed" measurements done if there is ever a question about new symptoms, large weight change or use of a medication known to interfere with thyroid hormone absorption or metabolism. Unless the clinician is investigating occult medication nonadherence or true malabsorption, there is no need to measure freeT4. At this stage, there are no reasons to ever measure freeT3 or thyroid antibodies.

2.2.10 HYPOTHYROIDISM: MAJOR SUMMARY POINTS

- The vast majority of hypothyroid disorders can be reliably diagnosed by measurement of TSH and, occasionally with combined freeT4 determination if pituitary disease is suspected.
- Interpretation of TSH and freeT4 follows an understanding of the normal feedback control of freeT4 effect upon TSH.
- Discordant or unexpected TSH and freeT4 pairs should be investigated by both the clinical chemist and a medical specialist/endocrinologist.
- There is no role for freeT3 or total T3 measurement in hypothyroidism.
- Outside of recurrent miscarriage in young women, there is very little value to routine measures of anti-TPO antibodies and no value to serial measures.
- Marginal elevations in TSH may not necessarily indicate the presence of a thyroid disease needing treatment; clinical chemists may play a large role in education of healthcare providers on this issue.

2.3 HYPERTHYROIDISM

2.3.1 CLINICAL PICTURE

As might be expected, the clinical picture of hyperthyroidism is essentially the opposite of what is seen in hypothyroidism and although it is less common than hypothyroidism, hyperthyroidism still accounts for a high proportion of endocrinology consultations. Depending on the severity, hyperthyroidism may present in the outpatient setting but severe and life-threatening manifestations (sometimes called "thyroid storm") may present to the emergency department and require intensive care unit admission.

The hallmark of all forms of hyperthyroidism is the appearance of sympathetic nervous system−related symptoms. Beyond that, there may be other clinical manifestations that are subtype specific (see above). In general, hyperthyroidism classically causes anxiety/nervousness, rapid heart rate, palpitations/cardiac arrhythmia, tremor, sweating, heat intolerance, and frequent bowel movements. When prolonged, it may also cause muscle weakness and substantial weight loss. When very severe, hyperthyroidism may also cause disorientation, coma, fever, and heart failure (features that typically define "thyroid storm" [47]). Evaluation of hyperthyroidism should not be delayed so that progression to a life-threatening form may be prevented.

2.3.2 CAUSES

Hyperthyroidism can be functionally divided into two major categories: transient, self-limited release of stored thyroid hormones (typically called thyroiditis) or the inappropriate, primary over-production of excess thyroid hormones.

2.3.2.1 Thyroiditis

It is probable that the most common cause of mild to moderate hyperthyroidism is that related to a self-limited thyroiditis. This term describes an acute, inflammatory condition of the thyroid wherein the stored contents of the thyroid follicles are released into the general circulation as a consequence of follicular cell injury. The hyperthyroid phase usually lasts between 4 and 6 weeks and then resolves spontaneously as the inflammatory condition passes and healing of the thyroid cells is completed. With the loss of all the stored thyroid hormone, it is common to then see the patient pass through a hypothyroid phase that may last several months, until thyroid hormone production and stores can be restored (see Section 2.1.2 for related details). Symptomatic patients may be treated with short-term beta-blocker drugs to blunt the effects of the sympathetic nervous system but use of antithyroid drugs such as methimazole or propylthiouracil (which inhibit thyroid hormone production) are not indicated and may provoke severe hypothyroidism during the later phase of the illness.

- *Subacute thyroiditis*—classically follows an upper respiratory tract infection and presents with severe anterior neck pain, fever, and mild to moderate transient hyperthyroidism.
- *Lymphocytic thyroiditis*—also known as "silent thyroiditis" or "post-partum thyroiditis" when it occurs in a woman after childbirth. This is typically mild to moderate, transient hyperthyroidism not associated with pain or fever. In some cases, persistent production of antithyroid antibodies and/or diffuse lymphocytic infiltration may progress to permanent hypothyroidism.

2.3.2.2 Primary thyroid hormone over-production

The following three causes of hyperthyroidism (Graves' disease, toxic thyroid adenomas, toxic multinodular goiter) may have either slow or abrupt onset and vary from very mild to life threatening in severity. In addition to the history and physical examination, functional nuclear medicine studies are often needed to demonstrate the continuous over-production of thyroid hormones, which characterize the diagnoses and define the treatment options. As a general rule, these causes of hyperthyroidism do not remit spontaneously although may not always be progressive in severity. In fact, where hyperthyroidism is shown to persist over 6 months, a diagnosis of thyroiditis is almost certainly excluded by definition. In all cases of thyroid hormone over-production, symptomatic relief may be achieved from short-term beta blocker drugs. For definitive therapy usually either antithyroid drugs, radioactive iodine thyroid ablation or surgical thyroidectomy is used. The treatment of life-threatening hyperthyroidism is beyond the scope of this text.

- *Graves' disease*—Typically affects women more than men perhaps related to its autoimmune nature. In this disease, auto-antibodies that simulate TSH are produced and interact with the TSH receptor resulting in enlargement of thyroid gland and over-production of thyroid hormone. Approximately 5% of affected patients develop Graves' orbitopathy, an inflammatory eye condition that may include severe bulging of the eyeballs known as proptosis. Very rarely, Graves' disease may also present with dermatologic manifestations known as pretibial myxedema on the lower legs.
- *Toxic thyroid adenoma*—Acquired, focal mutations in the TSH receptor of a thyroid gland may lead to local thyroid tumor growth with autonomous function. The tumor is typically benign, may be palpable, and is easily visualized on a nuclear thyroid scan.
- *Toxic multinodular goiter*—A multinodular thyroid is very common with ageing and as a chronic development among those born and raised in an iodine-deficient part of the world. On occasion and especially in concert with geographic relocation to an iodine sufficient locality, some of the nodules in the goiter develop the ability to autonomously produce thyroid hormone. In most cases, this is a slow, insidious process that may become relatively severe if not diagnosed for a long time.

> **KEY POINTS**
>
> The vast majority of simple hyperthyroidism can be divided into two types: self-limited thyroiditis and primary thyroid hormone over-production. The primary role of the clinical chemistry laboratory is to accurately define the presence and severity of thyroid hormone excess.

2.3.3 SPECIAL SITUATIONS: PREGNANCY, MEDICATIONS, CRITICAL ILLNESS

2.3.3.1 Pregnancy

Any type of hyperthyroidism may appear or be diagnosed during pregnancy. Untreated, hyperthyroidism is linked to a high rate of spontaneous miscarriage [48]. Because radioactive iodine thyroid ablation is contraindicated (to avoid destroying the fetal thyroid and/or cause fetal malformations), therapy of the primary thyroid hormone over-production requires expert attention with detailed discussion of the risks/uses of antithyroid drugs during gestation.

As mentioned in Section 2.2, there are physiologic pregnancy-related changes in thyroid hormones that may give the appearance of mild hyperthyroidism. This is typically referred to as "gestational thyrotoxicosis" and must be differentiated from other actual thyroid disease processes. Owing to its similar structure to TSH, when human chorionic gonadotropin (hCG) rises during the first 12 weeks of pregnancy, the hCG itself can function to stimulate the TSH receptor and increase the production of thyroid hormones. This TSH-independent means of raising thyroid production thus results in an expected drop in pituitary TSH production such that measured TSH may be slightly subnormal. In extreme cases, very high hCG can produce enough thyroid hormone that may make it difficult for the clinician to decide whether the patient has severe gestational thyrotoxicosis or early Graves' disease. However, gestational thyrotoxicosis spontaneously normalizes with the decline in hCG that occurs after 12 weeks gestation whereas Graves' disease may either persist or progress throughout pregnancy.

2.3.3.2 Medications

There are very few medications that actually cause hyperthyroidism. However, there are several drugs that can cause endocrine changes that give the appearance of mild biochemical hyperthyroidism. Amiodarone is by far the most common drug implicated and its effects causing thyrotoxicosis are complex [49]. Current thinking states that it may induce hyperthyroidism through a direct, toxic effect simulating a thyroiditis (also called type-1 amiodarone thyrotoxicosis) or alternately may cause an underlying abnormal thyroid (i.e., a previously euthyroid multinodular goiter) to become thyrotoxic by virtue of the massive iodine exposure from the high iodine content of the drug (also called type-2 amiodarone thyrotoxicosis). In some cases, both mechanisms may be operative.

Other medications that can cause a low TSH (hence the appearance of hyperthyroidism) include high-dose glucocorticoids/prednisone and dopamine.

The use of high-dose biotin supplementation has recently been reported to affect many immunoassays including TSH and freeT4; in some cases, the effect of such supplementation may cause both TSH and freeT4 to become abnormal simultaneously in a way that simulates biochemical hyperthyroidism [50]. Thus it is important to ask patients about biotin use (or any other "multi-vitamin") prior to interpretation of their thyroid laboratory results, especially if the results do not match the clinical picture.

2.3.3.3 Critical illness

As described in Section 2.1.3, serious medical illness is associated with changes in thyroid hormones' concentrations in blood that may mimic either mild hyperthyroidism or secondary hypothyroidism. TSH levels typically drop to slightly subnormal with acute illness and if severe, may be accompanied by a drop in freeT4 as well. A pronounced drop in both TSH and freeT4 may correlate with increased risk of mortality [51] but probably represent physiologic/adaptive changes to illness as opposed to thyroid pathology, hence the term "sick euthyroid syndrome" or "non-thyroidal illness." Some experts recommend against measurement of thyroid hormones in in-patients unless thyroid disease is truly suspected, in order to avoid having to deal with the confusion of separating "sick euthyroid" from mild thyroid diseases.

KEY POINTS

Gestational physiologic changes in thyroid hormones are very common in pregnant women and may appear to show mild hyperthyroidism that usually resolves as pregnancy progresses. Critical illness can also produce changes in TSH that appear to show hyperthyroidism but are more likely just adaptive to serious illness. Routine biochemical thyroid testing of in-patients is not recommended.

2.3.4 RARE SITUATIONS: EXOGENOUS THYROID HORMONE, ECTOPIC THYROID HORMONE

Very rarely, hyperthyroidism may be caused by the exogenous ingestion of thyroid hormones or meat products containing thyroid hormones [52]. Such situations may or may not be known to the patient who may thus deny the possibility when asked. Finally, in very rare settings, thyroid hormone can be produced outside of the thyroid gland such as in ovarian tumors. Both situations, while rare, may require expert assistance to diagnose but the main clue is persistent hyperthyroidism with negative nuclear thyroid imaging results.

2.3.5 DIAGNOSTIC APPROACH

2.3.5.1 It still starts with thyroid-stimulating hormone

As for the hypothyroid situation, diagnosis of hyperthyroidism is usually very straightforward and should begin with measurement of serum TSH using a third generation assay. As expected from physiology, the TSH level will be low, if not nearly undetectable in the presence of excess circulating thyroid hormone of any cause. Once critical illness is excluded, virtually all measures of TSH that are low will indicate a pathologic disease process although not always symptomatic.

2.3.5.2 FreeT4 can gauge the severity

Once the TSH is found to be low, the next question is usually to do with the degree of hyperthyroidism. When freeT4 is grossly elevated, this correlates with more severe disease.

2.3.5.3 FreeT3/total T3 may help differentiate between mild but overt hyperthyroidism and "subclinical hyperthyroidism"

On occasion, very mild hyperthyroidism (usually of the primary over-production variety) may present with preferential, isolated secretion of T3 rather than high levels of T4. Thus if a patient is found to have a low TSH with normal freeT4, one may consider measuring freeT3 or total T3. If high, the patient is deemed to have overt hyperthyroidism or "T3 Toxicosis." If both freeT4 and T3 are normal, the patient is typically diagnosed as having "subclinical hyperthyroidism." The clinical differentiation is such that most cases of T3 Toxicosis will be treated as active hyperthyroidism whereas subclinical hyperthyroidism is a more controversial indication for treatment.

In general, freeT3/total T3 measures do not usually add much to the diagnosis of hyperthyroidism outside of this setting and routine measurement is discouraged.

2.3.5.4 Nuclear imaging is often required to definitively diagnose the cause of hyperthyroidism

Beyond the scope of this text, either a technetium thyroid scan or I131 uptake will usually differentiate between transient thyroiditis versus thyroid hormone over-production. The relevant laboratory aspect has to do with the critical importance of having thyroid indices (particularly TSH) done in very close temporal proximity to the nuclear test since the interpretation of results are dependent upon TSH being suppressed at the time of testing.

2.3.5.5 Thyroid-stimulating hormone-receptor antibodies may be helpful as an adjunct or when nuclear studies cannot be done

A positive thyroid-stimulating hormone-receptor antibodies (TRAb) usually indicates a diagnosis of Graves' disease in the setting of hyperthyroidism with good specificity but poor sensitivity. For this reason, most endocrinologists will not order TRAb outside of pregnancy where nuclear studies cannot be performed. Some evidence exists to support measures of TRAb in late gestation among women with known or previous Graves' disease given a reasonable ability to predict fetal thyrotoxicosis amongst those with high antibody titer [53]. It is worth noting that there are two separate tests of this measure:

- TRAb—may cause either hyper or hypothyroidism
- thyroid-stimulating immunoglobulin (TSI)—specific for Graves' disease only

> **KEY POINTS**
>
> The first test for suspected hyperthyroidism should always be serum TSH. If the TSH is low, measurement of freeT4 and sometimes freeT3 can help quantitate the severity of hyperthyroidism. Determination of the exact cause of hyperthyroidism usually requires functional thyroid imaging but positive TRAb does suggest Graves' disease as the diagnosis.

2.3.6 DIAGNOSTIC DEFINITIONS

Fortunately the approach to low TSH is much less complex and controversial compared with the approach to high TSH. When measured in the sick inpatient, a low TSH usually indicates the presence of a "sick euthyroid" diagnosis which may be confirmed if a freeT4 is also low. However, more detailed thyroid work-up in the sick inpatient should probably be directed by an endocrinologist experienced in differentiating true thyroid disease in this setting.

For virtually all other nonpregnant outpatient measurements a low TSH reliably indicates the presence of hyperthyroidism.

- If the freeT4 and/or freeT3/total T3 are overtly high, the diagnosis is clearly hyperthyroidism.
- If the freeT4 and freeT3/total T3 are normal (albeit usually upper half of normal), a diagnosis of subclinical hyperthyroidism is more appropriate.
- Both thyroiditis and primary over-production diseases may present either overt or subclinical hyperthyroidism and so the degree of biochemical abnormality does not give diagnostic clarity.
- A freeT3 that is proportionally much higher than freeT4 may suggest a primary over-production situation owing to the excess production of T3 rather than release of small amounts of stored T3.

2.3.6.1 Pregnancy

Trimester-specific TSH reference ranges should be defined by laboratories based on their patient population. Unfortunately, this practice is not yet widely defined. With the hCG-driven decrease in TSH during the first trimester of normal pregnancy, it is common to see pregnant women with TSH measures of <1.0 mIU/L and as low as 0.03 mIU/L [31], which should not be construed as diagnostic of true thyroid disease.

Total T4 measures are preferred in pregnancy given the changes in both TBG and freeT4 with a normal upper limit suggested to be 1.5 times the nonpregnant reference range. Total T4 levels grossly above this adjusted range are more consistent with true hyperthyroidism/thyroid disease although a small subset of severe hyperemesis gravidarum patients may achieve slightly supranormal T4 levels as well [54]. In such cases a clinical assessment is required to determine whether other aspects of Graves' disease (goiter, eye disease, etc.) can clarify the diagnosis.

KEY POINTS

Outside of pregnancy or critical medical illness, a low TSH level signifies hyperthyroidism. If the accompanying freeT4 and freeT3 are normal, it is often called subclinical hyperthyroidism, which may still have associated health risks deserving of treatment in some cases. TSH levels drop below normal in most pregnancies during the first trimester. Total T4 is the preferred measure of thyroid hormones during gestation with an upper limit of normal around 1.5-fold the usual reference range.

2.3.6.2 The role of the chemistry laboratory in treatment/follow up of hyperthyroidism

2.3.6.2.1 Thyroiditis

Patients with self-limited thyroiditis do not usually require intensive biochemical follow up as the condition resolves spontaneously. It may be of value to check a TSH a few weeks after the initial presentation to determine if the patient is going into a hypothyroid phase. When this occurs, the hypothyroidism is usually mild and does not often require thyroid hormone replacement. Repeat TSH a few months later (or in response to progressive hypothyroid symptoms) can determine whether any postthyroiditis hypothyroidism is permanent or resolved. FreeT4 (and freeT3) measures are not usually needed in this setting.

2.3.6.2.2 Thyroid hormone over-production

In cases of subclinical hyperthyroidism, opinion is divided as to whether treatment is needed to restore a normal thyroid state [55]. However, if no treatment is given, it is reasonable to check TSH and freeT4 periodically every 3−6 months to determine if the disease is progressing to overt hyperthyroidism. It is important to measure both TSH and freeT4 in this setting to accurately assess the severity of the disease as serial-suppressed TSH measures cannot demonstrate change in disease severity.

When therapy such as antithyroid drugs (methimazole, propylthiouracil) or radioactive iodine is used, thyroid function may change quickly over a short period of time. The sequence of normalization upon treatment starts with normalization of freeT4, followed by freeT3 and lastly TSH. Depending on the severity and duration of the previous hyperthyroidism, normalization of TSH may take many months even after both freeT4 and freeT3 have normalized. In some cases, response to therapy can be so rapid that freeT4 and freeT3 actually become subnormal while the TSH is still suppressed. This emphasizes the importance of using serial tests of rapidly changing biochemical measures such as freeT4/freeT3 during initial drug therapy and with subsequent dose adjustments. Unlike with thyroid hormone replacement, dose changes in antithyroid drugs may cause relatively quick changes in free thyroid hormone levels and thus TSH measures at 6-week intervals may be insufficient to accurately titrate drug therapy without causing hypothyroidism. A general practice in endocrinology is to measure freeT4 every 3—4 weeks until it becomes normal after which one can usually simply follow the TSH to confirm ongoing euthyroidism.

The over-riding principle is to remember the relative rates of change in TSH versus freeT4 and freeT3 levels when interpreting changes in thyroid indices in concert with changes in therapy.

Serial measurements of TRAb are rarely of use outside of very select settings such as pregnancy complicated by Graves' disease.

KEY POINTS

In contrast to hypothyroidism where serial TSH measures are the cornerstone of follow up monitoring, a hyperthyroid state often undergoes more rapid change in thyroid hormones during treatment. Therefore treatment should be monitored and adjusted based on rapidly changing freeT4/T3 levels. TSH is measured during the chronic, stable phase and thereafter.

2.3.7 COMMON MISUSE OF LABORATORY TESTS IN HYPERTHYROIDISM

2.3.7.1 Measurement of the wrong thyroid auto-antibody

Many physicians know that autoimmune hyperthyroidism (Graves' disease) is a common cause of hyperthyroidism but very few recall exactly which antibody test is specific for the diagnosis. Care providers may simply write "thyroid antibodies" on a laboratory requisition form without specifying the exact antibody. This approach will not help the clinician with the diagnosis. For example, TPO is positive in a significant proportion of the population [28], and a "positive" TPO result in hyperthyroidism does not mean a definitive diagnosis of Graves' disease. Therefore physicians who see patients with thyroid disorders must be familiar with the diagnostic ability of thyroid auto-antibodies.

2.3.7.2 Reliance upon diagnostic imaging rather than clinical chemistry

Failure to use the laboratory when needed is a form of laboratory misuse. Many ultrasound reports of thyroid images may include a description of "thyroiditis" which is a radiologic term, not a biochemical or clinical term and should not be interpreted as indicative of hyperthyroidism. A diagnosis of hyperthyroidism cannot be seriously entertained without a suppressed TSH.

2.3.7.3 Failure to recognize surprisingly discordant pairs of TSH/freeT4 results

This is not common but can have serious clinical consequences when it occurs and is misinterpreted. In more than 99% of hyperthyroidism the TSH is low and the freeT4 is high or at least near the upper normal limit. In the case of a clinically hyperthyroid patient with high freeT4 but high or nonsuppressed TSH, one must consider the very rare possibility of a pituitary TSH-producing tumor causing hyperthyroidism. Similarly, low TSH with low freeT4 may indicate pituitary hypothyroidism or more commonly, the euthyroid sick syndrome in an ill patient. Discordant TSH/freeT4 results should never be ascribed to "laboratory error" without further investigation, especially in the symptomatic patient.

2.3.7.4 Over-use of freeT3/total T3 measures

The majority of hyperthyroid cases have an overtly high freeT4 and low TSH; the routine measurement of T3 for suspected hyperthyroidism is not necessary.

2.3.7.5 Failure to consider measurement of freeT3/total T3 when appropriate

Subclinical hyperthyroidism (low TSH, normal freeT4) is common and may not always require treatment. However, it is important to consider measurement of freeT3/total T3 in this particular setting to avoid missing T3 Toxicosis, especially in the symptomatic patient.

2.3.7.6 Failure to appreciate the diagnostic significance of persistently low thyroid-stimulating hormone

Modern TSH assays (third and fourth generation assays) are extremely sensitive and when TSH is low, it almost always indicates a hyperthyroid state. However, as indicated before, TSH alone cannot be used to assess the severity of hyperthyroidism or seriousness of the clinical condition as both mild/subclinical hyperthyroidism and severe hyperthyroidism have equally suppressed TSH. However, a low TSH should never be ignored. In some situations a healthcare provider may choose to repeat a measurement of TSH when a first test comes back showing suppression. If the repeat testing is delayed for some reason and continues to show suppression more than 8 weeks later, one must consider primary thyroid hormone over-production as most cases of thyroiditis will have resolved by that time.

2.3.7.7 Failure to appreciate the different rates of change of thyroid-stimulating hormone versus freeT4

In the setting of severe or longstanding hyperthyroidism, a serum TSH may remain suppressed for weeks or even months after restoration of a euthyroid state. The persistent suppression may be mistaken for failure of therapy, lack of patient compliance, or even as an indication to intensify antithyroid drug therapy. To avoid this error, it is better to follow the actual freeT4 levels when therapy is initiated; once freeT4 is normal, simple maintenance of therapy is usually all that is required for TSH to eventually normalize as well. In cases of life-threatening hyperthyroidism, there may even be a role for daily measures of freeT4/T3 to ensure therapy is being effective.

2.3.8 LABORATORY CONSIDERATIONS TO ASSESS THYROID DISORDERS

As indicated before, thyroid tests, like any other clinical tests, may be affected by different variables at preanalytical, analytical, and postanalytical stages producing erroneous results. Most of the factors that affect thyroid results in hypothyroid conditions can do the same in hyperthyroidism. Thus thyroid tests results are reliable only when the hypothalamic-pituitary—thyroid axis is intact and the impact of drugs, nonthyroidal diseases and heterophile antibodies is ruled out. TSH is an important marker for the diagnosis of hyperthyroidism. Currently, most of TSH assays used in clinical laboratories are highly sensitive (third or fourth generation assays). Basically, each generation of TSH assay represents a 10-fold (one log) improvement over the previous generation assay in their detection limits. For instance, first generation assays (mostly radioimmunoassays, RIA) had functional detection limits of $1-2$ mIU/L. Second generation assays had detection limits of approximately $0.1-0.2$ mIU/L. Third generation assays have a detection limits of $0.01-0.02$ mIU/L. Fourth generation assays have a detection limits of $0.001-0.002$ mIU/L. Third and fourth generation TSH assays are based on chemiluminescence technologies. With such sensitive assays, one can more easily distinguish hyperthyroidism from normal or hypothyroidism in all patient populations (e.g., hospitalized patients with euthyroid sick syndrome from those with hyperthyroidism) [6]. It is important that laboratories using these highly sensitive tests establish their own detection limits. As mentioned before, during pregnancy, TSH reference intervals should be established for each trimester by the laboratory using its own patient population.

FreeT4, and sometimes freeT3, is measured to confirm the diagnosis of hyperthyroidism. We mentioned earlier that both T4 and T3 are highly protein bound. That is, 99.97% of T4 and 97% of T3 are bound to TBG, albumin, and transthyretin and thus the concentration of the free thyroid hormones in circulation is <1 ng/dL, which presents great challenges to laboratorians for accurate measurement of these hormones. At present, almost all technologies used to measure freeT4 and T3 can provide an "estimate" of the actual concentrations. There are two reference methods, equilibrium dialysis and ultrafiltration. Both methods are time consuming, technically demanding and expensive; therefore not suitable for day-to-today clinical use. There are automated immunoassay techniques that can indirectly estimate the concentrations of freeT4 and T3. Fortunately, these assays provide reliable results accurate enough for diagnosis of most cases of thyroid disorders. However, under some rare conditions the reference methods may be used to provide more accurate results. Inappropriate use of reference methods for estimation of freeT4 and T3 is one of the most common cases misuse of the laboratory.

Clinical conditions that require measurement of freeT4 and T3 with the reference methods include familial dysalbuminemic hyperthyroxinemia with higher than normal albumin affinity for T4, some cases of pregnancies, congenital TBG deficiency, or excess and coadministration of medications that displace T4 or T3 from the binding protein such as heparin. The automated freeT4 and T3 immunoassays are not accurate enough to be used in the former conditions.

Serum or EDTA plasma (with or without gel separator) is the specimen of choice.

For more detailed effects of preanalytical errors please refer to Chapter 1, Endocrine laboratory testing: Excellence, errors, and the need for collaboration between clinicians and clinical chemists.

2.3.9 EXAMPLE CASE WITH DISCUSSION

A 34-year-old woman presents with a 5-month history of a 15 lb weight loss, tremors, palpitations, and anxiety. She also complains of a slight "bulging" appearance to her left eye, which has felt irritated and red over the past 2 months. On examination she is tachycardic, tremulous and has a diffusely enlarged thyroid gland approximately 2.5 times larger than normal. Her left eye shows mild episcleritis and appears more prominent than her right eye.

Question 1: What tests should the primary care provider order?

Response: This is a fairly classic picture of hyperthyroidism which can be confirmed by measurement of a TSH (expected to be very low/suppressed). With such a clear-cut presentation, we would automatically order the freeT4 along with the TSH in order to gauge the biochemical severity of the hyperthyroidism. Some endocrinologists might wish to also order a freeT3 but it is probably unnecessary at this point in light of the overt nature of the disease. Consideration could be given to checking for TRAb titer but is likely unnecessary given the long duration of illness and coappearance of ophthalmopathy which is virtually diagnostic of Graves' disease.

The blood tests are consistent with moderate hyperthyroidism and a subsequent technetium thyroid scan confirms the diagnosis of Graves' disease. Methimazole (an antithyroid drug) is given, along with a short-term beta blocker for symptom control.

Question 2: What is the role of the chemistry laboratory in her treatment and follow up for the next 6 months?

Response: Upon starting methimazole, her thyroid hormone levels will likely normalize within 3−4 weeks on average. However, the rate of normalization is dose dependent and may also depend upon the severity of disease at initiation. Thus it is possible for her thyroid indices to normalize in both a slightly shorter or even longer timeframe. The key will be to downtitrate the dose of her antithyroid drug in a timely fashion to avoid over-treatment causing hypothyroidism once she has achieved normal thyroid levels. In order to manage her care, one must rely upon clinical judgment and her free thyroid levels which will reflect the rapid changes. Her TSH might take weeks to become even low-detectable and thus cannot be used alone to make clinical decisions. We typically check freeT4 (and sometimes freeT3) every 3 weeks during the dose titration phase. Once both freeT4 and freeT3 have normalized on stable drug dosing, we then follow the TSH every 2−3 months to ensure long-term normality and stability. Owing to some diurnal variations in TSH secretion, we recommend that all the specimens be taken between 8:00 and 10:00 am.

The patient remains biochemically euthyroid for 12 months after which the antithyroid drug is stopped and her TSH continues to be normal thereafter, indicating disease remission. One year later, she becomes pregnant. She complains of fatigue, nausea, and vomiting with a 5 lb weight loss in the first 7 weeks of pregnancy. Her TSH is found to be 0.05.

Question 3: Is this a recurrence of her hyperthyroidism? What additional tests may help with the diagnosis?

Response: Perhaps but not necessarily. Graves' disease may occur de novo or as a recurrence during pregnancy but often preexisting Graves' disease may even enter a remission phase during early pregnancy. The marked rise in pregnancy-related hCG production in the first 12 weeks of pregnancy can act like TSH upon the thyroid thus lessening the need for pituitary TSH production; as a consequence, measured TSH levels fall often to subnormal and occasionally to undetectable levels. This hCG effect typically disappears as hCG levels fall after 12 weeks and TSH should then normalize. It may be worth measuring her total T4 and if it is very high then perhaps it is indeed recurrent Graves' disease. However, if the total T4 is within 1.5 times the usual upper reference limit, most endocrinologists would choose to follow the patient clinically and with serial biochemistry to see whether the apparent mild hyperthyroidism resolves or progresses to a more clear diagnosis of Graves' disease. Note that pregnancy is probably the only scenario where modern endocrinologists would choose to measure total T4 over freeT4.

Question 4: *What is the role of TRAb or TSI measurement in pregnancy?*

Response: Pregnancy is often an "immune-tolerant" state and TRAb titers, if positive, often fall or even normalize during late gestation. However, if TRAb titers are $>5\times$ the upper limit around 24 weeks, it may predict a higher risk of fetal thyrotoxicosis and should spur the endocrinologist to involve a high-risk obstetrical team or neonatologist to help follow the pregnancy [53]. Complex, severe, late pregnancy or fetal thyrotoxicosis requires very careful expertise in management.

2.3.10 MAJOR SUMMARY POINTS

- Measurement of TSH (low) and freeT4 (high) secures the diagnosis of hyperthyroidism.
- Milder forms with a normal freeT4 exist (subclinical hyperthyroidism)—treatment is controversial.
- Critical illness can show a low TSH that must be differentiated from hyperthyroidism although freeT4 and freeT3 are usually low, not high in this setting.
- Pregnancy changes can result in a low TSH that resolves spontaneously as pregnancy progresses; use of a pregnancy-adjusted total T4 measurement can help diagnose true hyperthyroidism in pregnancy.

2.4 THYROID CANCER
2.4.1 CLINICAL PICTURE

Thyroid cancer is one of the most common cancers and known to be increasing rapidly in incidence [56]. However, good data now show that the apparent increase in incidence may actually reflect an artifact of large scale over-diagnosis since the overall numbers of people dying from thyroid cancer are unchanged in decades despite a rapid rise in case detection [57]. The traditional presentation of thyroid cancer is the appearance of a palpable lump in the neck although many modern cases are now detected incidentally during imaging tests ordered for other reasons. The mass is typically painless and slow growing; it may be associated with palpably enlarged neck lymph nodes. Constitutional symptoms such as weight loss and night sweats are rarely present except in the very rare and highly aggressive forms.

2.4.2 CAUSES

Thyroid hormone—producing masses are virtually always benign and tend to present as hyperthyroidism (see prior section). Nonfunctional thyroid masses may be either benign or malignant although there are no biochemical tests that can differentiate the two. Diagnosis is usually made through fine needle aspiration cytology and/or diagnostic lobectomy for formal surgical pathology.

Thyroid cancer may be further divided into the well-differentiated type and the neuroendocrine and less differentiated types. The vast majority of thyroid cancers are of the differentiated type and most will have a generally good prognosis.

- Differentiated thyroid cancer—papillary thyroid carcinoma and follicular thyroid carcinoma are the most common, accounting for 85% and 12% of thyroid cancers, respectively [58]. Both have similar staging systems and prognostic scoring systems. It is worth noting that among patients <45 years of age, Stage 1 includes virtually all scenarios of primary tumor size and lymph node metastases with Stage 2 only diagnosed in the presence of solid organ metastases. Stages 3 and 4 only apply to persons over age 45, indicating the generally excellent prognosis for differentiated thyroid cancer in young people. Nonetheless both types of differentiated thyroid cancer may present with local recurrence, lymph node metastases or distant metastases, and long-term surveillance after primary surgery is the cornerstone of management. These cancers show growth-responsiveness to TSH and thus TSH suppression with thyroid hormone is a key part of treatment.
- Medullary thyroid cancer (MTC) arises from the C-cells of the thyroid and may be considered a neuroendocrine tumor. This type of tumor accounts for <5% of all thyroid cancer and many cases follow a more aggressive course than most differentiated thyroid cancers. MTC may be associated with multiple endocrine neoplasia syndromes which should always be considered after MTC diagnosis. MTC is not TSH-responsive and does not require TSH suppression. As well, this tumor type usually secretes calcitonin and rarely, other endocrine hormones such as adrenal corticotropic hormone (ACTH). Surgery is still the key in management although many cases present with lymph node metastases that can never be fully cleared and long-term surveillance with biochemistry and imaging is required. In individuals known to harbor an MTC-associated mutation in the re-arranged during transfection (RET) proto-oncogene, prophylactic thyroidectomy is commonly offered at ages as young as several months old, depending on which specific RET mutation is present [59].

KEY POINTS

The most common thyroid cancers are papillary and follicular thyroid carcinomas. MTC is a rarer thyroid cancer of neuroendocrine cell origin.

2.4.3 SPECIAL SITUATIONS: UNUSUAL CANCERS

Some rare subtypes of thyroid cancer such as Hurthle cell thyroid carcinoma or Tall cell thyroid carcinoma are still considered as differentiated thyroid cancers although often follow a more aggressive clinical course. As differentiated thyroid cancers, they are treated similarly to papillary and follicular thyroid cancers, including TSH suppression and serial measures of Tg. However, as a more aggressive and sometimes less well-differentiated tumor, these cancers may lose the ability to take up iodine or produce Tg which is usually an ominous development [60]. Anaplastic thyroid cancer is a highly malignant thyroid tumor with very high short-term mortality rates. As an undifferentiated, nonsecretory tumor, this tumor is predominantly diagnosed and followed with anatomical imaging rather than biochemistry.

2.4.4 ROLE OF THE CHEMISTRY LABORATORY IN DIAGNOSIS

2.4.4.1 High thyroid-stimulating hormone, low thyroid-stimulating hormone

For the patient presenting with a thyroid mass the first biochemical test should be a TSH measurement. If the TSH is suppressed, it raises the possibility that the thyroid mass is functional, which could be proved with thyroid scintigraphy. Functional thyroid masses are almost never malignant and so once confirmed, the functional thyroid nodule should not be biopsied [61]. On occasion a patient may present with Graves' disease and a coexistent nonfunctional thyroid mass, again proven through thyroid scintigraphy. There is some evidence that differentiated thyroid cancers that coexist with Graves' disease have a worse prognosis and so biopsy should be considered [62].

Although a high TSH reading is usually associated with autoimmune hypothyroidism, there is now evidence that in the setting of thyroid nodular disease, a coexisting high TSH may be a reasonable marker of malignancy with higher TSH levels predicting higher risk [63].

2.4.4.2 Molecular markers

Thyroid cytopathology from fine needle aspiration of thyroid nodules has a reasonably high sensitivity and specificity for malignancy but many aspirates are still nondiagnostic or labeled suspicious if they do not meet the criteria for diagnosis of thyroid cancer. Follicular thyroid tumors in particular cannot be diagnosed as adenoma or carcinoma by cytology alone since the definition of carcinoma depends upon the presence of local invasion which cannot be seen on aspiration biopsy. Traditionally, thyroid nodules with nondiagnostic or suspicious biopsy results have been treated with surgical hemithyroidectomy or careful clinical follow up depending on the clinical suspicion and patient preference. As a consequence, many hemithyroidectomies are performed for what ultimately turns out to be benign thyroid nodules.

Several factors affect the adequacy of thyroid aspirates. The most important is the skill of the person obtaining the aspirate. The nature of the aspirate, solid versus cystic, can also determine the success of a successful aspirate. On site evaluation of the aspirate, so called rapid on site evaluation (ROSE), can also greatly reduce inadequacy from 20%−30% to 5% or less.

An emerging approach to the problem involves the use of molecular markers that may be detected on thyroid aspirate material and now available commercially. The procedure involves RNA extraction and amplification from the aspirate material and then put through a gene expression classifier to produce either a "benign" or "suspicious" classification. Initial studies would appear to show that this step may reduce the uncertainty of nondiagnostic biopsies and thus reduce otherwise unnecessary hemithyroidectomies [64] however, other independent evaluations of this approach showed that it may not have sufficient accuracy to add much to patient management or outcomes [65].

2.4.4.3 Thyroglobulin

Tg, a protein produced, stored and secreted (in small amounts) from thyroid follicular cells is also produced by most differentiated thyroid cancers and often in proportion to the bulk of disease/metastases. The main clinical use of Tg measures relates to the follow up of differentiated thyroid cancer. Prior to thyroidectomy, Tg measures cannot determine the presence, absence, or bulk of thyroid cancer since both malignant and benign thyroid cells produce it. However, following thyroidectomy (and especially following subsequent radioactive iodine remnant ablation), Tg levels often drop to very low or even undetectable levels. Long-term serial measures of Tg in the postoperative, postablation patient may be used as an early marker of disease recurrence.

It is imperative that endocrinologists be familiar with the strength and weakness of thyroglobulin (Tg) assays used in their chemistry laboratories. One of the major problems with Tg assays is the presence of anti-TGs that are present in 20%−30% of patients with differentiated thyroid carcinomas [66]. These antibodies can interfere with Tg measurements resulting in false low or false high results, depending on the technologies used to measure Tg. Such discrepancies of results can have significant effect on patient management. That is, high false results in patients free of cancer may result in over-treatment and false-low result may result in under treatment. Currently, there are RIAs, immunometric assay, and liquid chromatography tandem mass spectrometry (LC−MS/MS) assays available for measuring Tg. Unfortunately, all of these assays are affected by the presence of TgAB and they behave differently based on the concentration of Tg and presence or absence of TgAB. In general, LC−MS/MS assays can provide the best results and should be used as the reference method when other methods fail to provide accurate results. RIA assays are the second most sensitive and specific assay for measuring Tg in TgAB positive cases. However, for most routine cases, current automated immunoassays offered by major vendors are acceptable. Both endocrinologists and clinical chemists should carefully watch for unusual discordant results and confirm their findings with LC−MS/MS or at least an RIA assay at a reputable laboratory.

There are new tests that are claimed to detect residual thyroid cancer in patients treated for thyroid cancer. These tests are specifically useful when Tg cannot be accurately measured due to the presence of TgAB. One of these tests measures Tg mRNA in circulation [67]. Although measuring Tg mRNA may provide some help as an early marker of relapse, the whole approach suffers from poor specificity and sensitivity. Investigators in this field are looking for new markers. Indeed, other markers such as mRNA for TSH receptors and thyroperoxidase and cytokeratin 20 detection by PCR are under investigation [68,69].

A rare occasional use of Tg may be seen in the investigation of the hyperthyroid patient in whom there is suspicion of undisclosed (or unknown) consumption of exogenous thyroid hormone . In cases of hyperthyroidism related to actual thyroid disease (either thyroiditis or Graves' disease), Tg levels are often high, in concert with the excess thyroid hormone release from the gland. However, if Tg levels are actually low despite hyperthyroidism, this suggests exogenous thyroid hormone ingestion.

2.4.4.4 Calcitonin

Calcitonin is a hormone of uncertain physiologic significance in humans secreted by the C-cells of the thyroid. The predominant clinical role is in the diagnosis and follow up of medullary thyroid cancer. This type of thyroid cancer is predominantly diagnosed in cytology laboratory by examining fine needle aspiration. Some experts recommend routine calcitonin measurement in all patients with thyroid nodular disease to ensure that the diagnosis is not missed [70]. However, given that MTC is a very rare thyroid cancer and not all calcitonin elevations indicate MTC, some current guidelines have not yet recommended the practice [62].

The one scenario where calcitonin measurements, either single or serial, might be useful in diagnosis of MTC is for patients known to have a RET mutation associated with multiple endocrine neoplasia type 2A or 2B (or patients related to known cases). This particular familial form of MTC requires aggressive monitoring and prophylactic thyroidectomy in childhood, depending upon the particular RET mutation. In such cases, serum calcitonin may alert the clinician to an undiagnosed MTC or may help define affected individuals where genetic testing is not available. Historically a calcium or pentagastrin-stimulated calcitonin was thought to have better sensitivity for early MTC but genetic testing, high-sensitivity ultrasound and prophylactic thyroidectomy have made such dynamic testing largely unnecessary.

Other neuroendocrine tumors have been reported to secrete calcitonin [71] and must be considered if investigating a patient with high levels. Very high calcitonin production may be associated with a clinical syndrome of flushing and diarrhea but the associated calcitonin-producing tumor is usually already known.

KEY POINTS

Papillary and follicular thyroid cancers are treated with surgery, radioactive iodine, and TSH suppression. Long-term follow up includes Tg monitoring. Medullary thyroid cancer is mostly treated surgically and is monitored through calcitonin levels.

2.4.5 ROLE OF THE CHEMISTRY LABORATORY IN FOLLOW UP

For patients with differentiated thyroid cancer (papillary/follicular subtypes), TSH is known to act as a growth factor for residual or recurrent disease. Long-term studies have shown a benefit to thyroid hormone replacement doses (since all such patients have had thyroidectomy) that are sufficiently high to result in a low or even slightly suppressed TSH [72]. The degree of TSH suppression should vary according to the risk category of the thyroid cancer. Low-risk patients may have a target TSH of <2.0 mIU/L whereas patients at higher risk of recurrence or known metastatic disease should have TSH suppressed to levels below 0.1 mIU/L [72].

Tg measurements are traditionally first measured when the patient presents for radioiodine thyroid remnant ablation following thyroidectomy for differentiated forms of thyroid cancer (papillary thyroid cancer (PTC) or follicular thyroid cancer (FTC)). Serum Tg is usually detectable before surgical thyroid remnants are radiated and the subsequent disappearance of measurable serum Tg during follow up is used to confirm complete ablation of any remaining thyroid cells.

Following surgery and radioiodine, Tg may be measured in either an "unstimulated" (random) state or as a "stimulated" measure following endogenous or exogenous TSH stimulation. Since TSH is known to stimulate thyroid cell growth and function, including Tg secretion, it represents a means of increasing the clinical laboratory's ability to detect any thyroid cells remaining (or, in the case of cancer, recurring). Traditionally the stimulated Tg is obtained by a random Tg measurement after 1 month of thyroid hormone replacement. However, this month long preparation results in clinical hypothyroidism for the patient which may be poorly tolerated. An alternate approach allows the measurement of serum Tg following two injections of recombinant human TSH (rhTSH) without need for discontinuation of thyroxine therapy. This method appears to be equally sensitive for detection of thyroid cancer recurrence [73,74] and does not adversely impact the patient's quality of life [75].

Whether Tg is measured in a "stimulated" or "unstimulated" fashion and at what frequency depends upon the extent/stage of disease and the level of concern of the clinician or patient. Patients without any nodal metastases and at low risk for local recurrence may have annual unstimulated Tg measures for a few years after diagnosis. Those with higher risk or suspected recurrence/metastases may undergo stimulated Tg on a more frequent basis in order to take advantage of the improved sensitivity for disease detection. In general a low-risk patient with undetectable stimulated Tg does not need repeated stimulated measures for at least 2−3 years thereafter, if ever [76].

There are several major limitations to the use of serum Tg in the follow up of thyroid cancer patients:

- Some cancers may be or may become less differentiated and less able to synthesize Tg at all. Therefore low Tg levels are not necessarily reassuring in the context of known metastatic disease.

- Tg levels do not always follow clear trends, especially at lower levels and therefore small apparent increases or decreases do not always mirror clinical disease activity.
- Thyroid/thyroid cancer cell death in response to radioiodine may be slow and thus Tg may also be slow to decrease and give the impression of disease resistance.
- Anti-TGs (as mentioned earlier) are common in the population and known to interfere with Tg measurements, giving false-low Tg determinations. Therefore, as a general rule, serum Tg should not be relied upon for clinical decision making in the context of positive anti-TGs. However, high titer anti-TGs themselves can sometimes function as a surrogate for actual Tg measures; a declining titer may correspond to disappearance of Tg-producing cells [77].

Calcitonin may be serially measured in the follow up of medullary thyroid cancer. Current guidelines now recommend use of serial calcitonin measures to calculate a "calcitonin doubling time" which appear to reflect the underlying tumor aggressiveness [78]. Those with a short doubling time should be viewed with more concern than those with a very long doubling time. Calcitonin measures do not require TSH stimulation since MTC is not a TSH-responsive tumor (due to the fact it does not derive from thyroid follicular cells). Other tests occasionally used to follow MTC include carcinoembryonic antigen (CEA) and alphafetoprotein [78].

2.4.6 COMMON MISUSE OF LABORATORY TESTS IN THYROID CANCER

2.4.6.1 Measurement of thyroglobulin in the investigation of malignant primary of unknown origin

Metastatic thyroid cancer is one possibility when investigating metastatic cancer of unknown primary origin. However, this diagnosis can only be made upon pathologic examination of the tissue. In the nonthyroidectomized patient, serum Tg can be highly variable and thus the presence or level of Tg does not yield any information about the presence or absence of thyroid cancer.

2.4.6.2 Failure to recognize that suppressed thyroid-stimulating hormone indicates a need to define a nodule as functional or nonfunctional before pursuing biopsy

Thyroid nodules are extremely common. However, functional (thyroid hormone producing) nodules are virtually never malignant although biopsies of such nodules may be hard to recognize as benign. Current guidelines recommend against biopsy of functional nodules. Thus if a thyroid nodule is present, the first step in investigation is a serum TSH. If the TSH is normal or elevated, it is reasonable to assume that the nodule is nonfunctional and biopsy may be pursued. However, if the TSH is subnormal, it becomes more likely that the nodule is functional and this can be confirmed with thyroid scintigraphy thus avoiding an unnecessary and sometimes misleading biopsy.

2.4.6.3 Failure to recognize that some thyroid cancer recurrences will have undetectable "unstimulated" thyroglobulin levels

Unstimulated Tg measures have significantly poorer sensitivity to detect residual or recurrent thyroid cancer following surgery and radioiodine remnant ablation. Some patients with anatomically visible thyroid cancer recurrences will continue to have undetectable Tg while taking levothyroxine replacement and therefore, in those with moderate to high risk of local recurrence, a random undetectable Tg cannot be taken as evidence of disease absence. A stimulated Tg measurement, either by rhTSH or thyroid hormone withdrawal is needed for maximum sensitivity in recurrence detection.

2.4.6.4 Failure to recognize that high titer antithyroglobulin antibodies preclude interpretation of serum thyroglobulin levels

Many laboratories recommend reflex testing of anti-TG with every request for serum Tg. When a high titer antibody is present, an appended comment reflecting the potential for positive or higher than reported Tg can help to ensure that the clinician is not falsely reassured by a low Tg level.

2.4.6.5 Reduction of thyroid hormone replacement doses in patients with thyroid cancer and low serum thyroid-stimulating hormone

Thyroid cancer patients will be a very small minority of patients receiving thyroid replacement in the average community practice. As such, many practitioners may be unfamiliar with the different targets of thyroid replacement compared with common primary hypothyroidism where the TSH is usually kept within a normal range. In moderate- to high-risk thyroid cancer the serum TSH should be kept at a slightly suppressed level to reduce the risk of recurrence and thus the low TSH should not result in a thyroid hormone dose reduction.

2.4.7 EXAMPLE CASE WITH DISCUSSION

A 28-year-old woman is found to have a 4 cm mass in her neck during routine annual examination. It is nontender and the remainder of her examination is unremarkable. Her health practitioner orders a serum TSH which is upper normal at 3.95 mIU/L. A fine needle aspiration biopsy is performed and a diagnosis of probable papillary thyroid cancer is given. At surgery she is found to have six lymph nodes in the central (level VI) compartment. Radioactive iodine adjuvant therapy is given. One year later, she takes two injections of rhTSH and a Tg measurement is done and found to be undetectable.

Question 1: Is there any other information needed to interpret this result?

Response: It would be helpful to know whether her postradioiodine scan showed any evidence of tumor outside the neck in order to know whether we are dealing with more advanced metastases although it will be unlikely without lateral neck involvement first. Secondly, it would be useful to know what her stimulated Tg level was when measured at the time of radioiodine ablation: a high level that subsequently becomes undetectable usually confirms successful ablation of all residual thyroid/thyroid cancer. Finally, it is imperative to know whether she has anti-TGs. If so, the "negative" Tg level cannot be assumed to be valid.

Question 2: Anti-TGs are measured and found to be present in high titer (1:160). How will this affect your management?

Response: The high titer anti-TGs essentially mean that we cannot rely upon the serum Tg to make clinical decisions. Careful neck ultrasonography is also a highly sensitive tool for the detection of the most common local thyroid cancer recurrences and should be included as a routine part of the patient follow up. If anti-TGs were present at ablation, a marked decrease in titer may be viewed as a surrogate measure for the expected decline in serum Tg posttherapy [77]. An alternate approach would be Tg measurement on a different, reliable RIA and/or LC−MS/MS that may be less impacted by the TGs.

2.4.8 MAJOR SUMMARY POINTS—ENDOCRINOLOGY

1. Differentiated thyroid cancer (papillary and follicular thyroid cancer) cannot be diagnosed biochemically although serum TSH may help in the process of risk assessment of thyroid nodules.
2. Differentiated thyroid cancers are treated with thyroid hormone replacement in doses often specifically chosen to suppress serum TSH.
3. The relatively rare medullary thyroid cancer may be diagnosed in known RET mutation carriers with calcitonin measurements.
4. Serum Tg (especially after TSH stimulation) is a cornerstone of long-term recurrence monitoring in patients with a history of differentiated thyroid cancer.
5. Serum Tg results in the presence of anti-TGs, generated by automated immunoassay analyzers should be interpreted with caution. If possible these specimens should be tested for both Tg and TGs using RIA or LC−MS/MS methods.

REFERENCES

[1] Rastogi MV, LaFranchi SH. Congenital hypothyroidism. Orphanet J Rare Dis 2010;5 (June 1):1.
[2] Vanderpump MP. The epidemiology of thyroid disease. Br Med Bull 2011;99 (September 1):39−51.

[3] Surks MI, Ortiz E, Daniels GH, Sawin CT, Col NF, Cobin RH, et al. Subclinical thyroid disease: scientific review and guidelines for diagnosis and management. JAMA. 2004;291(January 2):228−38.

[4] Vila L, Velasco I, González S, Morales F, Sánchez E, Torrejón S, et al. Controversies in endocrinology: on the need for universal thyroid screening in pregnant women. Eur J Endocrinol 2014;170(January 1):R17−30.

[5] Ogbera AO, Kuku SF. Epidemiology of thyroid diseases in Africa. Indian J Endocrinol Metab 2011;15(July 6):82.

[6] Demers LM, Spencer C. The thyroid: pathophysiology and thyroid function testing. In: Burtis CA, Ashwood ER, Bruns DE, editors. Tietz textbook of clinical chemistry and molecular diagnostics. 4th ed. Philadelphia, PA: Saunders; 2006.

[7] Biondi B, Cooper DS. The clinical significance of subclinical thyroid dysfunction. Endocrine Rev 2008;29(February 1):76−131.

[8] Garber JR, Cobin RH, Gharib H, Hennessey JV, Klein I, Mechanick JI, et al. Clinical practice guidelines for hypothyroidism in adults: cosponsored by the American Association of Clinical Endocrinologists and the American Thyroid Association. Thyroid 2012;22(December 12):1200−35.

[9] Jonklaas J, Bianco AC, Bauer AJ, Burman KD, Cappola AR, Celi FS, et al. Guidelines for the treatment of hypothyroidism: prepared by the american thyroid association task force on thyroid hormone replacement. Thyroid 2014;24(December 12):1670−751.

[10] Biondi B. Thyroid and obesity: an intriguing relationship. J Clin Endocrinol Metab 2010;95(August 8):3614−17.

[11] Hutfless S, Matos P, Talor MV, Caturegli P, Rose NR. Significance of prediagnostic thyroid antibodies in women with autoimmune thyroid disease. J Clin Endocrinol Metab 2011;96(June 9):E1466−71.

[12] Zimmermann MB, Jooste PL, Pandav CS. Iodine-deficiency disorders. Lancet 2008;372(October 9645):1251−62.

[13] Takasu N, Yamada T, Takasu M, Komiya I, Nagasawa Y, Asawa T, et al. Disappearance of thyrotropin-blocking antibodies and spontaneous recovery from hypothyroidism in autoimmune thyroiditis. N Engl J Med 1992;326(February 8):513−18.

[14] Soldin OP, Tractenberg RE, Hollowell JG, Jonklaas J, Janicic N, Soldin SJ. Trimester-specific changes in maternal thyroid hormone, thyrotropin, and thyroglobulin concentrations during gestation: trends and associations across trimesters in iodine sufficiency. Thyroid 2004;14(December 12):1084−90.

[15] De Groot L, Abalovich M, Alexander EK, Amino N, Barbour L, Cobin RH, et al. Management of thyroid dysfunction during pregnancy and postpartum: an Endocrine Society clinical practice guideline. J Clin Endocrinol Metab 2012;97(August 8):2543−65.

[16] Midgley JE, Hoermann R. Measurement of total rather than free thyroxine in pregnancy: the diagnostic implications. Thyroid 2013;23(March 3):259−61.

[17] Soldin OP. When thyroidologists agree to disagree: comments on the 2012 Endocrine Society pregnancy and thyroid disease clinical practice guideline. J Clin Endocrinol Metab 2012;97(August 8):2632−5.

[18] Smit BJ, Kok JH, Vulsma T, Briët JM, Boer K, Wiersinga WM. Neurologic development of the newborn and young child in relation to maternal thyroid function. Acta Paediatrica 2000;89(March 3):291−5.

[19] Lazarus JH, Bestwick JP, Channon S, Paradice R, Maina A, Rees R, et al. Antenatal thyroid screening and childhood cognitive function. N Engl J Med 2012;366 (February 6):493–501.

[20] Farwell AP. Nonthyroidal illness syndrome. Curr Opin Endocrinol Diabetes Obes 2013;20(October 5):478–84.

[21] Cohen-Lehman J, Dahl P, Danzi S, Klein I. Effects of amiodarone therapy on thyroid function. Nat Rev Endocrinol 2010;6(January 1):34–41.

[22] Crawford BA, Cowell CT, Emder PJ, Learoyd DL, Chua EL, Sinn J, et al. Iodine toxicity from soy milk and seaweed ingestion is associated with serious thyroid dysfunction. Med J Aust 2010;193(October 7):413–15.

[23] Abdulrahman RM, Verloop H, Hoftijzer H, Verburg E, Hovens GC, Corssmit EP, et al. Sorafenib-induced hypothyroidism is associated with increased type 3 deiodination. J Clin Endocrinol Metab 2010;95(August 8):3758–62.

[24] Stevenso HP, Archbold GPR, Johnston P, Young IS, Sheridan B. Misleading serum free thyroxine results during low molecular weight heparin treatment. Clin Chem 1998;44(5):1002–7.

[25] Howard D, La Rosa FG, Huang S, Salvatore D, Mulcahey M, Sang-Lee J, et al. Consumptive hypothyroidism resulting from hepatic vascular tumors in an athyreotic adult. J Clin Endocrinol Metab 2011;96(April 7):1966–70.

[26] Smithson MJ. Screening for thyroid dysfunction in a community population of diabetic patients. Diabetic Med 1998;15(February 2):148–50.

[27] Parle J, Roberts L, Wilson S, Pattison H, Roalfe A, Haque MS, et al. A randomized controlled trial of the effect of thyroxine replacement on cognitive function in community-living elderly subjects with subclinical hypothyroidism: the Birmingham Elderly Thyroid study. J Clin Endocrinol Metab 2010;95(August 8):3623–32.

[28] Rodondi N, Newman AB, Vittinghoff E, de Rekeneire N, Satterfield S, Harris TB, et al. Subclinical hypothyroidism and the risk of heart failure, other cardiovascular events, and death. Arch Intern Med 2005;165(November 21):2460–6.

[29] Hollowell JG, Staehling NW, Flanders WD, Hannon WH, Gunter EW, Spencer CA, et al. Serum TSH, T4, and thyroid antibodies in the United States population (1988 to 1994): National Health and Nutrition Examination Survey (NHANES III). J Clin Endocrinol Metab 2002;87(February 2):489–99.

[30] Rose SR, Brown RS. Update of newborn screening and therapy for congenital hypothyroidism. Pediatrics 2006;117(June 6):2290–303.

[31] Lee RH, Spencer CA, Mestman JH, Miller EA, Petrovic I, Braverman LE, et al. Free T4 immunoassays are flawed during pregnancy. Am J Obstet Gynecol 2009;200 (March 3): 260-e1.

[32] ACOG Practice Bulletin Number 148. Thyroid disease in pregnancy. Obstet Gynecol 2015;125:996–1005.

[33] Van den Boogaard E, Vissenberg R, Land JA, van Wely M, van der Post JA, Goddijn M, et al. Significance of (sub) clinical thyroid dysfunction and thyroid autoimmunity before conception and in early pregnancy: a systematic review. Hum Reprod Update 2011;17(September 5):605–19.

[34] Negro R, Formoso G, Mangieri T, Pezzarossa A, Dazzi D, Hassan H. Levothyroxine treatment in euthyroid pregnant women with autoimmune thyroid disease: effects on obstetrical complications. J Clin Endocrinol Metab 2006;91(July 7):2587–91.

[35] Spencer CA, Hollowell JG, Kazarosyan M, Braverman LE. National Health and Nutrition Examination Survey III thyroid-stimulating hormone (TSH)-thyroperoxidase antibody relationships demonstrate that TSH upper reference limits may be skewed by occult thyroid dysfunction. J Clin Endocrinol Metab 2007;92(November 11):4236−40.

[36] Vadiveloo T, Donnan PT, Cochrane L, Leese GP. The Thyroid Epidemiology, Audit, and Research Study (TEARS): the natural history of endogenous subclinical hyperthyroidism. J Clin Endocrinol Metab 2011;96(January 1):E1−8.

[37] Vadiveloo T, Donnan PT, Murphy MJ, Leese GP. Age-and gender-specific TSH reference intervals in people with no obvious thyroid disease in Tayside, Scotland: the Thyroid Epidemiology, Audit, and Research Study (TEARS). J Clin Endocrinol Metab 2013;98(January 3):1147−53.

[38] Rozing MP, Houwing-Duistermaat JJ, Slagboom PE, Beekman M, Frolich M, de Craen AJ, et al. Familial longevity is associated with decreased thyroid function. J Clin Endocrinol Metab 2010;95(November 11):4979−84.

[39] Spaulding SW, Taylor PN, Iqbal A. Why do patients with subclinical hypothyroidism get overtreated. Clin Thyroidol 2013;25:273−6.

[40] Boucai L, Hollowell JG, Surks MI. Thyroid dysfunction: hypothyroidism, thyrotoxicosis, and thyroid function tests: an approach for development of age-, gender-, and ethnicity-specific thyrotropin reference limits. Thyroid 2011;21(January 1):5.

[41] Newsome S, Hickman P. Thyroid. In: Kaplan LA, Pesce AJ, editors. Clinical chemistry: theory, analysis, correlation. 5th ed. St. Louis, MO: Mosby; 2010.

[42] Fedler C. Laboratory tests of thyroid function: pitfalls in interpretation. Contin Med Educ 2006;24(7):386−90.

[43] Bayer MF. Effect of heparin on serum free thyroxine linked to post-heparin lipolytic activity. Clin Endocrinol 1983;19:591−6.

[44] Demers LM, Spencer C. The thyroid: pathophysiology and thyroid function testing. Laboratory medicine practice guidelines: laboratory support for the diagnosis and monitoring of thyroid disease. Clin Endocrinol 2003;58(February 2):138−40.

[45] Villar HC, Saconato H, Valente O, Atallah AN. Thyroid hormone replacement for subclinical hypothyroidism. Cochrane Database Syst Rev 2007;3:CD003419.

[46] Walsh JP, Bremner AP, Bulsara MK, et al. Subclinical thyroid dysfunction as a risk factor for cardiovascular disease. Arch Intern Med 2005;165:2467−72.

[47] Carroll R, Matfin G. Review: endocrine and metabolic emergencies: thyroid storm. Therap Adv Endocrinol Metab 2010;1(August 3):139−45.

[48] Hamburger JI. Diagnosis and management of Graves' disease in pregnancy. Thyroid 1992;2:219−24.

[49] Cohen-Lehman J, Dahl P, Danzi S, Klein I. Effects of amiodarone therapy on thyroid function. Nat Rev Endocrinol 2010;6(January 1):34−41.

[50] Elston MS, Sehgal S, Du Toit S, Yarndley T, Conaglen JV. Factitious Graves' disease due to biotin immunoassay interference—a case and review of the literature. J Clin Endocrinol Metab 2016;101(June 9):3251−5.

[51] Plikat K, Langgartner J, Buettner R, Bollheimer LC, Woenckhaus U, Schölmerich J, et al. Frequency and outcome of patients with nonthyroidal illness syndrome in a medical intensive care unit. Metabolism 2007;56(February 2):239−44.

[52] Parmar MS, Sturge C. Recurrent hamburger thyrotoxicosis. Can Med Assoc J 2003;169(September 5):415−17.

[53] Peleg D, Cada S, Peleg A, Ben-Ami M. The relationship between maternal serum thyroid-stimulating immunoglobulin and fetal and neonatal thyrotoxicosis. Obstet Gynecol 2002;99(June 6):1040−3.

[54] Haddow JE, McClain MR, Lambert-Messerlian G, Palomaki GE, Canick JA, Cleary-Goldman J, et al. Variability in thyroid-stimulating hormone suppression by human chronic gonadotropin during early pregnancy. J Clin Endocrinol Metab 2008;93 (September 9):3341−7.

[55] Cooper DS. Approach to the patient with subclinical hyperthyroidism. J Clin Endocrinol Metab 2007;92(January 1):3−9.

[56] Pellegriti G, Frasca F, Regalbuto C, Squatrito S, Vigneri R. Worldwide increasing incidence of thyroid cancer: update on epidemiology and risk factors. J Cancer Epidemiol 2013;2013(May):965212.

[57] Ahn HS, Kim HJ, Welch HG. Korea's thyroid-cancer "epidemic"—screening and overdiagnosis. N Engl J Med 2014;371(November 19):1765−7.

[58] Mitchell I, Livingston EH, Chang AY, Holt S, Snyder WH, Lingvay I, et al. Trends in thyroid cancer demographics and surgical therapy in the United States. Surgery. 2007;142(December 6):823−8.

[59] Wells Jr SA, Asa SL, Dralle H, Elisei R, Evans DB, Gagel RF, et al. Revised American Thyroid Association guidelines for the management of medullary thyroid carcinoma: the American Thyroid Association Guidelines Task Force on medullary thyroid carcinoma. Thyroid 2015;25(June 6):567−610.

[60] Antonelli A, Fallahi P, Ferrari SM, Carpi A, Berti P, Materazzi G, et al. Dedifferentiated thyroid cancer: a therapeutic challenge. Biomed Pharmacother 2008;62(October 8):559−63.

[61] Haugen BR, Alexander EK, Bible KC, Doherty GM, Mandel SJ, Nikiforov YE, et al. 2015 American Thyroid Association Management Guidelines for adult patients with thyroid nodules and differentiated thyroid cancer: The American Thyroid Association guidelines task force on thyroid nodules and differentiated thyroid cancer. Thyroid 2016;26(January 1):1−33.

[62] Pellegriti G, Mannarino C, Russo M, Terranova R, Marturano I, Vigneri R, et al. Increased mortality in patients with differentiated thyroid cancer associated with Graves' disease. J Clin Endocrinol Metab 2013;98(January 3):1014−21.

[63] Haymart MR, Repplinger DJ, Leverson GE, Elson DF, Sippel RS, Jaume JC, et al. Higher serum thyroid stimulating hormone level in thyroid nodule patients is associated with greater risks of differentiated thyroid cancer and advanced tumor stage. J Clin Endocrinol Metab 2008;93:809−14. See more at: http://press.endocrine.org/doi/full/10.1210/jc.2011-2735#_i18.

[64] Alexander EK, Kennedy GC, Baloch ZW, et al. Preoperative diagnosis of benign thyroid nodules with indeterminate cytology. N Engl J Med 2012;367:705−15. See more at: http://press.endocrine.org/doi/full/10.1210/jc.2013-3584#sthash.Pwy0edSI.dpuf.

[65] McIver B, Castro MR, Morris JC, Bernet V, Smallridge R, Henry M, et al. An independent study of a gene expression classifier (Afirma) in the evaluation of cytologically indeterminate thyroid nodules. J Clin Endocrinol Metab 2014;99 (April 11):4069−77.

[66] Spencer C. Challenges of serum thyroglobulin (Tg) measurement in the presence of Tg autoantibodies. J Clin Endocrinol Metab 2004;89:3702−4.

[67] Takano T, Miyauchi A, Yoshida H, Hasegawa Y, Kuma K, Amino N. Quantitative measurement of thyroglobulin mRNA in peripheral blood of patients after total thyroidectomy. Br J Cancer 2001;85(July 1):102–6.

[68] Weber T, Lacroix J, Weitz J, Amnan K, Magener A, Hölting T, et al. Expression of cytokeratin 20 in thyroid carcinomas and peripheral blood detected by reverse transcription polymerase chain reaction. Br J Cancer 2000;82(January 1):157–60.

[69] Cheung K, Roman SA, Wang TS, Walker HD, Sosa JA. Calcitonin measurement in the evaluation of thyroid nodules in the United States: a cost-effectiveness and decision analysis. J Clin Endocrinol Metab 2008;93(June 6):2173–80.

[70] Nozières C, Chardon L, Goichot B, Borson-Chazot F, Hervieu V, Chikh K, et al. Neuroendocrine tumors producing calcitonin: characteristics, prognosis and potential interest of calcitonin monitoring during follow-up. Eur J Endocrinol 2016;174(March 3):335–41.

[71] Hovens GC, Stokkel MP, Kievit J, Corssmit EP, Pereira AM, Romijn JA, et al. Associations of serum thyrotropin concentrations with recurrence and death in differentiated thyroid cancer. J Clin Endocrinol Metab 2007;92(July 7):2610–15.

[72] Klubo-Gwiezdzinska J, Burman KD, Van Nostrand D, Mete M, Jonklaas J, Wartofsky L. Radioiodine treatment of metastatic thyroid cancer: relative efficacy and side effect profile of preparation by thyroid hormone withdrawal versus recombinant human thyrotropin. Thyroid 2012;22(March 3):310–17.

[73] Haugen BR, Pacini F, Reiners C, Schlumberger M, Ladenson PW, Sherman SI, et al. A comparison of recombinant human thyrotropin and thyroid hormone withdrawal for the detection of thyroid remnant or cancer 1. J Clin Endocrinol Metab 1999;84 (November 11):3877–85.

[74] Lee J, Yun MJ, Nam KH, Chung WY, Soh EY, Park CS. Quality of life and effectiveness comparisons of thyroxine withdrawal, triiodothyronine withdrawal, and recombinant thyroid-stimulating hormone administration for low-dose radioiodine remnant ablation of differentiated thyroid carcinoma. Thyroid 2010;20(February 2):173–9.

[75] Castagna MG, Brilli L, Pilli T, Montanaro A, Cipri C, Fioravanti C, et al. Limited value of repeat recombinant human thyrotropin (rhTSH)-stimulated thyroglobulin testing in differentiated thyroid carcinoma patients with previous negative rhTSH-stimulated thyroglobulin and undetectable basal serum thyroglobulin levels. J Clin Endocrinol Metab 2008;93(January 1):76–81.

[76] Kim WG, Yoon JH, Kim WB, Kim TY, Kim EY, Kim JM, et al. Change of serum antithyroglobulin antibody levels is useful for prediction of clinical recurrence in thyroglobulin-negative patients with differentiated thyroid carcinoma. J Clin Endocrinol Metab 2008;93(December 12):4683–9.

[77] Giraudet AL, Al Ghulzan A, Aupérin A, Leboulleux S, Chehboun A, Troalen F, et al. Progression of medullary thyroid carcinoma: assessment with calcitonin and carcinoembryonic antigen doubling times. Eur J Endocrinol 2008;158(February 2):239–46.

[78] Machens A, Ukkat J, Hauptmann S, Dralle H. Abnormal carcinoembryonic antigen levels and medullary thyroid cancer progression: a multivariate analysis. Arch Surg 2007;142(March 3):289–93.

FURTHER READING

Bjoro T, Holmen J, Kruger O, Midthjell K, Hunstad K, Schreiner T, et al. Prevalence of thyroid disease, thyroid dysfunction and thyroid peroxidase antibodies in a large, unselected population. The Health Study of Nord-Trondelag (HUNT). Eur J Endocrinol 2000;143(November 5):639−47.

Torosian L, Manrique G, Alvarez B, Lago G, Roca R, Belzarena C. Blood thyroglobulin and TSH receptor mRNA detection by RT-PCR in the follow-up of differentiated thyroid cancer patients. Rev Esp Med Nucl 2010;29(May−June 3):109−13.

Disorders related to calcium metabolism

3

Christopher Symonds, MD[1] and Joshua Buse, PhD[2]

[1]*Associate Professor, Department of Medicine, Division of Endocrinology,*
University of Calgary, Calgary, Alberta, Canada
[2]*Biochemistry Fellow, Department of Pathology and Laboratory Medicine,*
University of Calgary and Calgary Laboratory Services, Calgary, Alberta, Canada

CHAPTER OUTLINE

Endocrine Biomarkers. DOI: http://dx.doi.org/10.1016/B978-0-12-803412-5.00003-3

3.1 CALCIUM

3.1.1 OVERVIEW OF CALCIUM DISORDERS

Abnormalities in serum calcium are commonly sought by clinicians often as part of an investigation to unearth the cause of nonspecific symptoms. Medical school dogma recalls the uncommon presentation of hypercalcemia as the first indicator of an undiagnosed occult malignancy. Thus calcium is often part of a screening panel leading to the coincidental discovery of hypercalcemia or, less frequently, hypocalcemia. More often than not, these findings are asymptomatic and unrelated to the current complaints of the patient. Challenges arise in the interpretation of serum calcium particularly because of the influence of pH, fasting, nutrition, and both prescription and over the counter medications. The obligate intertwining of disorders of calcium with renal function, vitamin D, phosphate, and parathyroid function adds to these challenges. Therefore an understanding of the basic physiology of calcium and its allied hormones along with a systematic approach to ordering and evaluating serum calcium is imperative.

3.1.1.1 Epidemiology of selected calcium disorders

- Primary hyperparathyroidism (1HPTH), the most common cause of hypercalcemia, has an incidence of 1 in 500−1000 with the highest incidence found in women in their sixth and seventh decades [1].
- 1HPTH is usually asymptomatic and discovered incidentally. Kidney stones, fractures with severe osteoporosis, abdominal pain, and significant cognitive or behavioral impairment are distinctly uncommon in 21st century presentations.
- Malignancy is the most common cause of hypercalcemia in hospitalized patients.
- Granulomatous diseases such as sarcoidosis or tuberculosis, prolonged immobilization, medications, and vitamin D excess are most frequently encountered in a long list of nonparathyroid-mediated etiologies of hypercalcemia.
- Hypocalcemia may be autoimmune or congenital but is most commonly acquired following the inadvertent devascularization of all four parathyroid glands during thyroid surgery.

3.1.1.2 Basic physiology of calcium/calcitropic hormones including phosphate

The human body has an intricate system to maintain serum calcium within a very tight normal range (2.1−2.55 mmol/L). Understanding the interplay between calcium and its allied hormones is best accomplished by remembering the basic physiology starting with the parathyroid glands. While the parathyroids can uncommonly vary in number or location, there are usually four that reside at the superior and inferior poles of the thyroid gland bilaterally sharing some common blood supply but very little interaction with its larger thyroid neighbor. Each normal parathyroid gland is 4×6 mm in size and weighs 35−40 mg. Each parathyroid chief cell has a number of important receptors that evaluate, modulate, and maintain calcium homeostasis. Indeed the maintenance of normocalcemia via parathyroid hormone (PTH) is the primary function of the parathyroids, often at the expense of the skeletal or renal systems in disease states.

Calcium is a key hormone with obligate roles in cellular signaling, cardiac, neurologic, and muscular function. While serum phosphate (and its allied regulator fibroblast growth factor-23 (FGF-23)), calcitriol (1,25-dihydroxyvitamin D), and pH all play key roles in PTH secretion, serum calcium has the strongest influence through the calcium-sensing receptors (CaSR) on the parathyroid cell surfaces. Renal dysfunction, from mild chronic renal failure to end-stage renal disease requiring dialysis or transplant, may affect all of these hormones. Thus any abnormal serum calcium must be interpreted in light of the current serum creatinine or estimated glomerular filtration rate (eGFR).

Calcium physiology may be summarized as follows:

- Calcium, like many hormones, circulates in both free and bound forms. The major binding protein is albumin. Serum calcium measured in the laboratory should be thought of as "total" calcium. However, it is the ionized or "free" calcium that is responsible for the biological actions and therefore is of most clinical relevance. A simple correction factor to account for an abnormal albumin can be applied to derive a "corrected" or "free" calcium: corrected calcium = serum calcium + ((40-albumin (g/L)) × 0.02). Application of this "correction factor" (especially if ionized calcium measurement is not available) will permit the clinician to correctly diagnose calcium disorders even in the presence of abnormally low albumin/serum proteins.
- Measurement of ionized (or free) calcium may circumvent binding issues but come with challenges around availability, phlebotomy conditions, and expense.
- Phosphate, an anion, is an avid binder of the cation calcium. Therefore higher serum phosphate will result in a lower unbound or free calcium.
- Any *drop* in circulating free calcium is detected by the parathyroid CaSR and results in a rapid release of PTH. PTH has important subsequent effects on bone, kidney, and gastrointestinal tract all designed to reestablish normocalcemia:

- PTH will stimulate bone osteoclasts and osteocytes to resorb and release calcium into the circulation from the large skeletal calcium repository.
- PTH will increase the proportion of filtered calcium in the renal tubule that is reabsorbed into the circulation thus reducing the urinary loss of calcium. PTH will also reduce the reabsorption of filtered phosphate which will increase urinary phosphate loss. A lower serum phosphate means less binding of calcium resulting in a higher free component.
- PTH will increase the function of the one-alpha hydroxylase enzyme that mediates the conversion of 25-hydroxyvitamin D (the largest circulating form of vitamin D) to its active metabolite calcitriol. Calcitriol will in turn stimulate gastrointestinal cells to absorb a greater proportion of ingested calcium into the circulation than usual.

- Any *rise* in circulating free calcium will have the *opposite* effects via a drop in PTH release. This is another example of an intricate feedback loop to maintain calcium homeostasis.

3.1.2 HYPERCALCEMIA

3.1.2.1 Clinical picture

Hypercalcemia is frequently sought and not infrequently found as a part of multi-channel laboratory testing panels initiated by both physicians and insurance companies as screening tests for nonspecific symptoms or occult disease. An elevated serum calcium uncommonly causes direct symptoms when it is less than or equal to 0.25 mmol/L above the upper end of the reference range. The challenge to the clinician is that many patients with an incidental discovery of hypercalcemia will fall in this category yet have nonspecific, usually unrelated, symptoms.

However, patients certainly can present with symptoms directly related to high circulating serum calcium or its consequences and causes. Symptoms may include polyuria, polydipsia, and nocturia because of calcium-induced reduction in renal water reabsorption. Constipation, abdominal pain, bone pain, fatigue, cognitive dysfunction, muscle weakness, and joint pains are all well-described hypercalcemic symptoms but are nonspecific. Renal colic, known nephrolithiasis, low trauma fractures, and osteoporosis in patients with no good alternative causes are all more specific points on history that should lead to an evaluation for hypercalcemia.

Family history is important to note. Familial hypocalciuric hypercalcemia (FHH), the most common congenital cause of hypercalcemia (due to dysfunction of the CaSR), is inherited in an autosomal dominant fashion [2]. There are also rare but serious inherited autosomal dominant conditions such as multiple endocrine neoplasia types 1 and 2A that commonly present with hypercalcemia. A good medication history is crucial, particularly for current or recent ingestion of thiazide diuretics, lithium, or high-dose vitamins D and A. Usual supplemental doses of calcium or vitamin D rarely cause hypercalcemia in the face of normal renal function. A history of known malignancy especially with metastases seems obvious but may be overlooked. Similarly, hospitalized and/or immobilized patients, those with respiratory disease, TB exposure, and other endocrine conditions such as hyperthyroidism or adrenal insufficiency, may also present with hypercalcemia of various mechanisms.

The physical examination of a patient with asymptomatic hypercalcemia is often normal. Clinical signs of dehydration, band keratopathy, kyphosis, or an abnormal respiratory exam are important to evaluate. Parathyroid adenomas are rarely palpable in the neck—abnormal neck masses are more likely to be unrelated incidentally discovered thyroid nodules. Rarely a palpable parathyroid mass could represent a parathyroid carcinoma almost always in association with very high serum calcium and PTH levels.

3.1.2.2 Initial diagnostic approach

The approach to the differential diagnosis of a patient who presents with hypercalcemia starts with ensuring that the elevated serum calcium is corrected for albumin or is confirmed by ionized calcium: "true" hypercalcemia.

- The interpretation of "true" hypercalcemia always requires a concomitant PTH level: Is the PTH response appropriate or inappropriate for hypercalcemia?
- The physiologic response of normal parathyroid glands to hypercalcemia should always be a suppressed or low level of PTH. Confusion can arise when PTH measurements are toward the lower end of the reference range but not fully suppressed and therefore still detectable. Some PTH assays may detect circulating nonfunctional PTH fragments leading to this situation.
- Another common scenario arises when the PTH is within the upper half of the reference range when the calcium is high. The diagnosis is usually 1HPTH but clinicians can be misled by the "normal" PTH. In this situation the parathyroid glands are not acting physiologically and thus the PTH is deemed to be *inappropriately normal* for hypercalcemia. This inappropriate secretion of PTH despite the hypercalcemia indicates that the PTH secretion is excessive and autonomous and thus likely due to a parathyroid adenoma.

3.1.2.3 Causes of hypercalcemia

The two most important causes of hypercalcemia are 1HPTH and malignancy. However, practically, the most useful approach is to consider parathyroid-mediated and nonparathyroid-mediated etiologies.

3.1.2.4 Parathyroid mediated

- 1HPTH is the most common cause of hypercalcemia especially in the outpatient setting. 1HPTH results from the development of a usually solitary adenoma in one of the four parathyroid glands. However, double adenomas can be seen as well as an adenoma arising in an ectopic parathyroid location. 1HPTH may be associated with multiple endocrine neoplasia types 1 and 2A as well as other rare familial syndromes such as hyperparathyroidism-jaw tumor syndrome. Diffuse four gland hyperplasia is less common in apparently sporadic 1HPTH without current or past renal dysfunction.

- Secondary hyperparathyroidism (2HPTH) is *not* a cause of hypercalcemia but rather reflects the appropriate elevation in PTH to maintain normal serum calcium in response to renal failure or severe vitamin D deficiency. This is mentioned here for completeness in discussion of "hyperparathyroidism" but must be stressed that this physiologic compensatory PTH secretion does not appear with hypercalcemia and is not necessarily a disease process unto itself.
- Tertiary hyperparathyroidism (3HPTH): If the cause of 2HPTH remains untreated (i.e., chronic renal failure), prolonged parathyroid gland stimulation may lead to hypertrophy, hyperplasia and eventually autonomous and inappropriately high function causing a gradual and persistent rise in serum calcium. This latter progression may present with hypercalcemia despite renal transplant or vitamin D normalization.
- FHH is caused by an inactivating mutation in CaSR or one of its associated proteins. The "normal" set point for calcium rises to compensate for this autosomal dominant inherited desensitizing mutation and the urine calcium usually falls. Identifying FHH and differentiating it from 1HPTH is very important but can be challenging due to the biochemical similarity.

3.1.2.5 *Nonparathyroid mediated*

- Malignancy-associated hypercalcemia can arise via several mechanisms:
  - Most commonly the tumor produces pathologic amounts of PTH-related protein, a peptide physiologically highly similar to PTH with a common receptor and downstream effects. This form of malignancy-associated hypercalcemia is also called: humoral hypercalcemia of malignancy. PTH-related protein is normally produced in the female lactating breast to enrich milk calcium content. Solid tumors of the lung, head and neck, and kidney are the most common cancers associated with this cause of hypercalcemia but have been described in numerous other tumor types.
  - Some forms of non-Hodgkin's lymphoma may contain granulomatous tissue replete with one-alpha hydroxylase enzymatic activity causing a vitamin D (calcitriol)-mediated hypercalcemia.
  - Less frequently seen are: cancer-associated hypercalcemia from osteolytic bony metastases directly releasing calcium into the circulation or true ectopic PTH secretion from a tumor.

3.1.2.6 Less common other causes *of nonparathyroid-mediated hypercalcemia include but are not limited to*

- Sarcoidosis and other granulomatous diseases caused by tuberculosis, fungi, spirochetes, or connective tissue disorders.

- Milk-alkali syndrome can be seen more frequently in elderly patients who take alkaline calcium supplements in the face of dehydration and mild renal dysfunction.
- Immobilization hypercalcemia is an important cause in hospitalized and spinal-cord injured patients. Hypercalcemia results from unregulated and profound osteoclast-mediated bone resorption in the absence of the attenuating effects of gravity [3].
- Endocrine causes hyperthyroidism, adrenal insufficiency, pheochromocytoma, and vasoactive intestinal polypeptide (VIP) secreting lesions.
- Medications: thiazide diuretics, lithium, high-dose vitamin D or A, growth hormone, aminophylline, and many other agents implicated in case reports.

3.1.2.7 Special situations: pediatrics, pregnancy, and critical illness

Pediatrics: Pediatric causes of hypercalcemia will shift away from the acquired toward a heightened awareness of the congenital causes and those associated with other illnesses. In pediatric patients with parathyroid-mediated hypercalcemia the clinician needs to consider FHH, hyperparathyroidism-jaw tumor syndrome, MEN 1 & 2A, and familial-isolated hyperparathyroidism. Neonatal severe 1HPTH is a rare life-threatening form of hypercalcemia usually caused by homozygous inactivating mutations in CaSR [4]. The full list of nonparathyroid-mediated causes listed above will be more prevalent in children than adults and must be viewed through a pediatric lens.

Pregnancy: The complicated calcium-phosphate-parathyroid axis becomes more challenging to interpret in pregnant women and detailed discussion is beyond the scope of the current chapter. Fortunately, intact parathyroid function ensures calcium homeostasis in both mother and fetus in most situations. 1HPTH may be an exception to this rule and can provide challenges particularly when a patient with previously undiagnosed or untreated 1HPTH presents for her initial evaluation while already pregnant. Maternal hypercalcemia may result in significant suppression of the fetal parathyroid glands necessitating consideration of second trimester maternal parathyroid surgery.

Critical illness: The interpretation of calcium indices in critical illness requires knowledge of the patient's nutritional status, acid–base balance, and renal and hepatic function. All of these may shift binding of calcium such that the usual correction formula for total calcium may be unreliable. The ionized calcium may be the best test and is often routinely available on venous or arterial blood gas samples. Severe hypercalcemia may require intensive care unit (ICU) evaluation and treatment considering all of the above causes. Conversely, prolonged immobilization, particularly in younger patients or those who with acute spinal-cord injuries, may predispose to new hypercalcemia with often very high both serum and urine calcium concentrations. ICU staff should consider this diagnosis particularly in the face of unexpected polyuria, dehydration, delirium, or renal failure.

3.1.2.8 Diagnostic approach to parathyroid-mediated hypercalcemia

Once the cause of hypercalcemia is deemed to be parathyroid mediated, the differential diagnosis is usually limited to 1HPTH, FHH, or 3HPTH.

Differentiating 1HPTH and FHH can be challenging. As a general rule the following features would more strongly favor a diagnosis of 1HPTH but *not* FHH:

- Serum calcium >3.0 mmol/L.
- End-organ ramifications of hypercalcemia: kidney stones, nephrolithiasis, low bone density, fragility fractures, or radiographic evidence of heightened bone resorption such as osteitis fibrosa cystica.
- Fractional excretion of calcium (FeCa) > 1%, as calculated from concomitant serum and urine calcium and creatinine often requiring both a blood draw and a 24 h urine collection.
- No family history of hypercalcemia or hyperparathyroidism.

  In the absence of these differentiating features, genetic testing for mutations in the CaSR or its associated proteins such AP2S1 may be required. While expensive, the ramifications for ~50% of family members to be similarly affected with FHH may justify the cost. A definitive diagnosis may reduce or eliminate unnecessary investigations or surgery in first degree relatives of the individual diagnosed with FHH.
- 3HPTH by definition must have been preceded by 2HPTH commonly caused by progressive renal failure and end-stage renal disease requiring dialysis or transplantation. Severe, untreated vitamin D deficiency as may be seen in celiac disease or inflammatory bowel disease (IBD) can also cause substantial appropriate compensatory rises in PTH that may eventually lead to autonomous parathyroid function and hypercalcemia [5].

3.1.2.9 Diagnostic approach to nonparathyroid-mediated hypercalcemia

In cases where the low PTH response to hypercalcemia is deemed to be appropriate (i.e., suppressed), a detailed history and physical exam including a thorough medication/supplement review, family history, and travel/infectious-exposure history is needed to systematically address the list of nonparathyroid-mediated causes of hypercalcemia outlined above. Malignancy, for example, may often be strongly suspected after this initial clinical evaluation.

Other clinical tips for work-up include

- The most common way that malignancy causes hypercalcemia is through tumor production of PTH-related protein, also known as humoral hypercalcemia of malignancy. While PTH-related protein will not cross-react with modern PTH assays, both PTH and PTH-related protein interact with a common receptor on parathyroid, bone, kidney, and intestinal cells. Thus it would be common to see low or low-end normal serum phosphate, lower 25-hydroxyvitamin D, normal or only mildly elevated urine calcium and higher indices of bone resorption in *both* 1HPTH and PTH-related protein-associated malignancy.

- PTH-related protein can also be measured but practically speaking, a diagnosis of malignancy is often made before the PTH-related protein level is needed.
- Granulomatous diseases such as sarcoidosis or non-Hodgkin's lymphoma cause hypercalcemia by making inappropriate amounts of activated vitamin D: (calcitriol). Measuring 1,25-dihydroxyvitamin D may provide a clue particularly if high or normal, as low levels might be normally be expected in the presence of a low circulating PTH.
- Vitamin D excess may also promote increased phosphate reabsorption from the GI tract. A low PTH level will reduce urine phosphate losses. Thus serum phosphate may be high or at the upper end of the reference range in cases of vitamin D overdose.
- Marked hypercalciuria often >20 mmol/day may be seen in immobilization hypercalcemia or other non-PTH-mediated causes because PTH normally promotes the reabsorption of filtered calcium reducing urine calcium concentrations.
- Imaging studies starting with a simple chest X-ray may help identify granulomatous disease-like sarcoidosis. Mammography, fecal immunochemical test (FIT) colon cancer screening, CT imaging of chest, abdomen and pelvis, or total body bone scintigraphy may be selectively indicated if malignancy is suspected.

KEY POINTS

Hypercalcemia is most commonly discovered incidentally on routine blood tests.

- Serum calcium must be corrected for albumin or ionized calcium measured instead.
- PTH measurements must be interpreted in the context of concomitant circulating calcium levels.
- Parathyroid-mediated hypercalcemia is diagnosed when the PTH is inappropriately high for hypercalcemia. The differential diagnosis is narrow.
- Nonparathyroid-mediated hypercalcemia is diagnosed when the PTH is appropriately low for hypercalcemia. The differential diagnosis is broad.
- 1HPTH is the most common cause of hypercalcemia.

3.1.3 HYPOCALCEMIA

3.1.3.1 Clinical picture

Hypocalcemia is found less frequently than hypercalcemia yet may have serious or even fatal consequences. Clinical manifestations can be very heterogeneous and patient specific. Classic symptoms include peri-oral and fingertip numbness or paresthesias, neuromuscular irritability, abdominal pains, and hand or foot cramping. Worrisome and potentially life-threatening complications of severe hypocalcemia include laryngospasm, bronchospasm, and cardiac arrhythmias with a prolonged QTc (corrected Q-T interval) interval. In the long term, basal calcifications and associated seizure disorders are possible complications of untreated hypocalcemia. However, the symptoms and complications may be very dependent upon how long the hypocalcemia has been present. For example an acute drop in serum calcium to 1.7 mmol/L following thyroid surgery and inadvertent damage or removal of all four parathyroid glands will usually result in marked symptoms at the bedside. In contradistinction, with identically low serum calcium, patients with long standing hypocalcemia may remain symptom free. Hospitalized patients may have unique and often transient causes of hypocalcemia. Confirming true hypocalcemia with ionized calcium collected appropriately is very important particularly in patients with nutritional, renal, hepatic, or cardiac issues in hospital.

Clinicians should inquire about a personal and family history of hypocalcemia, autoimmune disease, and previous neck surgery, often a thyroidectomy. Renal disease, bowel surgery, malabsorption syndromes, cataracts, epilepsy, and liver disease are all important to note. A good medication and supplement history are also essential especially the intake of calcium, phosphate, vitamin D, and magnesium.

The physical examination may provide useful immediate information that may inform both the cause and urgency for treatment of hypocalcemia.

- A rapid assessment of airway, breathing, and circulation along with stridor and pulse rate/regularity is always the first step. With clinical stability assured a number of other findings should be sought.
- Acutely, look for neuromuscular irritability and signs of latent tetany. Generations of medical students have been taught to elicit Chvostek's sign by tapping over the facial nerve and Trousseau' sign by inflating the blood pressure cuff 10 mmHg above systolic and looking for the characteristic wrist flexion and finger extension in response to temporary ischemia.
- Practically speaking, while easy to perform, testing for Chvostek's sign has a low sensitivity and specificity for hypocalcemia.
- Trousseau's sign, while temporarily uncomfortable for patients, is much more reliable to detect those who may need more urgent treatment of hypocalcemia.
- Albright's hereditary osteodystrophy (AHO) is a constellation of physical features highlighted by short stature and one or more shortened metacarpals and/or metatarsals, often associated with obesity and a mild diminution in intellectual function. AHO is seen in the most common form of pseudohypoparathyroidism (PHP): type 1a.
- Anterior neck scars suggestive of past surgery, cataracts, vitiligo, and signs of hypothyroidism (autoimmune disorders) or chronic liver disease (severe vitamin D deficiency) may also give clinical clues as to the etiology of hypocalcemia.

3.1.3.2 Causes of hypocalcemia

Broadly speaking the causes of hypocalcemia can be divided into four groups:

1. Insufficient PTH production
   a. The most common etiology is iatrogenic hypoparathyroidism caused by parathyroid destruction, devascularization, or inadvertent removal during thyroid or parathyroid surgery.

    b. Autoimmune parathyroid damage or impairment of PTH action may be seen as an isolated condition or as part of autoimmune polyglandular syndromes (APS). APS type 1 is an autosomal recessive conditions caused by mutations in the AIRE gene [6]. It is most frequently diagnosed in childhood and is often seen with mucocutaneous candidiasis and adrenal insufficiency. Other more common autoimmune conditions such as type-1 diabetes mellitus or thyroid disease are less frequently associated with hypocalcemia.

    c. Direct destruction of all four parathyroid glands may also arise from external beam radiotherapy or infiltrative conditions such as hemochromatosis, metastatic tumors, or Wilson's disease.

    d. Any condition that interferes with normal parathyroid gland development, PTH release, or action may be diagnosed in neonates or children (congenitally). These conditions such as 11q22 syndrome (formerly called DiGeorge's sequence or syndrome) are quite uncommon.

    e. Uncommonly, autosomal dominant hypocalcemic hypercalciuria may arise from activating mutations in the CaSR causing a heightened sensitivity and reduced set point for PTH secretion with concomitant hypercalciuria. This condition is the "opposite" of FHH described in 3rd paragraph of Section 3.1.2.1.

    f. Hyper- or hypomagnesemia can cause hypocalcemia by impairing PTH release.

2. Impaired PTH action (PTH resistance)

    a. PTH resistance or PHP in which genetic defects in the PTH/PTH-related protein receptor or receptor action impair the ability to maintain normal serum calcium.

    b. Several forms of PHP exist: the most common is type 1a which can be diagnosed during physical examination by observing the features of AHO, as described above. Other forms such as type 1b have no phenotypic AHO features but are biochemically identical to type 1a with significant hypocalcemia and elevated PTH levels. PHP types 1c and 2 are less well characterized and may have overlap with type 1a or severe vitamin D deficiency.

    c. Confusingly, some patients may have classical AHO findings but completely normal calcium and PTH biochemistry—these patients are labeled as having pseudopseudohypoparathyroidism (pseudoPHP). PHP type 1a and pseudoPHP can occur in members of the same family suggesting the role of imprinting or epigenetics.

    d. Hypomagnesemia can impair PTH action.

3. Insufficient vitamin D or vitamin D action

    a. Vitamin D deficiency is extremely common yet rarely causes significant symptomatic hypocalcemia for two main reasons.

        i. Reference ranges for 25-hydroxyvitamin D are often set high such that, particularly in northern climates, a substantial proportion of the population may fall just below the reference or ideal range (>80 nmol/L in many labs). The clinical significance of mild vitamin D insufficiency remains unclear and does not lead to hypocalcemia (please see Section 3.4).

ii. Clinically relevant rises in PTH in response to vitamin D deficiency will act to maintain serum calcium in the normal range, thus preventing hypocalcemia: this is 2HPTH. Serum 25-hydroxyvitamin D levels need to be persistently below 50 nmol/L or often 25 nmol/L before 2HPTH occurs.

b. Malabsorption, gastric bypass, IBD, cirrhosis, and chronic renal failure are the most common causes of severe vitamin D deficiency. These patients may present with osteomalacia, profound proximal muscle weakness, hypophosphatemia, and hypocalcemia despite compensatory rise in PTH.

c. Vitamin D resistance can occur in two exceedingly rare forms usually presenting in childhood with rickets and hypocalcemia. Type 1 is caused by a mutation in the 1-alpha hydroxylase enzyme causing impairment in the conversion of 25-hydroxyvitamin D to the active form, calcitriol. Type 2 is true resistance to calcitriol caused by mutations in the vitamin D receptor.

4. Excess calcium binding or loss:
   - Hyperphosphatemia due to: acute or chronic renal failure, excess oral phosphate intake or enema phosphate load or tumor lysis syndrome following chemotherapy.
   - Excess bone calcium consumption after parathyroid surgery or thyroid surgery for hyperthyroidism is known as the "hungry bone" syndrome. Previous PTH or thyroid-mediated high bone turnover may result in significant prolonged needs for calcium and vitamin D replacement following surgery.
   - Acute pancreatitis likely due to excess binding or consumption of calcium in the inflamed pancreas.
   - Rhabdomyolysis, rapid blood transfusion, and chelation therapy.
   - Intravenous bisphosphonates treatment, foscarnet, and some chemotherapy agents such as imatinib can all cause acute hypocalcemia.

3.1.3.3 Special situations: pediatrics, pregnancy, and critical illness

Pediatrics: Suspecting hypocalcemia in neonates and children requires some diagnostic acumen. The causes will follow the above list for adults but congenital disorders (particularly those associated with other conditions and phenotypes) must be more strongly considered. For example, in patients with mucocutaneous candidiasis and hypocalcemia, clinicians must think of APS type 1 which can also present concomitantly or subsequently with life-threatening primary adrenal insufficiency. Children with the DiGeorge sequence can have associated cardiac defects, cleft palate, and immunodeficiency. The presence of rickets particularly once a child starts to walk necessitates a detailed work-up including evaluation of calcium, phosphate, and vitamin D.

Pregnancy: The most common clinical scenario involves a pregnant woman known to have preexisting hypoparathyroidism often because of past thyroid surgery. The doses of her usual maintenance oral calcium and activated vitamin D preparations may change during pregnancy and do require monitoring. Avoidance of maternal hypocalcemia is very important to prevent fetal 2HPTH. In lactating women, PTH-related protein levels will rise dramatically. This may result in a significant postpartum drop in the requirements for calcium and vitamin D treatment—indeed, these medications can often be temporarily discontinued. Close monitoring of calcium indices is needed because, as the woman weans her child, calcium and vitamin D therapy will need to be restarted.

Critical illness: Critical illness can commonly cause hypocalcemia in the ICU. Ionized calcium measurements should be routine to ensure true hypocalcemia is being identified. Causes are multifactorial but rapid identification and treatment with intravenous calcium infusions may be necessary.

3.1.3.4 Diagnostic approach (laboratory test focussed)

The diagnostic approach should focus on the most likely causes given the clinical picture elucidated from a detailed history and physical examination, as outlined earlier.

Important laboratory tests include

- Serum albumin and/or ionized calcium to confirm true hypocalcemia.
- Serum phosphate (often immediately available) gives an excellent clue as to the cause for hypocalcemia assuming normal renal function. Hypophosphatemia suggests appropriate or excess PTH secretion and action: this may lead the clinician to consider vitamin D deficiency or loss/binding of both calcium and phosphate. A high serum phosphate may indicate a loss of PTH secretion or action-most commonly, acquired hypoparathyroidism postsurgically or autoimmune damage. PHP also needs to be considered in patients with coexisting hypocalcemia and hyperphosphatemia.
- Measuring PTH with an intact assay will allow the clinician to separate conditions of parathyroid loss or damage (where the PTH is low) from disorders of PTH action such as PHP (that present with a high PTH).
- Serum creatinine and eGFR measurements are imperative. Progressive renal dysfunction will raise serum phosphate and PTH while impairing the activation of vitamin D. 2HPTH will ensue with severe hypocalcemia occurring uncommonly.
- Serum magnesium is also helpful: both hypomagnesemia and hypermagnesemia can cause hypocalcemia.
- Urine calcium measurements may provide useful information, often with a 24 h urine collection. PTH-deficient patients will have inappropriately normal or high urine calcium. Patients with severe vitamin D deficiency should have appropriately low urine calcium due to the physiologic rise in PTH. Interestingly, patients with PHP may have normal or low urine calcium

concentrations as the normal renal distal tubule calcium reabsorption response to high PTH may remain intact.

Urine cyclic amp measurements often following exogenous PTH administration can help differentiate between subtypes of PHP and other etiologies of a raised serum PTH [7].

KEY POINTS

Hypocalcemia is less common than hypercalcemia but can cause serious cardiac, respiratory, and neurologic problems if untreated.

- Inadvertent hypoparathyroidism following thyroid or parathyroid surgery is the most common cause of hypocalcemia in adults.
- Rare inherited disorders such as PHP will usually present in childhood but milder forms may go undetected until adulthood.
- The physical examination may provide useful immediate information that may inform both the cause and urgency for treatment of hypocalcemia: the features of AHO and eliciting Trousseau's sign should be reviewed.
- Long-term treatment of hypoparathyroidism should target serum calcium at the lower end or just below the reference range to reduce the risk of hypercalciuria and renal damage.

3.1.4 LABORATORY CONSIDERATIONS FOR ASSESSING HYPO- AND HYPERCALCEMIC DISEASES

The critical nature of abnormal calcemic states highlights the need to understand the (in)direct influence of various preanalytical, analytical, and postanalytical factors on each laboratory test in their differential diagnosis.

3.1.4.1 Total blood calcium

Calcium's biological transport within the circulation system relies upon the formation of a stable equilibrium between its ionized (free) and bound (albumin, salts, etc.) forms, with total blood calcium measuring both of these forms together; making it one of the most commonly ordered laboratory tests.

3.1.4.1.1 Preanalytical considerations

Inter- and intraindividual variables that influence the measurement of total calcium include

- Changes in the patient's posture, which have been shown to change both total and ionized calcium. Having the patient change from a standing to supine position increased plasma volume by 14.1% and directly reduced the measured concentration of calcium [8].
- Circadian rhythm, which results in cyclical changes in calcium concentration due to the direct influence of PTH. This influence produces an early morning

nadir and afternoon acrophase, with ionized calcium being more drastically affected; vary by as much as ± 0.13 mmol/L (0.5 mg/dL) [9].

- Albumin concentration, as approximately $1/3-1/2$ of plasma calcium being albumin bound and therefore a decrease in albumin results in a decrease in total calcium; ionized calcium concentrations are often normal. Therefore when a clinician encounters a patient with a depressed albumin level, it is preferable to rely on ionized calcium measurements.
- Underlying disease states can influence the concentration of both total and ionized calcium, with the most pronounced effect observed in those that induce either *in vivo* or *in vitro* hemolysis.

The impact of preanalytical specimen collection/processing on patient results is highlighted in Chapter 1, Endocrine laboratory testing: Excellence, errors, and the need for collaboration between clinicians and clinical chemists; however, specific considerations for a calcium measurement that does not correlate with the patient's clinical presentation include

- The use of blood collection tubes that contain anticoagulants (EDTA, citrate, and oxalate), which can reduce both the total and ionized calcium concentration. For plasma measurement of calcium, it is recommended that low concentration calcium-balanced heparin be used to reduce the error of measurement. Furthermore, use of the correct order of blood tube collection is important because the contact of anticoagulant tubes with the collection needle junction can transfer the anticoagulant to the next tube and may yield falsely low calcium values.
- Prolonged tourniquet use will increase the specimen's calcium concentration, with observed increases observed after as little as 1 min of tourniquet occlusion.

3.1.4.1.2 Analytical considerations

The analytical procedure for measuring total calcium can be performed by an array of techniques and instrumentation, including

- Spectrophotometric detection of calcium-dye-complexes.
  - The nonspecific nature of the calcium-dye complexation can allow for interference from divalent magnesium cations that falsely elevate calcium measurements.
  - Paraproteinemia, caused by an underlying disease state, leads to falsely elevate calcium measurements because they bind ionized calcium, shifting its equilibrium [10].
  - Contrast agents can lead to
    - falsely elevated calcium measurements if the contrast agent absorbs at the same wavelengths as the calcium-dye-complex,
    - falsely decreased calcium measurements because they act as powerful chelating agents (gadoversetamide and gadolinium) [11],

- critical that the contrast agents be cleared prior to collection of a specimen for calcium measurement. Contrast agent clearance typically requires up to 2 h, but is dependent on the patient's dosage and kidney function [12].
- Atomic absorption spectrophotometry has largely been replaced by Inductively coupled plasma mass spectrometry (ICP-MS) in the clinical laboratory, but both methods provide the most accurate and precise measurement of calcium and are not susceptible to same interferences as spectrophotometric detection of calcium-dye-complexes, with the exception of calcium chelating contrast agents.

3.1.4.1.3 Postanalytical considerations

The most important postanalytical consideration is the reference interval used. As such, appropriate use of a reference interval requires that it be generated from a cohort of "healthy" individuals, who are representative of the patient population encountered at your institution; it is vital that the reference intervals be instrumentation/method specific.

3.1.4.2 Ionized blood calcium

Ionized (free) calcium is the biologically active fraction of blood calcium and is a more accurate indicator of calcium status, which can reduce unnecessary work-up of a patient based upon a misleading total calcium result.

3.1.4.2.1 Preanalytical considerations

The contribution of an underlying disease state to an abnormal ionized calcium measurement occurs in a similar manner to total calcium. This primary consideration includes

- An underlying acidotic or alkalotic condition changes the concentration of ionized calcium inversely to the solution's pH; as such, it is directly related to hydrogen ion concentration. This means that with higher pH (i.e., the decreased H^+ ion concentration seen in respiratory alkalosis), albumin binding of calcium is increased and a lower concentration of ionized calcium is observed even though total serum calcium may remain normal.

  Calcium equilibrium is dependent on a number of variables and preanalytical standard operating procedures are put in place to maintain it. Thus while the measurement of total calcium, albumin, and total protein is available in standard laboratories, measurement of ionized calcium remains more difficult and is generally performed only in reference laboratories. Potential preanalytical errors include
- Changes in the pH of the whole blood specimen will modify the buffering action of the carbonate/carbon dioxide equilibrium formed in whole blood. Loss of carbon dioxide due exposure of the specimen to ambient air will increase the pH and decrease ionized calcium.

- Collection of the specimen using calcium-balanced heparin to prevent coagulation. The pretitration of the heparin with calcium reduces its capacity to bind ionized calcium within the whole blood specimen.
- Extended time between specimen collection and analysis can increase ionized calcium by up to 10 μmol/L/h (40 μg/dL/h) at room temperature. If the stability of ionized calcium needs to be extended it can be transported/stored at 4°C immediately after collection.
- Physical activity has been shown to induce an increase in total calcium and ionized calcium through the related increases in bicarbonate, lactate, and albumin concentrations [11].
- Massive blood transfusions depress ionized calcium because of the coadministration of citrate; ionized calcium rapidly returned to normal levels as citrate was redistributed and metabolized [13].

3.1.4.2.2 Analytical considerations

Confirmation of a spurious total calcium concentration is most easily achieved through ionized calcium measurements, which are performed by

- Indirect potentiometry, with a calcium selective electrode employing a selective membrane to measure the potential that is developed, with the calcium concentration being mathematically derived.
- Calculation of ionized calcium concentration can be performed by mathematically applying the measurement for total calcium, albumin, and pH. This is less accurate than the measurement of ionized calcium by indirect potentiometry [14].

3.1.4.2.3 Postanalytical considerations

It is important that the measured ionized calcium concentration be appropriately compared to an instrumentation/method-specific reference interval generated from a cohort of "healthy" individuals representative of the encountered patient population.

3.1.4.3 Urine calcium

Both random and 24 h urine collections are appropriate specimens for the measurement of calcium and can aid in assessing multiple aspects of calcium pathophysiology.

3.1.4.3.1 Preanalytical considerations

The calcium—creatinine ratio provides an accurate estimation of the daily urinary calcium excretion, with significant correlation between a random urine calcium and 24-h urine calcium [15]. A 24 h urine calcium, however, can reduce the influence of several preanalytical factors due to its averaging effect, including

- High/low calcium meals/diets directly influence the concentration of urinary calcium. A diet high in animal protein and salt leads to increased calcium excretion, while caffeine intake induces calcium loss due to its diuretic activity following consumption.

- Drugs or calcium supplements impact urinary calcium concentration; timing and dosage play an integral role. Thiazides reduce urinary calcium excretion, while high calcium supplements increase urinary calcium excretion [16].

Therefore a random urine specimen should be collected from a fasting patient and always normalized to creatinine excretion. The interindividual variability of creatinine levels is greatly impacted in specific populations, including

- males have a greater production of creatinine than females,
- athletes have a greater production of creatinine that is reflected in their larger muscle mass,
- renal impairment/chronic kidney disease (CKD) is a major confounder to interpretation of urine calcium excretion in direct accordance with the degree of decrease in GFR.

Although elevated urine calcium is a sign of an overactive parathyroid gland, this test is not reliable in differentiating or diagnosing hyperparathyroidism as approximately one-third of hyperparathyroid patients have normal urine calcium.

3.1.4.4 Parathyroid hormone

Regulation of calcium absorption/excretions by PTH provides a great indication for its measurement when differentially diagnosing the cause of calcium disorders because of its exquisitely sensitivity to both short- and long-term modifiers of calcium homeostasis.

3.1.4.4.1 Preanalytical considerations

Clinicians typically interpret PTH levels according to the concomitantly measured serum calcium but additional modifying influencers of PTH secretion should also be considered, including

- Seasonal variation, as a relative decrease in PTH being observed during summer months and a relative increase in PTH in the winter; the decrease in PTH concentrations is inversely related to vitamin D concentration [11].
- Circadian variation can provide an explanation for variation in sequential measurements of PTH. The presence of nocturnal and afternoon acrophase separated by a mid-morning and evening nadir should be accounted for during result interpretation; average amplitude of variation has been assessed to be 4.2 ng/L (4.2 pg/mL) [17].
- Consumption of food or supplements containing high concentrations of vitamin D have been linked to an acute, significant decrease in PTH concentration; with one study showing a difference of 15 ng/L (15 pg/mL) in the concentration of PTH between serum vitamin D levels <10 and >18 ng/mL [18].

The 5-min half-life of PTH typically maintains calcium within narrow equilibria, but also makes it susceptible to an array of preanalytical considerations, including

- Specimen selection, with both serum and plasma being acceptable for PTH measurement. Collection of a plasma specimen maximizes PTH stability through the elimination of coagulation time, increase in specimen volume, and reduction in coagulation-induced interferences. Both specimen types, however, benefit from the inclusion of protease inhibitors (aprotinin, leupeptin, or pepstatin) in the collection tube.
- Site of collection, with specimens collected from a peripheral vein can have a PTH concentration up to 30% greater than one collected from the central venous system. This consideration is critical for patients undergoing intraoperative PTH monitoring, where a central venous specimen is often collected and needs to be taken into account by the clinician; clinical monitoring should be based on specimen from the same site of collection. The elevated concentration of PTH in the centrally collected specimens may be a result of the central placement of the catheter in the vein immediately below the thyroid or the decreased concentration in the peripheral specimens may be caused by hormone degradation and hemodilution in the systemic circulation.

3.1.4.4.2 Analytical considerations

Accounting for all of the factors that can influence the concentration of intact, active PTH within a specimen is difficult and further complicated by interassay variability. Several generations of PTH assays have been developed, with second and third generation assays currently being employed for testing [19].

- Second generation assays were purported to react only with the bioactive form of PTH, amino acid sequence 1–84. However, utilization of the assay has demonstrated that this generation also reacts with large C-terminal and N-terminal PTH fragments that contain partially preserved N-terminal structure. This method is represented within the literature as the measurement of "intact parathyroid hormone" and is the most widely used methodology, with numerous vendors providing suitable methodology for its measurement.
- Third generation assays were developed to be more selective for biologically active PTH although reactivity is still observed with N-terminal. This assay is able to assess the ratio of active PTH and PTH fragments. The specificity of the assay is believed to be superior for the whole PTH; however, limited availability of the methodology to two manufactures has reduced its implementation.
- Standardization of the PTH assays is currently ongoing, with a world health organization international standard PTH 1–84 molecule being selected. This recombinant and commutable standard has been deemed suitable for both the third generation immunoassay and LC–MS/MS methods, but not the second

generation due to the difference in cross relativities between the measurement of intact and whole PTH.

- Both second and third generation assays also detect PTH that has undergone posttranslational modification to inactivate it; oxidation of methionine 8 and 18. The importance of this is observed in patients who undergo extensive oxidative stress, where the measurement of PTH does not match the clinical observations.
- Furthermore the methodology employed by various second and third generation assays utilizes a biotin−streptavidin interaction for isolation of the detected PTH. The concentration of the analyte of interest will be measured by either a competition assay with a labeled antigen competing for antigen binding sites or being bound by a second labeled antibody in a sandwich assay. In both instances, high levels of biotin in patient specimens can significantly interfere with the assay resulting in falsely low or high results. *Further information on this subject can be found within* Chapter 1, *Endocrine laboratory testing: Excellence, errors, and the need for collaboration between clinicians and clinical chemists.*

3.1.4.4.3 Postanalytical considerations

The action of PTH in the regulation of various ions' absorption, storage, and excretion means that changes in its concentration do not solely influence calcium concentrations nor are they regulated in a singular manner. For example, pregnant women have a significant decrease in PTH concentrations during their first and second trimesters, modulating calcium concentrations to allow for its optimal delivery to the developing fetus. Therefore awareness of these influences on the level of circulating PTH can prevent confusion for both physicians and/or laboratory personnel when evaluating the apparent changes in sequential parathyroid measurements. Furthermore the use of an appropriate reference range will minimize inappropriate result interpretation. The identification of the reference range for PTH requires well-developed inclusion/exclusion criteria for the selection of a population with "normal" PTH levels. For example, if a healthy patient population has a relatively low serum 25-hydroxyvitamin D concentration than the population used to establish a normal PTH concentration should also have a similarly low concentration of 25-hydroxyvitamin D. In instances where a reference range was developed that excluded individuals with low 25-hydroxyvitamin D concentrations, the PTH reference range was up to 35% lower than the upper normal limit reported for the assay and have led to some individuals recommending that the season, latitude and vitamin D status be taken into account when developing a reference range of PTH. This consideration may be most relevant at the population level, as opposed to interpretation of individual PTH−calcium pairings [11]. Similarly, hyperparathyroidism has been assessed to be more common in individuals with a higher body mass index associated with morbid obesity. Such significant influences on PTH concentration demonstrates that reference ranges should consider the influence of renal function, age, dietary calcium intake, and overall health when assessing how to establish a reference range or considerations to ask laboratory personnel when investigating a PTH measurement that does not correlate with the patient's clinical presentation.

3.1.4.5 Parathyroid hormone-related protein

The structural similarities between PTH and PTH-related protein result in both interacting with the PTH receptor, with PTH-related protein having a stimulating, blocking, or neutral effect on the receptor.

3.1.4.5.1 Preanalytical considerations

Very low levels of circulating PTH-related protein in healthy individuals are a result of its normal mechanism of action in autocrine or paracrine physiology. Therefore it is important to know which physiological states lead to elevated and sustained over-production of PTH-related protein, including pregnancy, lactation, some nonmalignant diseases, and ectopic secretion of PTH-related protein by tumors. This ectopic PTH-related protein production is most commonly seen in carcinomas of breast, lung, head, neck, kidney, bladder, cervix, uterus, and ovary. In many cases, measurement of PTH-related protein is not needed once a characteristic malignancy is found in the setting of hypercalcemia and suppressed PTH.

As with any protein, conservation of structural features and inhibition of degradation are paramount to achieving an accurate measurement of their concentration, with use of a protease inhibitor preventing its denaturation and cleavage. Specimens collected in EDTA tubes and stored on ice or frozen are also suitable for analysis as the rapid separation of cells and/or clots from the plasma further reduce the rate of peptide cleavage.

3.1.4.5.2 Analytical considerations

Measurement of the three most common PTH-related protein isoforms that result in hypercalcemia is achieved using immunoassays with specificity for a wide range epitopes, including N-terminal, midregion, or C-terminal epitopes. However, variability in the possible in vivo sequences and structures of PTH-related protein produce measurement that can differ by quite a bit between methods. Such variation is further exacerbated by the lack of a common calibration standard, preventing the comparison of different PTH-related protein assays to one another, preventing the comparison/use of patient values from different vendor's instruments. Therefore it is important to measure both PTH and PTHrP and discuss with the laboratory the potential for each analyte to interfere within the assays employed.

3.1.4.5.3 Postanalytical considerations

The role of PTH-related protein as a prognostic indicator remains unclear, while the lack of a common calibration standard and variability in its detection limits its routine use [20]. This highlights the need for a gold-standard method to measure

PTH-related protein so as to ensure the comparative specificity and sensitivity of the methods available.

3.1.4.6 Cyclic adenosine monophosphate

Although an increase in cAMP levels occurs in response to exogenous PTH administration, it is rarely used as a marker of biologic PTH activity during the investigation of suspected pseudohyperparathyroidism (PTH resistance); any results must be interpreted in conjunction with other tests.

3.1.4.6.1 Preanalytical considerations

Preanalytical considerations for the measurement of cAMP are minimal due to the 24 h collection of the urine specimen, which eliminates the influence of diurnal variation. However, considerations of dietary factors are required, with excessive caffeine being associated with an increase in cAMP due to the inhibition of phosphodiesterase.

3.1.4.6.2 Analytical considerations

Measurement of cAMP in a urine specimen collected over a 24 h period of time can be attained using an array of laboratory assays, including

- Enzymatic colorimetric assays have high sensitivity and excellent throughput capacity, but lack of sufficient specificity.
- Liquid chromatography—mass spectrometry offer the best specificity and sensitivity.

3.1.4.7 Albumin

Information regarding a patient's disease state can be obtained by the measurement of various proteins. Albumin, a serum and extravascular protein, makes up more than half of the total protein in serum and regulates the colloid osmotic or oncotic pressure that regulates the passage of water and diffusible solutes through capillaries.

3.1.4.7.1 Preanalytical considerations

As nearly everything within the blood can be bound to albumin, it acts in the vascular transport of analytes and regulates their unbound, available concentration at any period of time. A decrease in albumin levels is caused by intestinal malabsorption syndromes, advanced liver disease, protein-losing states, and malnutrition along with almost any acute illness.

3.1.4.7.2 Analytical considerations

Albumin is most commonly measured following its complexation with a dye, with the wavelength at which the dye absorb light differing depending upon whether it is complexed with albumin or free in solution. This method may overestimate albumin due to the dye binding to other proteins and can influence the calculation ionized calcium and total globulin fraction; accounting for this overestimation can be difficult.

Electrophoresis is the most common means of further fractionating serum proteins, through the use of either capillary or gel techniques. Both processes separate proteins in a stationary phase and provides the most accurate measurement of the albumin fraction of serum protein.

3.1.4.7.3 Postanalytical considerations

Although the reference measurement of serum albumin system is well defined and ensures equivalence of patient results, it is critical to apply the appropriate reference range for both the patient and methodology employed and validating it against the gold-standard method.

Phosphate, magnesium, and vitamin D The measurement of phosphate, magnesium, and vitamin D is useful for the diagnosis of endocrine disorders that influence calcium. However, discussion of these analytical measurements and preanalytical, analytical, and postanalytical variables will be covered in *laboratory considerations and interpretational issues for assessing an abnormal phosphate/magnesium/vitamin D state.*

KEY POINTS

A patient-specific history is critical for assessing the influence of inter- and intraindividual variable on test outcome

- Discussion with the laboratory regarding optimal specimen collection, transport, and storage techniques will maintain optimal specimen integrity.
- Utilization of the gold-standard methods for analysis will provide a superior result; however, an understanding of its limitations will ensure correct clinical interpretation.
- Inadequate laboratory performance and a lack of an appropriate reference range for each analyte can result in an incorrect interpretation and decision by clinicians, with serious consequences for clinical practice, for the health care, and, ultimately, for the patient.

3.1.5 MISUSE OF LABORATORY TESTING IN CASES OF HYPERCALCEMIA

- Failure to identify spurious elevations in serum calcium-mild or unexpectedly discovered high calcium levels should be verified by simply repeating the test under the appropriate conditions prior to embarking upon an expensive detailed investigation.
- Failure to correct serum calcium for albumin or confirm a borderline elevation with ionized calcium.
- Drawing ionized calcium incorrectly or in a setting without the necessary quality control.

- Failure to measure calcium and PTH concomitantly: PTH cannot be interpreted without knowing the current ambient serum calcium.
- Labeling 1HPTH as definitely "ruled out" when a normal range PTH is obtained in the face of hypercalcemia. A "normal" PTH level in the setting of sustained hypercalcemia is usually evidence of 1HPTH.
- Labeling 1HPTH as definitely "ruled in" when a low-end normal range PTH is obtained in the face of hypercalcemia. Although PTH levels should be suppressed in non-PTH causes of hypercalcemia, the diagnosis may be difficult and/or a second cause present when PTH is low but still detectable. Careful additional diagnostic tests or consultation with endocrinology is advised.
- Ignoring a persistent often mild elevation in serum calcium. Although hypercalcemia may be mild or asymptomatic, its presence may still herald an important underlying medical disorder.
- Over-frequent measurement and over-interpretation of abnormal calcium levels in hospitalized patients. Shifts in volume status, renal function, and acid−base balance may complicate interpretation of calcium levels; such confounders must be considered before proceeding to expensive additional investigations.
- Inappropriate ordering of expensive PTH-related protein testing prior to completion of a thorough history and physical supplemented by readily available laboratory testing and imaging studies.
- Repeated measurement of serum PTH is patients with an established diagnosis of 1HPTH: absolute PTH levels do not often correlate to the degree of hypercalcemia and have little clinical relevance in patients with normal renal function. In patients with 1HPTH who are being observed, an annual serum calcium and albumin may suffice [21].
- Assumption that hypercalcemia is due to malignancy in patients with known malignancy; in actuality, such patients may just as well have hyperparathyroidism or other causes and mis-diagnosis of the cause of hypercalcemia may lead to inappropriate therapies.

3.1.6 MISUSE OF LABORATORY TESTING IN HYPOCALCEMIC DISORDERS

- Failure to correct serum calcium for albumin or confirm a borderline decrease with ionized calcium.
- Overutilization of serum calcium in cases of established asymptomatic hypocalcemia of known etiology.
- Repeated measurement of serum PTH is patients with an already established cause for hypocalcemia.
- Measurement of calcitriol (1,25-dihydroxyvitamin D) in the diagnosis of hypocalcemia outside of suspecting rare vitamin D resistance syndromes.

Otherwise the concentrations of calcitriol in blood (pmol/L) can be difficult to interpret and circulate in 1/1000th the concentration of 25-hydroxyvitamin D (nmol/L). Thus 1,25 vitamin D levels are virtually never needed to investigate the vast majority of hypocalcemia cases.

3.1.7 EXAMPLE WITH DISCUSSION (HYPERCALCEMIA)

A 65-year-old healthy woman presents to her family doctor for her annual review. She feels well and has no chronic medical problems. She takes a 500 mg calcium supplement and 1000 IU of vitamin D daily but no other supplements or prescription medications. She is concerned about osteoporosis because her 95-year-old mother had a hip fracture last year. The patient herself has not lost height nor has she had any fractures as an adult. Her family doctor orders a routine blood panel.

Question 1: Should serum calcium be a part of "routine" blood testing?

Answer: No. Serum calcium should be considered selectively in the correct clinical situation. These scenarios may include: declining renal function, a new kidney stone, symptoms of possible hypercalcemia (new polyuria, nocturia, constipation, bone pain, or unexpected cognitive dysfunction), osteoporosis or low trauma fracture, or diagnoses associated with hypercalcemia (malignancy, sarcoidosis, tuberculosis, immobilization, etc.). Clinical gestalt that "something is going on" cannot be ignored—serum calcium may be a very reasonable and inexpensive test to consider.

Question 2: Serum calcium is ordered and the result is: 2.6 mmol/L (reference range 2.1−2.55 mmol/L). Are any other laboratory tests needed initially?

Answer: Most labs routine measure total serum calcium which includes both bound and free components. Because binding proteins (mostly albumin) as well as pH can vary in many patients, it is important to measure a serum albumin concomitantly and mathematically "correct" the calcium, if the albumin is abnormal, in order to get an accurate estimation of the biologically active ionized calcium level. If albumin was not ordered on the original test, the laboratory may have saved some serum to run this test without the need to send the patient for another phlebotomy. If another phlebotomy is required, some clinicians may choose to order ionized calcium if offered by their lab. In addition, given the initial elevated calcium, adding PTH, and serum creatinine (with estimated GFR) to the repeat calcium test may also be worthwhile. Sending the patient to have the repeat test while fasting can sometimes make a difference but the literature is not clear on this as a necessity. Transient, postprandial hypercalcemia may theoretically occur but should be easily recognized.

Question 3: *Should the family physician advise the patient to stop her low-dose calcium and vitamin D supplements prior to repeating the calcium test?*

Answer: No. Low-dose calcium and vitamin D supplements will have little influence on serum calcium levels. Furthermore, some studies suggest that stopping these supplements may actually exacerbate the problem if the underlying diagnosis is 1HPTH. Vitamin D deficiency is a strong stimulus for PTH release with consequent stimulation of bone resorption. In cases of 1HPTH, stopping supplements may thus lead to an increase in bone resorption without really improving the serum calcium level. High-dose supplements without a strong indication should be reduced to bone health prevention doses: combined diet and supplements no more than 1200 mg calcium and 800−2000 IU vitamin D per day [22].

Question 4: *Serum calcium is repeated along with other suggested basic tests.*

Results: Calcium 2.63 mmol/L (2.1−2.55 mmol/L), albumin 40 g/L (33−48 g/L) (corrected calcium is thus also 2.63 mmol/L), creatinine 55 (eGFR 90 (> 60)) and PTH 35 ng/L (7−37 ng/L).

What is the most likely cause of this patient's incidentally discovered hypercalcemia?

Answer: The most likely and most common cause is 1HPTH. She is in a higher incidence age group with women three times more likely than men to develop 1HPTH. In the face of normal renal function and clear-cut, albeit mild, hypercalcemia, the serum PTH is inappropriately at the upper end of the reference range. This is therefore *parathyroid-mediated hypercalcemia* for which the differential diagnosis is narrow. Nonparathyroid-mediated causes such as malignancy need not be considered and the patient should be reassured. The only other diagnosis to consider is a germline mutation in the CaSR, also known as FHH. In the absence of any family history of hypercalcemia or other inherited endocrine syndromes in a woman in her seventh decade, FHH is unlikely. However, a 24 h urine calcium and creatinine collection will allow the calculation of the FeCa. FeCa > 2% will exclude FHH in this situation whereas most cases of FHH will have FeCa of <1% [23]. An adequate urine collection is important to prevent falsely low urine calcium results. If results are persistently equivocal (FeCa, 1%−2%), serum calcium testing of first degree relatives should be considered first because FHH is inherited in an autosomal dominant fashion and similar hypercalcemia in a first degree relative will markedly raise the possibility of a CASR mutation presence. Genetic testing for mutations in the CaSR and its associated proteins such as AP2S1 would allow a definitive FHH diagnosis in the few selected cases needed.

Question 5: With a confirmed diagnosis of 1HPTH in an asymptomatic person, what are the options for management now?

Answer: The only proven treatment for 1HPTH is surgical removal of the parathyroid adenoma (solitary in 80%−85% of cases). This should be performed by an expert high volume parathyroid surgeon and may be guided by preoperative imaging studies such as ultrasound and sestamibi parathyroid nuclear scans. The latter two tests are not indicated for diagnosis as they lack the appropriate sensitivity. However in an asymptomatic otherwise healthy patient with normal renal function, observation without intervention may be a very appropriate option. Natural history studies suggest that only ∼1/3 of patients will progress to develop a surgical indication when observed over a 10-year period [24]. Thus ∼2/3 of patients may remain stable and not develop any complications from 1HPTH. Surgical indications include the development of renal stones, renal dysfunction, serum calcium >0.25 mmol/L over the local lab's upper end reference range, and clinical or bone density evidence for osteoporosis. Patients with significant hypercalciuria >10 mmol/day may also be considered for surgery as may patients under the age of 50 who are young enough that they may inevitably progress within their expected lifespan. These recommendations, however, should be considered as guidelines to inform the discussion between physician and patient at the bedside [21]. For example, many elderly patients have osteoporosis but no other strong indications for surgery in 1HPTH. Medical therapy for osteoporosis with close observation may be a valid option [22].

Question 6: After a detailed discussion with the patient, a decision is made to observe and monitor 1HPTH. What tests should be considered at baseline and then for monitoring? How often should these tests be performed?

Answer: At baseline, if not recently performed, a 24 h urine collection for calcium and creatinine, renal ultrasound or X-ray, spinal X-ray, and bone density evaluation is recommended. If no evidence for hypercalciuria, nephrolithiasis, nephrocalcinosis, vertebral fractures, or low bone mass is discovered, then observation can proceed. If patients experience a new kidney stone or low trauma fracture, they should be re-evaluated as possible surgical candidates. Laboratory monitoring should include an annual corrected serum calcium and creatinine with eGFR. Repeat serum PTH tests are not needed and will not influence any change in management. Bone density tests could be repeated every 2 years. Follow-up spinal X-rays or renal ultrasound is only done if clinically indicated. Patients should be advised to remain well hydrated and avoid the use of thiazide diuretics.

3.1.8 EXAMPLE CASE WITH DISCUSSION (HYPOCALCEMIA)

A 31-year-old man presents to his family doctor with complaints of irritability, intermittent numbness in his lips and fingertips along with periodic uncomfortable cramping of his hands. His past medical history is unremarkable except for thyroid surgery 1 month ago for a multinodular thyroid gland that proved to be benign. He takes no medications apart from the occasional calcium carbonate 500 mg which was suggested upon his discharge from hospital. He has no family history of calcium or parathyroid disorders. The physical examination reveals only a healing thyroidectomy scar. The patient reports his right hand cramped when the nurse took his blood pressure.

Question 1: What is the most likely diagnosis and how should it be confirmed?

Answer: Hypocalcemia due to temporary or permanent parathyroid injury during his recent thyroid surgery is the most likely diagnosis. The patient should be sent for a calcium and albumin or ionized calcium test to confirm.

Question 2: Given the high clinical suspicion for hypocalcemia, would any other laboratory tests be helpful to obtain right now?

Answer: Serum phosphate, PTH, and creatinine + eGFR would all be helpful now. In this scenario, one might expect the serum phosphate to be at least mildly elevated and the PTH to be low or inappropriately at the lower end of the normal range. Serum creatinine + eGFR should be normal in this healthy young man. Unless the patient was taking diuretics or had diarrhea, serum magnesium would not likely be useful at this stage. 24 h urine calcium or 25-hydroxyvitamin D may not provide a lot of further information right now and should not be routine, given the fairly obvious clinical history.

Results: Serum calcium 1.68 mmol/L with albumin 38 g/L (corrected calcium = 1.72 mmol/L). Phosphate 1.8 mmol/L (reference range 0.8−1.5 mmol/L), PTH 8 ng/L (reference range 7−37 ng/L), creatinine 93 μmol/L, and eGFR 76. A diagnosis of postoperative hypoparathyroidism is made. The patient is treated with calcium carbonate 1000 mg tid and calcitriol 0.25 mcg bid. An activated vitamin D preparation such as calcitriol is important to prescribe. Traditional vitamin D supplements will be inconsistently converted to the active form in the absence of PTH.

Question 3: One year later the patient presents to his physician for follow-up. He feels well with no complaints and all of his presenting symptoms rapidly disappeared with treatment. He continues to take the same doses faithfully. Is laboratory testing indicated now? If so, which tests should be requested?

Answer: Calcium, albumin, phosphate, creatinine/eGFR and 24 h urine for calcium and creatinine

Results: Calcium 2.5 mmol/L, albumin 39 g/L (corrected calcium = 2.52 mmol/L), phosphate 1.2 mmol/L, creatinine 80 μmol/L/eGFR

88, 24 h urine calcium 12 mmol/day (expect normal 24 h urine calcium to be <0.1 mmol/kg/day—in this patient who weighs 80 kg: <8 mmol/day), urine creatinine 13 mmol/day (confirming an adequate 24 h urine collection).

Question 4: Should any adjustments be made to his calcium and calcitriol doses? If so, why?

Answer: His serum calcium is at the upper end of the reference range. The greater the amount of calcium filtered by the glomerulus, the higher the urine calcium given the absence of PTH that normally would stimulate significant tubular reabsorption back into the circulation. This combined with the measured hypercalciuria will increase the patient's risk for kidney stones and nephrocalcinosis. His calcium carbonate dose should be reduced perhaps to 500 mg tid-qid before retesting calcium and albumin in 1–2 weeks. The calcitriol dose could be reduced as a second step as need be, may be down to 0.25 mcg once daily. Persistent hypercalciuria may necessitate the addition of a thiazide diuretic to promote renal calcium reabsorption. Target serum calcium in patients with permanent hypoparathyroidism: lower end of the reference range or just below. (In our lab, target: 2.0–2.2 mmol/L (reference range 2.1–2.55 mmol/L).) The major exception to this rule is in pregnant women in whom mid-upper normal range calcium levels are targeted to reduce the risk of fetal 2HPTH from any maternal source hypocalcemia.

3.2 PHOSPHATE DISORDERS

3.2.1 OVERVIEW

Phosphate is a ubiquitous component of the Western diet. Phosphate deficiency through nutritional deprivation of phosphate-containing foods alone is uncommon outside of eating disorders and alcoholism. Phosphate plays crucial roles in oxygen transport, cellular membrane function and signaling, DNA structure, cardiac function, and musculoskeletal integrity and strength. ATP and 2,3DPG are the two key phosphate-containing molecules without which life could not exist. Serum phosphate levels depend upon intake and gastrointestinal absorption, pH, circulating binders, renal function for excretion, and three key-regulating hormones—PTH, calcitriol, and FGF-23.

- PTH will promote renal phosphate excretion resulting in less phosphate to bind calcium and thus higher "free" calcium.
- Calcitriol will increase the intestinal absorption of phosphate and calcium.
- FGF-23, secreted primarily by osteocytes in bone in response to hyperphosphatemia, will increase renal phosphate losses in the urine and dampen the production of calcitriol, such that the net effect is to lower serum phosphate levels.

Progressive deterioration in renal function impairing the ability to excrete phosphate is the most common cause of hyperphosphatemia. The high serum phosphate may both transiently reduce serum-free calcium stimulating PTH release and directly stimulate PTH release resulting in 2HPTH.

3.2.2 HYPOPHOSPHATEMIA

3.2.2.1 Clinical picture

Hypophosphatemia is often multifactorial and can be seen in up to 5% of hospitalized patients [25]. The majority of these resolve as the patient recovers and resumes normal nutrition, activity, and renal function. Most mild cases of hypophosphatemia do not cause significant morbidity or symptoms and thus do not require treatment apart from addressing the underlying illness. In acute situations with rapid drops in serum phosphate, tissue hypoxia and both cardiac and skeletal muscle impairment may occur. Chronic hypophosphatemia can lead to muscle weakness, osteomalacia, bone pain, and fractures. The complex interplay of calcium, PTH, and vitamin D with phosphate must be considered—disorders involving these hormones can result in hypophosphatemia and overlap with disorders of calcium. Genetic disorders while uncommon can lead to significant isolated hypophosphatemia often diagnosed in childhood. However, the heterogeneity of these conditions may mask discovery until patients are into adulthood. A family history is thus very important in the work-up of patients presenting with hypophosphatemia. A detailed medication history should be sought for prescription drugs such as bisphosphonates, diuretics, and glucocorticoids. Over the counter supplements particularly phosphate binders such as calcium carbonate or citrate and aluminum- or magnesium-containing antacids should be directly queried.

The physical examination may be unremarkable. However, particular attention should be paid to any signs of rickets, kyphosis, or diminution in muscular strength. Subcutaneous masses or bony prominences in cases of suspected tumor-induced hypophosphatemic osteomalacia (TIO) are very important to note.

3.2.2.2 Causes of hypophosphatemia

The causes of hypophosphatemia can be divided into three categories: reduced phosphate intake or GI absorption, increased renal phosphate losses, and shifts from the extracellular to the intracellular compartment.

- *Reduced or impaired phosphate intake/GI absorption* can be seen in sick hospitalized patients with inadequate nutritional replacement. Patients with eating disorders and alcoholism are also predisposed. More commonly, malabsorption disorders, bariatric or other stomach or bowel surgery, celiac disease or IBD can lead to reduced GI absorption of ingested phosphate. Vitamin D deficiency, if severe, will reduce the calcitriol-mediated GI uptake of both phosphate and calcium. Oral phosphate binders particularly calcium salts or other antacids taken in excess will reduce the dietary phosphate absorption.

- *Increased renal phosphate excretion* is a frequent etiology of chronic hypophosphatemia in outpatients—it may be congenital or acquired. Hyperparathyroidism is a common cause of a high urine phosphate/low serum phosphate and may also be accompanied by a calcium disorder. Primary over-expression of FGF-23 will also promote phosphaturia as well as impair the protective reflex rise in calcitriol seen with other forms of hypophosphatemia. TIO can occur when small, occult mesenchymal tumors in bone, skin, or muscle may produce high amounts of FGF-23. There are a number of inherited forms of hypophosphatemia that results in a higher circulating concentration or effectiveness of FGF-23 and its consequent renal phosphate losses, inappropriately normal or low calcitriol and osteomalacia. The most common of these disorders is X-linked hypophosphatemia (XLH). Medications such as diuretics, glucocorticoids, or bicarbonate can also promote urine phosphate loss. Antiviral medications such as tenofovir are increasingly recognized as causes of mild to severe hypophosphatemia [26].
- *Intracellular shifts* can be seen most frequently in tumor lysis syndrome following chemotherapy, during the treatment of diabetic ketoacidosis, the hungry bone syndrome after parathyroid or hyperthyroidism surgery, severe infections or the refeeding syndrome. This is usually a transient situation seen in severely ill in-patients.

3.2.2.3 Special situations

In many cases, acute hospitalization, concomitant illness, or obvious GI and nutritional disorders will explain the low serum phosphate which resolves upon recuperation or treatment of the underlying condition. In otherwise well outpatients with chronic hypophosphatemia, the clinician must exclude 1HPTH and vitamin D deficiency. Only then should a hunt proceed for less common conditions such as TIO or an adult presentation of a rare renal phosphate wasting disorder, like XLH.

Children and neonates have higher serum phosphate concentrations. This can cause confusion particularly if the laboratory does not report age-specific reference ranges. For example, in our lab the adult reference range for serum phosphate is: 0.80–1.50 mmol/L. If this range was applied to any child younger than 15 years, a serum phosphate of 0.90 mmol/L might be passed off as normal rather than low—thus a potentially pathologic cause of hypophosphatemia could be missed. The index of suspicion for genetic conditions presenting with hypophosphatemia must be heightened in the pediatric population. Lower limb bowing and other signs of rickets may be unapparent in XLH until the child is consistently walking and weight bearing.

3.2.2.4 Diagnostic approach (laboratory test focussed)

- Confirm persistent low serum phosphate, particularly in hospitalized patients.
- Stop any medications or supplements that may be contributing to hypophosphatemia and optimize nutrition.

- Measure concomitant serum calcium and albumin (or ionized calcium), creatinine, PTH, and 25-hydroxyvitamin D. These results should identify common causes: 1HPTH, 2HPTH, and vitamin D deficiency.
- Isolated renal phosphate wasting disorders usually have a normal calcium and PTH. In this case a 24 h urine collection for phosphate or spot urine with creatinine to calculate a fractional excretion of phosphate may be needed. With a coexisting low serum phosphate, it is inappropriate for the 24 h urine phosphate to be more than 3.2 mmol/day or the fractional excretion of phosphate to be >5% [27]. Please note that urine phosphate excretion is highly dependent on phosphate intake which can vary greatly from day to day. To account for this, measuring the renal tubular maximum reabsorption rate of phosphate to GFR (TmP/GFR) may be useful. A low TmP/GFR (<0.8) in the face of a low serum phosphate suggests pathologic urine phosphate wasting [28].
- Calcitriol measurement may also help in this situation as FGF-23-mediated phosphaturic disorders such as TIO and XLH will have an inappropriately normal or low 1,25-dihydroxyvitamin D when the serum phosphate is low. FGF-23 can be sent away to reference labs at some expense—the result must be interpreted in light of the phosphate, calcium, PTH, and vitamin D levels at the time.

3.2.3 HYPERPHOSPHATEMIA

3.2.3.1 Clinical picture

Because the kidneys possess a great capacity to excrete significant oral phosphate loads, persistent hyperphosphatemia is unusual in patients with normal renal function. Clinical manifestations may relate more to concomitant hypocalcemia (e.g., with hypoparathyroidism) with the neuromuscular irritability, paresthesias and hand cramping, and characteristic physical findings described in Section 3.1.3. Normal or high serum calcium in conjunction with chronic hyperphosphatemia may lead to organ or blood vessel calcification from the deposition of calcium-phosphate complexes. When presented with a patient found to have hyperphosphatemia, a detailed history searching for renal disease risk factors and predisposing conditions such as diabetes mellitus and hypertension is imperative. The use of laxatives and enemas that contain high amounts of phosphate should be queried [29]. Sick hospitalized patients receiving chemotherapy, total parenteral nutrition (TPN), and those with critical illness particularly sepsis, acidosis, and rhabdomyolysis are all prime candidates to have high serum phosphate. In patients for whom the clinical picture does not fit with pathologic hyperphosphatemia, laboratory analytic issues need to be considered to exclude pseudohyperphosphatemia [30].

3.2.3.2 Causes

The ability of the kidneys to excrete phosphate and prevent hyperphosphatemia can be overwhelmed in three types of situations: reduced renal excretion, excessive phosphate load, and extracellular shifts. Often times these causes can overlap—for example, a patient discovered to have a very high serum phosphate while hospitalized on the oncology unit. Following chemotherapy, he is dehydrated causing prerenal failure, develops the tumor lysis syndrome and requires TPN.

- *Reduced renal excretion*: this can be as a result of any prerenal, renal, or postrenal cause but is most prevalent in patients with progressive renal failure due to chronic diseases such as hypertension or diabetes. Any cause of a low PTH such as hypoparathyroidism following thyroid surgery will reduce the renal excretion of phosphate leading to higher serum phosphate and is usually obvious from the concomitant calcium abnormality. Acromegaly is an uncommon but important cause of mild hyperphosphatemia [31].
- *Excessive phosphate load*: laxatives- or enemas-containing phosphate, TPN, or other intravenous phosphate treatment.
- *Extracellular shifts*: most commonly seen in hospitalized patients often with critical illness. Hyperthermia, tumor lysis syndrome, acidosis, sepsis, and rhabdomyolysis will all cause acute usually transient elevations in phosphate.

3.2.3.3 Rare situations

Familial tumoral calcinosis is a rare, inherited syndrome that can cause calcium phosphate crystal deposition in joints and soft tissues [32].

3.2.3.4 Diagnostic approach (laboratory test focussed)

- In the asymptomatic patient with no predisposing risk factors for hyperphosphatemia, ask the laboratory to help rule out analytic interference due to hyperlipidemia, hemolysis, or high levels of protein or bilirubin.
- Serum creatinine and eGFR are the most helpful tests as the majority of cases of hyperphosphatemia are related to renal dysfunction.
- Correcting the underlying cause and supportive care with careful hydration and removal of the phosphate source may result in rapid normalization of the serum phosphate.
- Serum calcium and albumin or ionized calcium: hypocalcemia may be the presenting clinical feature and take priority for investigation and treatment. If the corrected calcium is low, measure PTH to assess for defects in parathyroid secretion (PTH low or lower end reference range) or action (PTH high) as in PHP. Serum phosphate may be lower in PHP as PTH-mediated phosphate excretion may be maintained.
- A high serum calcium and high serum phosphate are important to identify as it will increase the risk for metastatic calcification. The product of serum calcium and serum phosphate has been evaluated. The risk of soft tissue calcification is quite high if the calcium-phosphate product exceeds $70 \text{ mg}^2 \text{ dL}^2$ ($5.6 \text{ mmol}^2 \text{ L}^2$). Guidelines suggest aiming for a product below $55 \text{ mg}^2 \text{ dL}^2$ ($4.4 \text{ mmol}^2 \text{ L}^2$) in patients with stages 4 and 5 CKD [33].

3.2.4 LABORATORY CONSIDERATIONS TO ASSESS HYPO- AND HYPERPHOSPHATEMIC DISORDERS

3.2.4.1 Blood phosphate measurement

The modulation of phosphate storage, exchange, uptake, and excretion relies upon a tightly controlled relationship that influences the pathophysiological states of the gastrointestinal tract, kidneys, and bone, which relates to the concentration of phosphate in the blood.

3.2.4.1.1 Preanalytical considerations

Phosphate's equilibrium within blood is formed between inorganic phosphate and organically bound phosphoric acid, with changes to one component shifting the equilibrium of the system. This can be influenced by

- Anticoagulants that can reduce the concentration of phosphate within plasma specimens and has made serum specimens the most commonly accepted specimen type for the measurement of phosphate. For example, plasma specimens collected within a lithium heparin tube have been found to reduce the concentration of phosphate by 0.1 mmol/L (0.3 mg/dL).
- Hemolysis, which results in the release of intracellular red blood cell phosphate stores.
- Delayed separation of red blood cells from plasma/serum has been associated with leakage of phosphate occurring due to the imperfect barrier of the membrane and leakage of ions down their concentration gradients.

In addition the influence of abnormal phosphate measurements can result in the incorrect diagnosis of hyper- and hypophosphatemia due to

- increased phosphate load (food intake, vitamin D intoxication, renal failure), which requires specimens collected for routine phosphate measurements occurring after a patient has been fasting for 12−14 h.
- decreased intestinal absorption due to excessive antacid use.

3.2.4.1.2 Analytical considerations

The analytical procedures for measuring phosphate can be performed by an array of methods, including

- Complexation of phosphate with colorimetric dyes. However, these methods can be susceptible to interference from:
  - elevated phospholipids due to abnormal phospholipid metabolism, can lead to interference with the colorimetric method of detection
  - paraproteinemia can result in a pseudohyperphosphatemia due to an increase in specimen turbidity.
  - contrast agents that absorb at the same wavelengths as the phosphate-dye-complex.

- Although atomic absorption spectrophotometry is a more specific method for the quantification of phosphate, its use of specialized instrumentation and extensive sample preparation reduces its application in the clinical laboratory.

3.2.4.1.3 Postanalytical considerations

Appropriate reference intervals for phosphate need to be applied among healthy populations stratifications encountered by the lab. For example the age of an individual can influence phosphate levels, with children's and males >40 having decreasing phosphate levels.

3.2.4.2 Urine phosphate measurement

3.2.4.2.1 Preanalytical considerations

Urine phosphate concentrations suffer from many of the same issues as serum/plasma as described earlier. Measurement of 24 h urine phosphate is almost exclusively performed in the investigation of hypophosphatemic disorders with virtually no indication to perform this test in other clinical scenarios.

3.2.4.2.2 Postanalytical considerations

The influence of physiological variables on serum phosphate concentration can result in erroneous results and highlight the need of appropriate reference intervals. The ratio of the maximum rate of tubular phosphate reabsorption (TmP) to the GFR, or [TmP/GFR], corresponds to the theoretic lower limit of plasma phosphate. At any concentration lower than this value, all filtered phosphate would be reabsorbed. It may still be useful in assessing renal reabsorption of phosphorus in a variety of pathological conditions associated with hypophosphatemia even though direct measurements of PTH have replaced much of the utility of TmP/GFR measurements.

3.2.4.3 Fibroblast growth factor-23

FGF-23 is a bone-derived hormone secreted in response to dietary phosphate intake and serum phosphate levels, regulating phosphate levels through its action on the kidneys to increase urinary phosphate excretion. Furthermore, it acts as an inhibitor of 25-hydroxyvitamin-D3-1-α-hydroxylase. Its use as a marker of CKD is especially important due to the limited specificity and sensitivity of current CKD markers (serum creatinine and albuminuria) and its response to states of phosphate excess. At the present time, however, the routine measurement of FGF-23 in CKD patients is not considered in the first line of diagnostic tests; measurement of FGF-23 is increasingly being used for the investigation of hypophosphatemic disorders [34].

3.2.4.3.1 Preanalytical considerations

The measurement of FGF-23 is a more accurate biomarker of body phosphate accumulation because its increase begins early in kidney dysfunction, preceding serum phosphate concentrations that are considered beyond the upper reference interval. However, detection of these incremental increases requires

- stabilization of hormone, as both serum and plasma specimens collected without a protease inhibitor were found to be unsuitable for analysis of FGF-23 due to the degradation over time
- standardization of the time of collection, as plasma FGF-23 concentrations undergo a substantial diurnal variation [34].

Furthermore the serum concentrations of FGF-23 progressively increase as renal function declines, requiring the patient's renal function being accounted for during interpretation of results. Furthermore the interpretation of FGF-23 results requires serum phosphate levels in a similar manner to the requirement of serum calcium when interpreting PTH levels. There is a reasonable evidence to support that it is a progressive, chronic attempt to increase phosphaturia in CKD although some discussions exist on whether these increases represent a response to maintain normal phosphate levels or reflect an end point in its production.

3.2.4.3.2 Analytical considerations

Variability in the measurement of FGF-23 can be observed in choice of antibodies specific for it intact or C-terminal regions; however, current methods for its measurement show poor analytical agreement and cannot be used interchangeably. Although such differences are mainly due to differences in calibration, discordance between the measurement of intact and C-terminal FGF-23 is observed. Therefore harmonization of available assays using a common international standard would facilitate more meaningful interpretation of data, providing the ability to

- convert results between assays
- understand each assay's accuracy and reproducibility
- produce a standardized reference interval for each test.

Furthermore it is unclear if the measurement of FGF-23 through the use of immunoassays for intact or C-terminal regions of its structure may be more relevant.

3.2.4.3.3 Postanalytical considerations

Although the importance of a clearly defined reference interval is critical for clinically applying FGF-23, population-based studies that define cut-off values for clinical risk assessment are scarce; those reference ranges which are currently available refer to specific assays and overlook many of the variables that can influence its concentration. Therefore before these assays are applied in clinical practice, a consideration of the influence the biological variability of FGF-23 has on the reference ranges and validated cut-off values for diagnostic thresholds are required; the exclusion of patients afflicted with phosphate wasting diseases (X-linked dominant hypophosphatemic rickets/

osteomalacia, autosomal dominant hypophosphatemic rickets/osteomalacia, and tumor-induced rickets/osteomalacia) is paramount to ensuring the reference range accuracy.

3.2.5 MISUSE OF LABORATORY

- Over-investigation with multiple tests prior to a careful history and physical. Many phosphate disorders particularly in hospitalized, critically ill patients or those with multiple medical problems can be addressed by treating and supporting the underlying illness. Expensive assays such as calcitriol, FGF-23, or genetic tests are usually unnecessary in these situations.
- Twenty-four hour urine collections are unpopular with patients and nursing staff. Furthermore they are subject to both over- and under-collection making results difficult to interpret, particularly in patients who are in hospital.
- Over-diagnosing relatively mild vitamin D insufficiency. While severe vitamin D deficiency is an important cause of hypophosphatemia, 25-hydroxyvitamin D values that fall between 50 and 80 mmol/L are unlikely to stimulate PTH above the reference range and will have minimal effects on phosphate metabolism.
- Trying to put together a diagnosis with important laboratory tests all measured at different times. Phosphate, calcium, and vitamin D concentrations in the serum can change rapidly so need to be drawn to together to make meaningful inferences.
- Urine phosphate is highly dependent upon phosphate ingestion which can vary greatly from day to day.
- Failure to consider false elevations in serum phosphate due to hemolysis, abnormal lipids, circulating proteins, or high bilirubin.
- Failure to apply age-specific reference ranges for serum phosphate may lead to either over-diagnosis or under-diagnosis of pathologic phosphate disorders.

3.2.6 EXAMPLE WITH DISCUSSION

A 40-year-old woman presents to her family physician with 1 year history of mild but progressive fatigue, reduced exercise capacity, and difficulties climbing the stairs in her three-story walk-up apartment. She has no chronic health problems or past medical history. She takes no regular medications or supplements. She has a brother with type-1 diabetes. The physical exam is normal. Her dentist noted one of her fillings had become loose and suggested she get her calcium, phosphate, glucose, and TSH measured at the laboratory prior to seeing the family doctor.

Results: Calcium 2.12 mmol/L (2.1−2.55 mmol/L), phosphate 0.75 mmol/L (0.8−1.5 mmol/L), glucose 4.1 mmol/L, TSH 1.7 mU/L (0.4−4.0 mU/L).

Question 1: Do the results explain her symptoms? Are other tests needed?

Answer: Mild hypophosphatemia does not usually cause symptoms. Her serum calcium is at the lower end of the reference range but we do not know her albumin. The screening tests for diabetes and thyroid disease are normal. These tests are very appropriate for this fatigued patient with a history of autoimmune disease in her brother. The calcium needs to be repeated with an albumin or ionized calcium. Phosphate should be repeated along with creatinine + eGFR and PTH.

Results: Calcium 2.11 mmol/L, albumin 35 g/L (corrected calcium: 2.21 mmol/L), phosphate 0.74 mmol/L, creatinine 35 (eGFR 95), PTH 130 ng/L (7−37 ng/L).

Question 2: What is the laboratory diagnosis?

Answer: Mild hypophosphatemia with a serum calcium at the lower end of the reference range. Secondary hyperparathyroidism in a patient with normal renal function. This suggests that the primary problem here is something that is forcing an excessive secretion of PTH in order to maintain serum calcium. The low serum phosphate is likely due in part to this higher PTH level.

Question 3: What other test(s) should be ordered next?

Answer: 25-Hydroxyvitamin D. Severe vitamin D deficiency would explain the low phosphate, "lowish" calcium and high PTH. It is the most common cause of 2HPTH in patients with normal renal function. Given her nonspecific symptoms and the family history of autoimmune disease, celiac disease should be screened for.

Results: 25-Hydroxyvitamin D: 13 nmol/L (> 80 nmol/L), celiac screen: positive (confirmed by duodenal biopsy).

Question 4: Should the patient receive oral phosphate supplements in addition to commencing a dietitian-supervised gluten-free diet and high-dose vitamin D supplements?

Answer: No. The mildly low phosphate is not the cause of her symptoms but has been very helpful clue to make the diagnosis. Oral phosphate preparations are poorly tolerated in many patients because of gastrointestinal side effects. Furthermore phosphate tablets may bind her serum calcium putting her at risk for symptomatic hypocalcemia and worsen her 2HPTH.

Question 5: Will the 2HPTH be reversed with the appropriate treatment of her celiac disease and vitamin D deficiency?

Answer: Yes—in most cases but it may take some time. As her gluten-free diet promotes improved small intestinal function, vitamin D doses can be gradually weaned over time.

KEY POINTS

Phosphate disorders are commonly seen in hospitalized patients and can be of a multifactorial etiology.

- Serum phosphate is strongly influenced by gastrointestinal absorption, renal function, and three key-regulating hormones: PTH, calcitriol, and FGF-23.
- Most mild cases of hypophosphatemia do not cause significant problems or symptoms.
  - No treatment apart from addressing the underlying illness and comorbidities.
- Hyperparathyroidism and severe vitamin D deficiency are the most common acquired pathologic causes of hypophosphatemia.
  - XLH is the most common congenital cause.
- Renal failure is the most common cause of hyperphosphatemia.

3.3 MAGNESIUM DISORDERS

3.3.1 INTRODUCTION

Magnesium is the second most prevalent cation in the body, the vast majority of which resides within cells and bone. It serves key roles in neuromuscular function, cellular energy regulation, enzymatic activity, and membrane transport. Only 1% circulates in the extracellular compartment with about 30% being protein bound. Dietary deficiency of magnesium is uncommon outside of starvation or alcoholism. GI absorption through the small intestine may be accelerated by calcitriol (1,25-dihydroxyvitamin D). Magnesium excretion is primarily regulated by the kidney through glomerular filtration and tubular reabsorption. Magnesium deficiency may be seen together with hypocalcemia, hypercalcemia, phosphate disorders, and vitamin D deficiency. The clinician who is presented with one disorder of mineral metabolism may thus be obligated to seek out and treat another disorder concomitantly.

3.3.1.1 Hypomagnesemia

3.3.1.1.1 Causes

Mild hypomagnesemia (0.5–0.7 mmol/L) may be commonly seen in hospitalized patients. The etiology can be multifactorial. Patients are often asymptomatic and do not require treatment. Moderate to severe magnesium deficiency <0.5 mmol/L can be divided into three main categories: reduced GI absorption or intestinal loss, increased renal loss, and acute shifts from the extracellular to the intracellular compartments.

Common causes in each category include:

- Reduced GI absorption or intestinal loss:
  - excess vomiting or diarrhea
  - malabsorption syndromes and small bowel diseases
  - bariatric or small intestinal surgery

- vitamin D deficiency
- use of proton pump inhibitors [35]
- Increased renal loss:
  - congenital renal syndromes that impair tubular magnesium reabsorption such as Gitelman's and Bartter's syndromes.
  - acquired renal parenchymal and tubular diseases particularly with diabetic nephropathy [36]
  - medications—most commonly diuretics and some antibiotics
- Acute shifts from the extracellular to the intracellular compartments:
  - treatment of diabetic ketoacidosis
  - "Hungry bone" syndrome postparathyroidectomy

3.3.1.1.2 Clinical picture

Patients may be asymptomatic or present with symptoms of neuromuscular fatigue or irritability. There may be an overlap with the symptoms of hypocalcemia such as paresthesias and muscular cramping; indeed, hypocalcemia may also be present and needs to be treated. Hypomagnesemia may impair both PTH action and secretion. Electrocardiographic abnormalities and symptomatic arrhythmias can be seen.

3.3.1.1.3 Diagnostic approach

Clinical suspicion for hypomagnesemia to trigger serum magnesium laboratory testing is most important when presented with patients who may have nonspecific symptoms particularly when they have known GI or renal disease. Refractory hypocalcemia is a classic example where patients may fail to respond to calcium infusions without addressing the coexisting magnesium deficiency. Similar to potassium, deficiency in total body stores of magnesium may not be adequately reflected in the serum magnesium level.

- 25-Hydroxyvitamin D testing may be helpful as true vitamin D deficiency may cause or exacerbate hypomagnesemia.
- A detailed review of patient medications may reveal the cause of low magnesium.
- Measure calcium, phosphate, creatinine/eGFR, and PTH.
- Inappropriate renal magnesium loss is the most common cause of hypomagnesemia—measurement of an inappropriately high fractional excretion of magnesium on a random or 24 h urine sample may confirm this etiology. A fractional excretion of magnesium of >4% suggests inappropriate urinary magnesium loss [37].
- Urine calcium excretion may also help: some congenital disorders of magnesium wasting may be associated with either hypocalciuria or hypercalciuria.

3.3.2 **TREATMENT**

- Address the underlying cause.
- Correct any coexisting calcium and/or vitamin D deficiencies.
- For serum magnesium <0.5 mmol/L, consider either oral or IV replacement.
- Oral magnesium preparations may be dose limited due to symptomatic diarrhea.

3.3.2.1 *Hypermagnesemia*

High serum magnesium is distinctly less common than hypomagnesemia.

- Hypermagnesemia is often seen in the setting of parenteral IV magnesium treatment such as in the treatment of eclampsia by obstetricians.
- High-dose magnesium-containing cathartics or enemas may also deliver a large magnesium load to the intestine that cannot be rapidly excreted by the kidneys.
- Often there is some concomitant renal disease that may impair the ability to deal with a magnesium load.
- Mild hypermagnesemia has been reported in FHH caused by an inactivating mutation in the CaSR.
- High magnesium may also impair PTH synthesis and cause hypocalcemia.
- Hypermagnesemia may cause hypotension through vasodilation, neuromuscular impairment, constipation, and electrocardiographic abnormalities/arrhythmias.
- The approach to disorders of high magnesium is usually to identify and remove the exogenous magnesium source. Hydration and dialysis, if needed, will then ensure adequate clearance.

The identification and/or confirmation of magnesium levels that exist in the hyper- or hypostates outlined above relies upon its measurement within biological specimens, with differential diagnosis of the various underlying disease states requiring the measurement of additional analytes, including calcium, phosphate, PTH, and 25-hydroxyvitamin D. The close association between magnesium, calcium, phosphate, and vitamin D makes their analysis in either blood or urine specimens integral to differential diagnosis of the aforementioned disease states.

3.3.2.2 *Laboratory considerations to assess hypo- and hypermagnesemia diseases*

3.3.2.2.1 Blood magnesium measurement

Preanalytical considerations As serum magnesium comprises ~0.3% of total body magnesium concentrations, it poorly correlates with the concentration of magnesium in most tissues and is further influenced by

- diet, as serum magnesium concentrations are higher in vegetarians/vegans in comparison to omnivorous [38]

- supplements and drugs that contain magnesium can lead to elevated levels in the blood
- exercise, with high exertion exercise leading to elevated serum levels in comparison to endurance exercises
- the pregnancy trimester, with the third trimester being associated with a marked increase in serum magnesium concentration [39].

In addition to inter/intraindividual variability, numerous laboratory-associated preanalytical factors have been associated with erroneous magnesium measurements, including

- Specimen collection tube, with some tubes using a silicone compound that gives falsely high magnesium values. Some values may be twice the real value and even higher, requiring consultations with the laboratory to understand which specimen collection tubes are appropriate for magnesium measurement.
- Hemolysis, as the concentration of magnesium with red blood cells can be up to three times greater than in serum.
- Bilirubin, which has been shown to interfere in the measurement of magnesium.

These factors indicate that the measurement of ionized magnesium requires defined sample handling, collection and transport procedures to reduce the occurrence of false results.

Analytical considerations The analytical procedures for measuring magnesium can be performed by an array of methods, including

- Complexation of magnesium with colorimetric dyes can be susceptible to interference from:
  - inorganic phosphate that can inhibit the formation of the colorimetric dye complex
  - paraproteinemia can interfere as it increases the solutions turbidity
  - contrast agents that absorb at the same wavelengths as the magnesium-dye-complex
- The specialized instrumentation and extensive sample preparation associated with atomic absorption spectrophotometry have limited its application in the clinical laboratory for the measurement of magnesium.

3.3.2.2.2 Urine magnesium measurement

Preanalytical considerations The determination of the concentration of magnesium within urine is susceptible to interference from the individuals circadian rhythm, which underlies renal magnesium excretion. Therefore it is important to collect a 24-h urine specimen to assess magnesium excretion and absorption accurately.

Postanalytical considerations Interpretation of urinary magnesium concentration results requires an associated serum measurement to be of use.

3.3.2.3 Misuse of laboratory tests in magnesium disorders

- Repeating serial serum magnesium testing after initial result is either normal or falls just below the lower end of the reference range. Important hypomagnesemia generally involves markedly low levels, not borderline low levels.
- Over-treatment of mild hypomagnesemia (0.5−0.7 mmol/L) in the asymptomatic patient.
- Stool magnesium measurements likely of little routine utility. If the fractional excretion of magnesium in the urine is appropriately low, then the cause of hypomagnesemia is usually GI.
- Serum calcitriol levels are usually unhelpful—vitamin D status is best evaluated with a 25-hydroxyvitamin D measurement.

3.3.3 EXAMPLE CASE WITH DISCUSSION

A 43-year-old man has recently immigrated to Canada from Sudan. He works on the night shift in a meat packing plant. Recently, he has noticed some worsening fatigue, muscle weakness, irritability, and occasional palpitations. His family physician examines him and finds him to have mild but persistent hypertension over several visits. The patient is prescribed hydrochlorthiazide 25 mg/day. He returns to his doctor 4 weeks later at which time his blood pressure has normalized but his symptoms are worse. He is now having more palpitations and troubles working given the heavy lifting required. ECG shows frequent premature atrial contractions.

Question 1: *What laboratory tests might be appropriate at this time?*

Answer: Electrolyte panel, creatinine/eGFR, complete blood count (CBC), and TSH are ordered. Results are all normal with the exception of mild hypokalemia: 3.3 mmol/L.

Question 2: *Does the hypokalemia fully explain the patient's symptoms?*

Answer: While hypokalemia is a common side effect of diuretic medication, the patient clearly had milder symptoms that predated the hydrochlorthiazide prescription. Oral potassium supplements are prescribed that improves his serum potassium to 3.6 mmol/L after 2 weeks.

Question 3: *What other tests should be considered?*

Answer: Serum magnesium should be drawn given the known association between diuretic use and renal magnesium wasting. Furthermore hypokalemia may be more difficult to treat in the face of concomitant hypomagnesemia.

Results: Serum magnesium 0.4 mmol/L (reference range 0.65−1.05 mmol/L). Hydrochlorthiazide is replaced with perindopril and his potassium tablets are stopped. The patient is started on oral magnesium preparations but continues to have symptoms and is experiencing more frequent looser stools. Repeat serum magnesium is 0.5 mmol/L. Potassium is now normal: 4.1 mmol/L.

Question 4: *Will any other laboratory tests provide help?*

Answer: Yes. Given his darker skin and night job with little daytime sun exposure, he is quite predisposed to vitamin D deficiency which can impair the intestinal absorption of magnesium. Measuring calcium, albumin, and PTH may also be useful.

Results: 25-Hydroxyvitamin D 15 nmol/L (severe deficiency range), PTH 80 ng/L (7−37 ng/L), calcium 2.12 mmol/L (2.1−2.55 mmol/L), albumin 38, corrected calcium 2.16 mmol/L.
- He has clear-cut vitamin D deficiency with 2HPTH.

The patient is started on vitamin D 4000 IU daily and oral magnesium is continued at a low dose such that he has no further diarrhea. Six weeks later his 25-hydroxyvitamin D is 60 nmol/L and his serum magnesium is 0.8 mmol/L. His symptoms have all resolved. Oral magnesium is stopped and vitamin D 2000 IU/day is continued.

Discussion: This patient had preexisting vitamin D deficiency and symptomatic hypomagnesemia which was exacerbated by a thiazide diuretic causing further urinary magnesium loss. Vitamin D replacement, oral magnesium supplements, and discontinuing the diuretic resulted in a resolution of symptoms and normalization of his serum magnesium. This case illustrates the interplay and codependence between many bone-related minerals and hormones.

KEY POINTS

Magnesium disorders require clinical suspicion to consider measuring serum magnesium.

- Renal wasting and GI malabsorption/intestinal loss are the most common causes of hypomagnesemia.
- Inappropriately high urine magnesium or fractional excretion of magnesium in the face of hypomagnesemia strongly suggests a renal cause.
- Consider coexisting hypocalcemia, vitamin D deficiency, or hypokalemia when investigating and treating hypomagnesemia.
- Hypermagnesemia is much less common than hypomagnesemia and is usually caused by intravenous or GI magnesium loads that may overwhelm the capacity for renal excretion.

3.4 VITAMIN D DISORDERS

3.4.1 VITAMIN D INSUFFICIENCY AND DEFICIENCY

3.4.1.1 Introduction

The definition, evaluation, and treatment of vitamin D disorders have been the subject of much recent controversy. The almost ubiquitous nature of lower vitamin D levels particularly in northern countries during the wintertime has led many researchers, clinicians, and patients to try and connect this phenomenon to many disease conditions. Indeed, association studies can link lower vitamin D levels to an increased incidence of cancer, diabetes, and multiple sclerosis [40]. However, despite diligent study, vitamin D supplementation has not yet been shown to reduce the incidence or severity of any of these conditions. Even in the skeleton, where a very plausible mechanism for the essential role of vitamin D is established, vitamin D treatment for those found to have a mild to moderate insufficiency has not shown consistent effects on fracture reduction [41]. Despite the lack of evidence, many patients and healthcare professionals recommend vitamin D often in high doses as a disease-preventing supplement. A recent report from the Institute of Medicine (IOM) challenges the concept that low vitamin D levels are indeed truly pathologic [42].

Part of the problem lies in the definition of vitamin D insufficiency and true deficiency. In 2011 the IOM in the United States concluded that true deficiency of vitamin D should only be defined as 25-hydroxyvitamin D levels falling below 25 nmol/L and insufficiency in patients with values between 25 and 50 nmol/L [43]. That same year the Endocrine Society (ES) came out with guidelines defining deficiency below 50 nmol/L and insufficiency as those having 25-hydroxyvitamin D levels between 50 and 75 nmol/L [44]. One of the key pieces of evidence that both groups cited was the inverse relationship between population levels of vitamin D and PTH. The IOM does not believe that a meaningful reflex rise in PTH (2HPTH) ever occurs at 25-hydroxyvitamin D levels of 50 nmol/L or higher. The ES, on the other hand, cites a few studies purporting clinically important PTH rises below 75 nmol/L. Not surprisingly the two organizations have quite different recommendations for empiric vitamin D supplement doses which lead to confusion amongst both patients and clinicians.

Studies and meta-analyses of studies looking at the independent effect of vitamin D supplementation on fractures and falls are heterogeneous and difficult to interpret. Many trials look at combinations of different doses of vitamin D and calcium in different patient populations and varied community or institutionalized settings. Most clinicians feel that doses of vitamin D 800–1000 IU/day combined with a total calcium daily dose of 1000 mg (preferably from the diet with supplement as need be) is safe and reasonable in patients with osteoporosis or those at risk for same.

Finally the controversy outlined above for possible disease associations, conflicting 25-hydroxyvitamin D targets and supplement doses for vitamin D, has led to a significant push by the population and healthcare providers for the routine screening of blood vitamin D levels. This overutilization has resulted in a substantial financial burden. In our lab, policy changes instituted in early 2015 requiring a special form to accompany requisitions for this blood test resulted in a significant drop in testing and meaningful cost savings.

3.4.1.2 Clinical picture

The clinical relevance of vitamin D insufficiency continues to be debated whether one defines this by either the IOM or ES criteria. However, true vitamin D deficiency that the IOM defines as those having 25-hydroxyvitamin D levels <25 nmol/L may have substantial clinical ramifications that may respond very well to inexpensive and safe vitamin D supplementation. Clinical or true vitamin D deficiency is less common in the 21st century with better nutrition and reduced air pollution improving sun exposure.

The skeletal ramifications of severe vitamin D deficiency are different in children with growing bones and active growth plates (commonly called "rickets") as opposed to an adult whose growth plates have closed and is simply known as osteomalacia (impaired bone mineralization). Improvements in nutrition, hygiene, education, and pollution have all resulted in a marked fall in childhood rickets and adult osteomalacia. However, clinicians must be cognizant of this condition particularly in patients from developing countries and those whose customs mandate covering up from the sun many of whom may also have darker skin and thus less vitamin D production from sunlight. Medical conditions, drugs, and surgeries can also predispose to severe vitamin D deficiency.

Classic features to seek on history and physical exam include

- bone pain especially in the ribs and sternum in children
- delayed growth and development in children who may fall off the normal curves of growth charts
- infants presenting with seizures and hypocalcemia
- proximal muscle weakness with difficulty climbing stairs or rising from a chair
- low trauma fractures or unexpected low bone mass in patients with no obvious risk factors for osteoporosis
- bony deformities such as the rachitic rosary over the ribs and sternum or bowing of lower extremities upon weight bearing in toddlers
- mothers with darker skin who may also cover much of their skin and face with clothing are at high risk for vitamin D deficiency as are their children
- patients with known hepatic or renal disease
- patients with malabsorption syndromes or small bowel disorders such as celiac or Crohn's disease
- patients with past gastric or small bowel surgery, including bariatric surgery [45]
- long standing anticonvulsant therapy
- in adults, signs and symptoms of hypocalcemia would be uncommon presenting features of vitamin D deficiency given the reflex protection of the parathyroid glands 2HPTH

3.4.1.3 Causes of vitamin D deficiency

- Insufficient oral vitamin D intake often in the setting of malnutrition or other illnesses.
- Limited sunlight exposure due to climate, clothing habits, or darker skin color.

- Reduced gastrointestinal absorption of vitamin D in patients with malabsorption disorders, gastric, bariatric, or small intestinal surgeries, celiac disease, and IBD.
- Liver disease impairing the conversion of both skin and GI-derived vitamin D to the stable storage form: 25-hydroxyvitamin D.
- Renal dysfunction impairing renal one-alpha hydroxylation of 25-hydroxyvitamin D to its active form, calcitriol.
- Medications that promote rapid breakdown and metabolism of vitamin D such as the anticonvulsants phenytoin, carbamazepine, and phenobarbital.
- Rare congenital forms often presenting with hypocalcemia in neonates:
  - type-1 vitamin D resistance caused by a defect in 1-alpha hydroxylase enzyme
  - type-2 vitamin D resistance caused by a partial or complete defect in the vitamin D receptor

3.4.1.4 Diagnostic approach (laboratory test focussed)

Readers are referred to Section 3.1.1.2 for an overview of vitamin D physiology which should be remembered as a model by which to consider the appropriate vitamin D metabolite to be measured for any specific clinical context.

A high index of suspicion when identifying the above clinical features should lead to the measurement of the stable serum form, 25-hydroxyvitamin D. X-rays in children may show classic findings at growth plates or changes of rickets. Pseudo-fractures can be seen in vertebrae or longer bones. Adults may be mistakenly labeled as having osteoporosis based upon a low bone mineral density (BMD) result. Clinicians should consider osteomalacia in patients with low BMD for whom there is not a plausible explanation especially younger men and premenopausal women. The definitive diagnosis of rickets or osteomalacia can only be made on a bone biopsy showing widened osteoid seams and a defect in bone mineralization. Practically speaking, a bone biopsy is uncommonly needed when the clinical picture and biochemistry support the diagnosis.

Vitamin D deficiency may be supported by the expected laboratory findings:

- low 25-hydroxyvitamin D usually <25 nmol/L (<12.5 nmol/L in severe cases)
- reflex secondary rise in PTH (2HPTH) that will increase bone resorption and promote calcium efflux from the skeleton
- mild hypocalcemia can be seen in severe cases but in adults the 2HPTH will often keep the serum calcium at the lower end of the reference range
- 1,25-dihydroxyvitamin D (calcitriol) is not usually helpful as 2HPTH may keep this in the reference range; even in severe vitamin D deficiency (with skeletal consequences), the serum 1,25 vitamin D levels are usually normal
- low or low-end normal range serum phosphate as vitamin D normally promotes GI phosphate absorption
- low urine calcium because of both low vitamin D and high PTH

3.4.2 VITAMIN D EXCESS

Uncommonly, patients will present with hypercalcemia due to vitamin D excess. As described in Section 3.1.2, granulomatous diseases such as sarcoidosis or tuberculosis or malignancy such as lymphoma can cause high serum calcium mediated by an unregulated production of calcitriol in the pathologic granulomatous or cancerous tissue. 1,25-Dihydroxyvitamin D (calcitriol) levels may helpful in these unique situations. Treatment of the underlying disease, hydration, and adjuvant corticosteroid therapy will often rapidly normalize the calcium.

Intentional or unintentional vitamin D toxicity is well described although the threshold 25-hydroxyvitamin D level above which hypercalcemia may occur is not well known [46]. The IOM suggests caution when values are >200 nmol/L. The etiology is usually readily identified and remedied with hydration and supplement discontinuation. However, some alternative medication or supplements may not be well labeled or patients not well informed about high-dose vitamin D therapy for which there are some enthusiasts. Kidney stones, nephrocalcinosis, and hypercalciuria with low or suppressed PTH may all be seen with prolonged high-dose vitamin D use.

3.4.3 RENAL BONE DISEASE

3.4.3.1 Clinical picture

Renal bone disease is a complex and clinically challenging disorder. The older term renal osteodystrophy has been replaced by CKD-mineral bone disorder (MBD) [47]. Renal osteodystrophy is now used to describe the histopathology. CKD-MBD covers a spectrum that can range from high to low bone turnover states with different mechanisms and potential approaches. CKD-MBD denotes perturbations in renal, bone, parathyroid, gastrointestinal, and vascular function. Marked advancements in understanding the pathophysiology of CKD-MBD have been made. However, limitations in the interpretation of circulating hormone and mineral levels and noninvasive assessments of bone quality and quantity continue to impair the ability of physicians to consistently reduce the serious cardiovascular and musculoskeletal morbidity and mortality of patients with kidney disease.

3.4.3.2 Hormone and mineral abnormalities in chronic kidney disease-mineral bone disorder

- *FGF-23*: Produced by osteocytes and osteoblasts, FGF-23 increases in the early stages of CKD often prior to any significant measured changes in serum phosphate, calcium, PTH, or vitamin D. FGF-23 will promote renal phosphate loss and impair the activation of 25-hydroxyvitamin D to its active form calcitriol leading to reduced GI absorption of both phosphate and calcium. It may therefore be viewed as a potentially protective or compensatory mechanism for the phosphate retention of CKD.

- *Hypocalcemia*: As renal function declines, serum calcium will fall as a result of diminishing calcitriol production from both the renal parenchyma and inhibition by FGF-23.
- *Hyperphosphatemia*: Rises in serum phosphate, despite the rise in FGF-23 as glomerular and tubular function is impaired, will lead to more calcium-phosphate binding and thus less free or ionized calcium.
- 2HPTH: Lower calcium and lower calcitriol plus hyperphosphatemia will all stimulate the parathyroid glands to increase PTH production. 2HPTH will increase bone resorption more than formation leading to a high bone turnover state with loss of bone quantity and quality. Patients with CKD have a higher risk for fractures.
- *Vitamin D deficiency*: Reduced calcitriol and 25-hydroxyvitamin D (from the malnutrition associated with chronic illness) may have deleterious effects on bone, muscle, and possibly immune function.

Nephrology guidelines, best described for those with end-stage renal disease on dialysis, outline strategies for maintaining normal serum phosphate and calcium with a permissive mild to moderate PTH elevation [48]. With the goal of reducing the risk of vascular and soft tissue calcifications, fractures and cardiovascular mortality, these strategies include:

- reducing dietary phosphate intake
- using oral phosphate binders to reduce GI absorption: calcium carbonate, calcium acetate, sevelamer, lanthanum, and newer iron-containing binders all have shown utility to lower the phosphate
- activated vitamin D analogues including calcitriol and paricalcitol to inhibit PTH production but may also undesirably increase GI calcium and phosphate absorption
- calcimimetics such as cinacalcet will act directly on parathyroid chief cells to reduce PTH

Unfortunately, there is not yet compelling data that demonstrates the efficacy of one regime or combination over the other for reducing morbidity and mortality. Significant patient and disease heterogeneity with lack of bone histomorphometry availability may be the main factors limiting larger scale clinical trials. Furthermore many of these agents are expensive and add to an already substantial "pill burden" for the patient with kidney disease.

3.4.3.3 Spectrum of renal bone disorders

- *High bone turnover disease*, predominantly due to uncontrolled severe hyperparathyroidism, leads to the classic bone pathology of osteitis fibrosa cystica. PTH and bone turnover markers may be very high.
- *Low bone turnover disease* may be caused by adynamic bone disease or osteomalacia. These patients have lower PTH levels that may be in the usual adult reference range. Over-judicious efforts to control phosphate and suppress

PTH above along with aluminum or bisphosphonate exposure may all be predisposing factors.

- *Adynamic bone disease* is usually associated with virtual absence of either osteoclastic or osteoblastic activity and may significantly impair the uptake of calcium and phosphate into the skeleton. This may lead to hypercalcemia and a higher calcium-phosphate product with its inherent risk of vascular and metastatic soft tissue calcification [33,49].
- Hypogonadism, corticosteroid therapy, prolonged heparin exposure, and vitamin D deficiency can also coexist and lead to patients with an admixture of high and low bone turnover disease.
- Osteoporosis, particularly in older patients, may predate the onset of kidney disease further complicating the approach to evaluation and therapy to reduce the risk for fracture.

3.4.3.4 Diagnostic approach (laboratory test focussed)

Ideally, all patients with CKD would be evaluated with a bone biopsy to better guide where they fall in the spectrum of low to high bone turnover disease. Unfortunately, bone biopsies are time-consuming, invasive, resource intensive and not widely available. The physician is thus usually left with evaluating clinical, biochemical, and diagnostic imaging surrogates to guide monitoring and therapy.

Laboratory goals for patients with CKD (KDIGO) [48]:

- calcium: maintain in the normal range
- phosphate: maintain in the normal range or toward normal range in dialysis patients
- PTH: ideal value is unknown but aim for 2−9 times the upper end of the reference range for dialysis patients

Bone density testing is of limited use in patients with CKD due to a lack of good data-relating BMD to fracture risk in this population. Furthermore high, low, and mixed turnover bone disease may all have low or normal bone mass and thus these subtypes cannot be differentiated.

KEY POINTS

Renal bone disease is a complex and clinically challenging disorder.

- CKD-MBD is a spectrum that can range from high to low bone turnover states.
- Laboratory goals for patients with CKD include keeping serum calcium and phosphate in the reference range.
- The ideal value for PTH is unknown but aim for 2−9 times the upper end of the reference range for dialysis patients.
- Bone density testing is of limited use in patients with CKD due to a lack of good data-relating BMD to fracture risk in this population.
- Physicians should involve a nephrologist early in the care of all patients with CKD.

3.4.4 **LABORATORY CONSIDERATIONS TO ASSESS VITAMIN D–ASSOCIATED DISEASES**

3.4.4.1 *Serum 25-hydroxyvitamin D*

The natural existence of 25-hydroxyvitamin D as two different structural isomers results in a difference in their biological activities and prevents the measurements of 25-hydroxyvitamin D being indicated for differential diagnosis of vitamin D–associated disorders.

3.4.4.1.1 Preanalytical considerations

Environmental conditions influence 25-hydroxyvitamin D synthesis including the amount of sunlight an individual receives, geographical region they inhabit, and the time of year. The impact of these environmental conditions has not resulted in a call for increased testing, but instead an increase in the usage of vitamin D supplementation if the individual is at a risk for being deficient. A clinical understanding of the minimal utility of assessing free 25-hydroxyvitamin D has led to a reduction in the number of vitamin D tests ordered. In addition, many preanalytical factors associated with the type of specimen drawn and storage/transport conditions have been found to not influence the measured concentration.

3.4.4.1.2 Analytical considerations

The long half-life and low interday variability of 25-hydroxyvitamin D make it the preferred analyte for measuring total body or nutritional vitamin D status by either

- Immunoassays (both competitive and noncompetitive), which do not equally differentiate the two isomers and lead to an underestimation of 25-hydroxyvitamin D concentrations in comparison to liquid chromatography methods.
- Liquid chromatography, employing mass spectrometry or absorbance detectors, is considered the reference method for the measurement of 25-hydroxyvitamin D because of its capability to chromatographically separate each structural isomer and metabolite. This eliminates the overestimation of different 25-hydroxyvitamin D isomers and metabolites caused by immunoassay cross reactivity.
  - The lack of standardization of liquid chromatographic methods introduces variability between laboratories that require the development of custom reference ranges to reduce the implication of measurement bias on the differential diagnosis of a disease state.
  - The LC–MS/MS methods have been able to detect the C3 epimer of 25-hydroxyvitamin D, which has a lower affinity for binding the vitamin D receptor, but has the same capacity to suppress PTH secretion. Therefore the C3 epimer is hypothesized to have a higher metabolic stability as

compared in comparison to 25-hydroxyvitamin D. It should be noted that discrepant findings among these studies on the level of the C3 epimer of 25-hydroxyvitamin D necessitate further investigation into its prevalence across various populations.

3.4.4.1.3 Postanalytical considerations

A patient's clinical presentation that prompts the measurement of 25-hydroxyvitamin D is quite diverse and special consideration is required for the selection of a normal population for development of a reference range. Identification of the reference range for 25-hydroxyvitamin D requires well-developed inclusion/exclusion criteria for the selection of a population that takes into account both disease states and seasonal variation.

Environmental conditions influence 25-hydroxyvitamin D synthesis and should be taken into account when creating a reference range, including:

- Seasonality, as blood levels during the winter months (at high latitudes) being significantly lower than those measured during the summer months, with a relative increase observed in the spring and decrease observed in the fall [52]. The capacity to understand seasonality as an important preanalytical factor, however, allows for its use in the interpretation of results.
- Regional influence, which is important for proper result interpretation as the proportion of individuals who have low or deficient 25-hydroxyvitamin D levels occurs much more readily in regions above 37° north or below 37° south [52].

3.4.4.2 (1,25)-Dihydroxyvitamin D

The structural similarities between 25-hydroxyvitamin D and 1,25-dihydroxyvitamin D not only allow for its measurement by either immunoassay or liquid chromatography but also result in it susceptibility to the same preanalytical and analytical factors. As the physiological most active form of vitamin D, 1,25-dihydroxyvitamin D is produced by the kidney and is useful for evaluating hypercalcemia, vitamin D intoxication, granulomatous disease, lymphomas, and renal failure. These clinical presentations, however, would only utilize the measurement of 1,25-dihydroxyvitamin D for confirmation of a differential diagnosis that was based upon the previous measurement of calcium, PTH, and other clinical indicators. In fact, many clinicians feel that the measurement of 1,25-dihydroxyvitamin D is clinically unnecessary in most cases since 1,25-dihydroxyvitamin D–related disorders are rare and usually diagnosed through other clinical features. The enzyme that produces 1,25-dihydroxyvitamin D is localized in the kidney, thus the measurement of 1,25-dihydroxyvitamin D is sometimes of interest in patients on dialysis or with end-stage kidney disease.

3.4.4.3 (24,25)-Dihydroxyvitamin D and CYP24A1 mutations

Vitamin D regulates bone and mineral homeostasis by acting as signaling hormone. However, the free circulation of 1,25-dihydroxyvitamin D can lead to a hypercalcemic state if it is not properly regulated. As with most hormones, it is regulated through a negative feedback loop, but the undue delay in the action of this pathway requires an even faster method of reducing 1,25-dihydroxyvitamin D concentrations, namely 25-hydroxyvitamin D−24-hydroxylase or CYP24A1. The function of 25-hydroxyvitamin D−24-hydroxylase is to catabolize 1,25-dihydroxyvitamin D into its inactive metabolite form, with its transcription being mediated by a positive vitamin D response element within its promoter region that is designed to protect cells from excess vitamin D.

The influence of the 25-hydroxyvitamin D−24-hydroxylase in the regulation of 1,25-dihydroxyvitamin D is readily apparent in individuals that have a mutation within this region. Loss of 25-hydroxyvitamin D−24-hydroxylase function is the most common mutation and predominantly prevents inactivation of 1,25-dihydroxyvitamin D or conversion of 25-hydroxyvitamin D to 24,25-dihydroxyvitamin D, leading to an overabundance of active vitamin D. For example, missense mutations in the 25-hydroxyvitamin D−24-hydroxylase gene have led to an increase in vitamin D sensitivity in patients with idiopathic infantile hypercalceima; some have inferred that it is the loss of the heme binding site which prevents enzymatic catabolism [50]. Some of these mutations in 25-hydroxyvitamin D−24-hydroxylase have resulted in elevated concentrations of 1,25-dihydroxyvitamin D, which is exacerbated through vitamin D intake and sunlight exposure. The incidence of such mutations has been observed in individuals diagnosed with osteopenia, corneal calcifications, chondrocalcinosis, CKD, nephrocalcinosis, and calcium-containing stone formation across a wide racial and geographic distribution [51]. An evaluation of individuals suffering from mutations in 25-hydroxyvitamin D−24-hydroxylase is recommended for the differential diagnosis of some hypercalcemic states, most notably hypercalciuric nephrolithiasis.

3.4.5 MISUSE OF LABORATORY WITH VITAMIN D TESTING

- Screening patients for vitamin D with a routine blood test in the absence of any risk factors, clinical features, or radiographic findings that suggest a predisposition for low vitamin D. Population vitamin D screening has never been recommended and is unnecessary outside of very select clinical conditions.
- Repeat vitamin D testing when a previous test result was in the reference range.
- Repeat vitamin D testing when a previous test result was in the insufficient range (IOM: 25−50 nmol/L, ES: 50−75 mmol/L). If the patient then starts appropriate vitamin D supplements, there should be no reason to remeasure or "follow" the vitamin D levels since virtually all patients will respond to

simple supplementation. Measuring 1,25-dihydroxyvitamin D (calcitriol) is rarely indicated except during the investigation of hypercalcemia. Even in severe vitamin D deficiency, calcitriol levels may fall within the reference range and thus have no clinical utility in this situation.

- Considering vitamin D deficiency to be *ruled out* if the serum calcium is normal. 2HPTH will usually result in maintaining serum calcium in the lower third of the reference range.
- Failure to consider CKD-MBD as a clinically relevant diagnostic possibility in CKD patients who still have a serum phosphate within the reference range.
- Failure to realize that a mild to moderate PTH elevation may be an appropriate goal of the nephrologist in patients with CKD. Inappropriate normalization or suppression of PTH in CKD may lead to low bone turnover or a dynamic bone disease.
- Not correcting serum calcium for albumin or using the ionized calcium.
- Inappropriate ordering of expensive laboratory tests such as FGF-23 or sclerostin. While both clearly play a key role in the development of CKD-MBD, there is not currently sufficient information as how to best relate the serum values to pathophysiology and intervention strategies for individual patients.

3.4.6 EXAMPLES WITH DISCUSSION

A 42-year-old woman presents to her family physician with progressively worsening fatigue and some nonspecific muscles aches and pains. She has unintentionally lost 5 kg in the past 6 months. She is previously well, continues to have regular menses, and takes no medications. Her family history is unremarkable apart from a diagnosis of osteoporosis in her mother when she was in her early seventies. The physical exam is normal although the physician notes she is a bit slow to get out of her chair to sit up on the examining table. Her family doctor orders a routine blood panel.

> *Question 1*: Should serum 25-hydroxyvitamin D be a part of "routine" blood testing?

> *Answer*: No. Unless there are clinical features or identifiable risk factors for vitamin D deficiency, the high chance of finding mild vitamin D insufficiency in the average person does not justify screening and the global expense of such a practice could be very significant. Low-dose routine vitamin D supplementation may be a better strategy for those at risk for mild vitamin D insufficiency.

> *Results*: CBC, electrolytes, creatinine/eGFR, TSH, calcium, and albumin all come back normal. The patient is reassured but asked to return in 1 month if she is not feeling any better.

One month later, she comes back to see her family doctor. In the interim time, she has lost another 2 kg of weight. Her stools are "a bit loose" and she is now finding that climbing the stairs in her home takes a bit longer than usual.

The doctor now orders a chest X-ray, mammogram, urinalysis, FIT test and sends her stool away for culture, ova, and parasites. All of these tests come back normal. Despite her complaints, the patient continues to work full time as a lawyer and look after her two school-aged children and husband.

Question 2: *Are other tests indicated at this time?*

Answer: Not necessarily. Her symptoms are nonspecific and she is still highly functioning at work and home. A basic malignancy and infection screen are both normal. Her general practitioner (GP) wants to see her again in 2 weeks.

Her mother had convinced her own doctor to order a bone density test on her daughter given the family history. The result shows her lumbar spine T-score to be quite low at −3.5 and total hip T-score also low at −3.1 (see Chapter 4, Bone metabolism, for interpretation details).

Question 3: *Does the patient have osteoporosis? Is this the cause of her weakness, fatigue, and weight loss?*

Answer: Osteoporosis can be defined a few different ways. The best definition is evidence of a low trauma or fragility fracture often supported by low bone mass on BMD. However, the BMD definition of a T-score <−2.5 is entrenched in the minds of many practitioners. In this case, finding a very low BMD in this previously well premenopausal woman is unusual and suggests the need for a work-up for other causes of low bone mass besides osteoporosis. In the absence of fractures, her symptoms are unlikely to be directly explained by the low bone density alone.

Question 4: *What might account for her low bone mass and how might the family doctor evaluate this?*

Answer: There is no history to support an eating disorder, hypogonadism, or malignancy. The normal calcium and TSH exclude 1HPTH and hyperthyroidism as secondary causes of low bone mass. With her progressing proximal muscle weakness, weight loss, and low bone mass, measuring 25-hydroxyvitamin D is now certainly indicated.

Result: 25-Hydroxyvitamin D = 10 nmol/L indicating severe vitamin D deficiency

Question 5: *What other tests are needed to determine the cause of her severe vitamin D deficiency?*

Answer: With the weight loss and loose stools, GI malabsorption or liver disease seems possible. The physician orders a celiac screen and a full panel of liver transaminases and liver function tests. In addition, calcium, albumin, PTH, and serum phosphate are requested. A GI consult is also requested.

Results: Celiac screen: negative. Liver transaminases normal. Calcium 2.05, albumin 28, phosphate 0.7 PTH 100.

Endoscopy and biopsy reveals IBD: Crohn's disease.

Diagnosis: Severe vitamin D deficiency due to GI malabsorption from IBD, causing presumed osteomalacia with concomitant 2HPTH.

Question 6: *What is the treatment?*

Answer: In patients found to be vitamin D deficient because of GI malabsorption often higher doses of supplemental vitamin D 2000−5000 IU/day may be needed initially with 25-hydroxyvitamin D monitoring. Treatment of her IBD is also essential after which the vitamin D supplement dose may be safely reduced to 1000 IU/day. Improvement in muscle strength and bone mass would be anticipated but may take some time.

3.4.7 EXAMPLE 2 WITH DISCUSSION

A 45-year-old woman presents to her family physician with symptoms of fatigue and a reduced appetite. Her weight has been stable. Her past medical history is unremarkable except for a history of bladder and ureter surgery as a child. She takes no regular medications and continues to have regular monthly menses. The physical examination is normal. A laboratory panel is ordered.

Results: Electrolytes, ferritin, and hemoglobin A1C are all normal. Hemoglobin 110 with normal mean corpuscular volume (MCV).

Creatinine 170 μmol/L, eGFR 28 mL/min/1.73 m^2

Question 1: *What is the most likely cause of her renal dysfunction?*

Answer: Previous damage perhaps from vesico-ureteral reflux as a child and surgery seems the most likely cause. However, an evaluation of possible prerenal, renal, and postrenal causes is indicated.

Question 2: *What other tests should be ordered now?*

Answer: To start, a urinalysis looking for blood and active sediment is needed. A kidney and bladder ultrasound looking for hydronephrosis, stones, and renal size is also indicated.

Results: Apart from postsurgical changes and possible renal parenchymal scarring, these tests are normal. A nephrology consult is requested.

Question 3: *What other laboratory tests may be helpful now in this patient with stage 4 CKD?*

Answer: Calcium, albumin, phosphate, PTH, and 25-OH (25-hydroxy) vitamin D

Results: Calcium: 2.20 mmol/L (2.1−2.55 mmol/L), albumin: 38 g/L (33−48 g/L), corrected calcium: 2.24 mmol/L, phosphate: 2.0 mmol/L (0.8−1.5 mmol/L), PTH: 370 ng/L (7−37 ng/L), and 25-OH vitamin D: 40 nmol/L

Question 4: How would you interpret these results?

Answer: She has 2HPTH due to:

- Reduced vitamin D and almost certainly low calcitriol due to:
    - loss of functioning renal parenchyma
    - elevated FGF-23 inhibiting 1-alpha hydroxylase activation of 25-OH vitamin D
    - reduced vitamin D stores
- Reduced free or ionized calcium because of calcium-phosphate binding. The corrected calcium has normalized because of the effects of a high PTH.
- Hyperphosphatemia

The patient is at substantial risk for CKD-MBD with high bone turnover state most likely.

Question 5: What should be the approach to 2HPTH in this case?

Answer: Phosphate binding with inexpensive calcium carbonate taken three times daily with meals with be the first step. Goals are to normalize the phosphate and reduce the PTH. Vitamin D supplementation 1000−2000 IU/ day should also be considered. If goals are not being met, the addition of calcitriol would be the next step. The patient's care should be supervised by a nephrologist.

KEY POINTS

The definition, evaluation, and treatment of vitamin D disorders have been the subject of much recent controversy.

- Vitamin D plays a crucial role in bone quality and mineral metabolism.
- Severe vitamin D deficiency, best defined as <25 nmol/L, is associated with rickets and osteomalacia.
- Purported links to other medical conditions and disease prevention through vitamin D supplementation have not yet been substantiated.
- Routine screening of patients with serum 25-hydroxyvitamin D is not recommended in the absence of any risk factors, clinical features, or radiographic findings that suggest a predisposition for low vitamin D.

REFERENCES

[1] Marcocci C, Cetani F. Primary hyperparathyroidism. N Engl J Med 2011;365:2389−97.
[2] Vargas-Poussou R, Mansour-Hendili L, Baron S, Bertocchio JP, Travers C, Simian C, et al. Familial hypocalciuric hypercalcemia types 1 and 3 and primary hyperparathyroidism: similarities and differences. J Clin Endocrinol Metab 2016;101(5):2185−95.
[3] Cheng C-J, Chou C-H, Lin S-H. An unrecognized cause of recurrent hypercalcemia: immobilization. South Med J 2006;99(4):371−4.

[4] Arnold A, Marx S. Familial primary hyperparathyroidism (including MEN, FHH, and HPT-JT). Prim Metab Bone Dis 2013. Available from: http://dx.doi.org/10.1002/9781118453926.ch69.

[5] Pitt SC, Sippel RS, Chen H. Secondary and tertiary hyperparathyroidism, state of the art surgical management. Surg Clin North Am 2009;89(5):1227−39.

[6] Eisenbarth GS, Gottlieb PA. Medical progress: autoimmune polyendocrine syndromes. N Engl J Med 2004;350(20) 2068−79.

[7] Mantovani G. Pseudohypoparathyroidism: diagnosis and treatment. J Clin Endocrinol Metab 2011;96(10):3020−30.

[8] Lippi G, Salvagno GL, Lima-Oliveira G, Brocco G, Danese E, Guidi GC. Postural change during venous blood collection is a major source of bias in clinical chemistry testing. Clin Chim Acta 2015;440:164−8.

[9] Jubiz W, Canterbury JM, Reiss E, Tyler FH. Circadian rhythm in serum parathyroid hormone concentration in human subjects: correlation with serum calcium, phosphate, albumin, and growth hormone levels. J Clin Invest 1972;51(August 8):2040−6 [Internet] [cited 2016 Dec 12]. Available from: http://www.ncbi.nlm.nih.gov/pubmed/5054463.

[10] Elfatih A, Anderson NR, Fahie-Wilson MN, Gama R. Pseudo-pseudohypercalcaemia, apparent primary hyperparathyroidism and Waldenström's macroglobulinaemia. J Clin Pathol 2007;60(April 4):436−7. [Internet] [cited 2016 Dec 12]. Available from: http://www.ncbi.nlm.nih.gov/pubmed/17405982.

[11] Pasco JA, Henry MJ, Kotowicz MA, Sanders KM, Seeman E, Pasco JR, et al. Seasonal periodicity of serum vitamin D and parathyroid hormone, bone resorption, and fractures: the geelong osteoporosis study. J Bone Miner Res 2004;19(August 5):752−8. [Internet] [cited 2016 Dec 12]. Available from: http://doi.wiley.com/10.1359/jbmr.040125.

[12] Löwe A, Balzer T, Hirt U. Interference of gadolinium-containing contrast-enhancing agents with colorimetric calcium laboratory testing. Invest Radiol 2005;40(8):521−5. [Internet] [cited 2016 Dec 12]. Available from: http://www.ncbi.nlm.nih.gov/pubmed/16024990.

[13] Kahn RC, Jascott D, Carlon GC, Schweizer O, Howland WS, Goldiner PL. Massive blood replacement: correlation of ionized calcium, citrate, and hydrogen ion concentration. Anesth Analg 1979;58(4):274−8. [Internet] [cited 2016 Dec 12]. Available from: http://www.ncbi.nlm.nih.gov/pubmed/36816.

[14] Takano S, Kaji H, Hayashi F, Higashiguchi K, Joukei S, Kido Y, et al. A calculation model for serum ionized calcium based on an equilibrium equation for complexation. Anal Chem Insights 2012;7:23−30. [Internet] [cited 2016 Dec 12]. Available from: http://www.ncbi.nlm.nih.gov/pubmed/22837641.

[15] Gökçe Ç, Gökçe Ö, Baydinç C, İlhan N, Alaşehirli E, Özküçük F, et al. Use of random urine samples to estimate total urinary calcium and phosphate excretion. Arch Intern Med 1991;151(August 8):1587. [Internet] [cited 2016 Dec 12]. Available from: http://archinte.jamanetwork.com/article.aspx?doi = 10.1001/archinte.1991.00400080083015.

[16] Brickman AS, Massry SG, Coburn JW. Changes in serum and urinary calcium during treatment with hydrochlorothiazide: studies on mechanisms. J Clin Invest 1972;51(April 4):945−54. [Internet] [cited 2016 Dec 12]. Available from: http://www.ncbi.nlm.nih.gov/pubmed/4552338.

[17] Fuleihan GE-H, Klerman EB, Brown EN, Choe Y, Brown EM, Czeisler CA. The parathyroid hormone circadian rhythm is truly endogenous—a general clinical research center study 1. J Clin Endocrinol Metab 1997;82(January 1):281−6. [Internet] [cited 2016 Dec 12]. Available from: http://press.endocrine.org/doi/10.1210/jcem.82.1.3683.

[18] Steingrimsdottir L, Gunnarsson O, Indridason OS, Franzson L, Sigurdsson GPL, et al. Relationship between serum parathyroid hormone levels, vitamin D sufficiency, and calcium intake. JAMA 2005;294(November 18):2336. [Internet] [cited 2016 Dec 12]. Available from: http://jama.jamanetwork.com/article.aspx?doi = 10.1001/jama.294.18.2336.

[19] Cavalier E, Plebani M, Delanaye P, Souberbielle J-C. Considerations in parathyroid hormone testing. Clin Chem Lab Med 2015;53(January 12):1913−19. [Internet]. [cited 2016 Dec 12]. Available from: http://www.ncbi.nlm.nih.gov/pubmed/26035114.

[20] Donovan PJ, Achong N, Griffin K, Galligan J, Pretorius CJ, McLeod DSA. PTHrP-mediated hypercalcemia: causes and survival in 138 patients. J Clin Endocrinol Metab 2015;100(May 5):2024−9. [Internet]. [cited 2016 Dec 12]. Available from: http://www.ncbi.nlm.nih.gov/pubmed/25719931.

[21] Bilezikian JP, Brandi ML, Eastell R, Silverberg SJ, Udelsman R, Marcocci C, et al. Guidelines for the management of asymptomatic primary hyperparathyroidism: summary statement from the fourth international workshop. J Clin Endocrinol Metab 2014;99(10):3561−9.

[22] Marcocci C, Bollerslev J, Khan AA, Shoback DM. Medical management of primary hyperparathyroidism: proceedings of the fourth international workshop on the management of asymptomatic primary hyperparathyroidism. J Clin Endocrinol Metab 2014;99(10):3607−18.

[23] Eastell R, Brandi ML, Costa AG, D'Amour P, Shoback DM, Thakker RV. Diagnosis of asymptomatic primary hyperparathyroidism: proceedings of the fourth international workshop. J Clin Endocrinol Metab 2014;99(10):3570−9.

[24] Silverberg S, Shane E, Jacobs T. A 10-year prospective study of primary hyperparathyroidism with or without parathyroid surgery. N Engl J Med 1999;341(17):1249−55.

[25] Halevy J, Bulvik S. Severe hypophosphatemia in hospitalized patients. Arch Intern Med 1988;148(1):153−5.

[26] Mateo L, Holgado S, Mariñoso ML, Pérez-Andrés R, Bonjoch A, Romeu J, et al. Hypophosphatemic osteomalacia induced by tenofovir in HIV-infected patients. Clin Rheumatol 2016;35(May 5):1271−9.

[27] Ruppe MD, de Beur SMJ. Disorders of phosphate homeostasis. Primer on the metabolic bone diseases and disorders of mineral metabolism.. Hoboken, NJ, USA: John Wiley & Sons, Inc; 2009. p. 317−25, Chapter 69.

[28] Payne R. Renal tubular reabsorption of phosphate (TmP/GFR): indications and interpretation. Ann Clin Biochem 1998;35(Pt 2):201−6.

[29] Domico M, Huynh V, Anand S, Mink R. Severe hyperphosphatemia and hypocalcemic tetany after oral laxative administration in a 3-month-old infant. Pediatrics 2006;118(5):e1580−3.

[30] Larner A. Pseudohyperphosphatemia. Clin Biochem 1995;28(4):391−3.

[31] Grunenwald S, Tack I, Chauveau D, Bennet A, Caron P. Impact of growth hormone hypersecretion on the adult human kidney. Ann Endocrinol (Paris) 2011;72(6):485−95.

[32] Ramnitz MS, Gourh P, Goldbach-Mansky R, Wodajo F, Ichikawa S, Econs MJ, et al. Phenotypic and genotypic characterization and treatment of a cohort with familial tumoral calcinosis/hyperostosis-hyperphosphatemia syndrome. J Bone Miner Res 2016;31(October 10):1845−54.

[33] Massry SG, Coburn JW, Chertow GM, Hruska K, Langman C, Malluche H, et al. K/DOQI clinical practice guidelines for bone metabolism and disease in chronic kidney disease. Am J Kidney Dis 2003;42(4 Suppl. 3):S1−201.

[34] Wesseling-Perry K. FGF23: is it ready for prime time? Clin Chem 2011;57(11):1476−7.

[35] Cundy T, Dissanayake A. Severe hypomagnesaemia in long-term users of proton-pump inhibitors. Clin Endocrinol (Oxf) 2008;69(August 2):338−41.

[36] Pham P-CT, Pham P-MT, Pham SV, Miller JM, Pham P-TT. Hypomagnesemia in patients with type 2 diabetes. Clin J Am Soc Nephrol 2007;2(February 2):366−73.

[37] Elisaf M, Panteli K, Theodorou J, Siamopoulos KC. Fractional excretion of magnesium in normal subjects and in patients with hypomagnesemia. Magnes Res 1997;10(December 4):315−20.

[38] Vormann J. Magnesium: nutrition and metabolism. Mol Aspects Med 2003;24(1):27−37.

[39] Standley CA, Whitty JE, Mason BA, Cotton DB. Serum ionized magnesium levels in normal and preeclamptic gestation. Obstet Gynecol 1997;89(January 1):24−7. [Internet]. [cited 2016 Dec 12]. Available from: http://www.ncbi.nlm.nih.gov/pubmed/8990431.

[40] Bouillon R, Carmeliet G, Verlinden L, Van Etten E, Verstuyf A, Luderer HF, et al. Vitamin D and human health: lessons from vitamin D receptor null mice. Endocr Rev 2008;29(6):726−76.

[41] Rosen CJ. Vitamin D insufficiency. N Engl J Med 2011;364(January 3):248−54.

[42] Manson J, Brannon P, Rosen C. Vitamin D deficiency—is there really a pandemic? New Engl J 2016;375(19):1817−20.

[43] Ross A, Manson J, Abrams S, Aloia J, Brannon P, Clinton S, et al. The 2011 report on dietary reference intakes for calcium and vitamin D from the Institute of Medicine: what clinicians need to know. J Clin Endocrinol Metab 2011;96(1):53−8.

[44] Holick MF, Binkley NC, Bischoff-Ferrari HA, Gordon CM, Hanley DA, Heaney RP, et al. Evaluation, treatment, and prevention of vitamin D deficiency: an endocrine society clinical practice guideline. J Clin Endocrinol Metab 2011;96(July 7):1911−30.

[45] Madan A, Orth W, Tichansky D, Ternovits C. Vitamin and trace mineral levels after laparoscopic gastric bypass. Obes Surg 2006;16(May 5):603−6.

[46] Hathcock JN, Shao A, Vieth R, Heaney R. Risk assessment for vitamin D 1, 2. Am J Clin Nutr 2007;85:6−18.

[47] Moe S, Drüeke T, Cunningham J, Goodman W, Martin K, Olgaard K, et al. Definition, evaluation, and classification of renal osteodystrophy: a position statement from Kidney Disease: Improving Global Outcomes (KDIGO). Kidney Int 2006;69(11):1945−53.

[48] Ketteler M, Elder GJ, Evenepoel P, Ix JH, Jamal SA, Lafage-Proust M-H, et al. Revisiting KDIGO clinical practice guideline on chronic kidney disease—mineral and bone disorder: a commentary from a kidney disease: improving global outcomes controversies conference. Kidney Int 2015;87(3):502−28.

[49] Bover J, Ureña P, Brandenburg V, Goldsmith D, Ruiz C, DaSilva I, et al. Adynamic bone disease: from bone to vessels in chronic kidney disease. Semin Nephrol 2014;34(6):626−40.

[50]   Ji H-F, Shen L. CYP24A1 mutations in idiopathic infantile hypercalcemia. N Engl J Med 2011;365(November 18):1741−3. [Internet] [cited 2016 Dec 12]. Available from: http://www.nejm.org/doi/abs/10.1056/NEJMc1110226.

[51]   Dinour D, Beckerman P, Ganon L, Tordjman K, Eisenstein Z, Holtzman EJ. Loss-of-function mutations of CYP24A1, the vitamin D 24-hydroxylase gene, cause long-standing hypercalciuric nephrolithiasis and nephrocalcinosis. J Urol 2013;190(August 2):552−7. [Internet] [cited 2016 Dec 12]. Available from: http://www.ncbi.nlm.nih.gov/pubmed/23470222.

[52]   Bolland M, Grey A, Ames R, Mason B, Horne A, Gamble G, Reid I. The effects of seasonal variation of 25-hydroxyvitamin D and fat mass on a diagnosis of vitamin D sufficiency. Am J Clin Nutr 2007 Oct [cited 2016 Dec 12]; 86(4):959−964.

FURTHER READING

Fawcett WJ, Haxby EJ, Male DA. Magnesium: physiology and pharmacology. Br J Anaesth Br J Anaesth 1999;83(83):302−20.

Bone metabolism

4

Gregory Kline, MD[1], Dennis Orton, PhD[2] and Hossein Sadrzadeh, PhD[3]

[1]*Clinical Professor, Department of Medicine, Division of Endocrinology, University of Calgary, Calgary, Alberta, Canada*
[2]*Clinical Biochemist, Dr. CJ Coady Associates. Surrey Memorial Hospital, Surrey, British Columbia, Canada*
[3]*Professor, Department of Pathology and Laboratory Medicine, and Graduate Studies, University of Calgary and Section Chief of Clinical Biochemistry, Calgary Laboratory Services, Calgary, Alberta, Canada*

CHAPTER OUTLINE

Endocrine Biomarkers. DOI: http://dx.doi.org/10.1016/B978-0-12-803412-5.00004-5
© 2017 Elsevier Inc. All rights reserved.

4.1 OVERVIEW

Osteoporosis and other metabolic bone diseases form a large group of disorders of bone metabolism. Since bone loss is a natural consequence of aging, the development of age-associated low bone mass with subsequent fragility fractures of the spine or hip is a major factor in both patient morbidity and healthcare costs among societies with increasingly aged population. Owing to the sheer size of the population older than the age of 60, "routine" osteoporosis dominates the majority of clinical bone issues but there are a number of other metabolic bone problems that must be remembered and investigated during the care of the patient who suffers a fragility fracture. And, while bone fracture remains the most obvious clinical consequence of a metabolic bone problem, other symptoms such as bone pain, bone deformity, and mineral metabolism disorders may be a manifestation that leads to clinical investigation.

4.1.1 CLINICAL EPIDEMIOLOGY OF BONE DISEASES (OSTEOPOROSIS)

- Loss of bone mass is universal following cessation of estrogen production at any age but particularly after menopause.
- The prevalence of low bone density depends upon the way in which it is defined (see later) and the sex and age of the population in question.
- Using bone density definitions, osteoporosis is more common in women than men and increases with age such that more than 70% of women older than the age of 70 will have low bone density/osteoporosis [1].
- Given the close association between age and bone loss, it makes more sense from an epidemiological standpoint to report the incidence of osteoporosis-associated fragility fracture rather than simply the incidence of low bone density.
- "Lifetime risk of fracture" is a commonly quoted statistic in both women and men but is of very limited use for defining the problem since fracture risk is not constant over time.

- Global hip fracture rates vary greatly, from over 500/100,000 population in some Nordic countries to a low of <25/100,000 in some African and Middle Eastern countries [2].
- It is more clinically relevant to speak about age range–specific fracture risks in order to better describe the important links between ageing, frailty, and fracture.
- Similarly the 10-year risk of a major osteoporotic fracture in women older than the age of 65 varies from 5% to 25% according to individual country data [3]. This emphasizes the importance of considering ethnicity in all osteoporosis assessments.
- Other metabolic bone diseases such as nutritional rickets may be very rare in developed nations but quite common in the developing world.
- In addition to pain and acute management complications, osteoporotic fractures are sometimes associated with permanent deformity, depression, institutionalization, and increased short- and long-term mortality rates [3].

KEY POINTS

The most common metabolic bone disorder is age-associated loss of bone mass. It is largely asymptomatic until it results in fracture which accounts for the majority of the associated morbidity and healthcare costs. Given the high rate of osteoporotic fracture in some countries, a strategy to reduce fractures may be a very important public health measure.

4.1.2 BASIC BONE PHYSIOLOGY

A comprehensive overview of microscopic bone structural anatomy is well beyond the scope of this chapter. In brief, bone is comprised of both organic components (largely the collagen matrix structure and bone cells) and inorganic components (essentially calcium hydroxyapatite crystals embedded in the bone matrix). Bone serves many purposes including body structure, protection of vital organs, hematopoiesis, a source of endogenous calcium reserve, and acid buffering. The two major types of bone are cortical/compact bone (as seen in long bones like the femur) and cancellous/trabecular "spongey" bone (as seen in vertebral bodies). Both types are vascular, especially trabecular bone.

Bone is not an inert structure as it may seem upon gross inspection. It is a highly dynamic metabolic environment that is able to respond to increasing/repetitive force loads, it is able to grow during development and able to self-repair after fracture. The metabolic bone processes are crucial to its ability to store or release calcium as needed. The cells and processes of metabolic bone activity are largely derived from marrow elements; therefore it is predominantly trabecular bone (e.g., in the spine) that is most affected, either positively or negatively, by metabolic bone changes.

An extremely important concept in the discussion of metabolic bone disease is "remodelling." This speaks to the ongoing natural process by which old bone tissue is targeted for removal (resorption), to be replaced by new bone formation that follows the same architectural format. Healthy, normal bone will thus show continuous, targeted resorption followed by formation. This is necessary for growth and for the healing of fractures.

The basic sequences and mediators of bone remodeling are as follows:

- Cells trapped within bone structure (*osteocytes*) initiate the targeted remodeling in response to strain, load, or fracture which is sensed by their long dendritic processes that course through bone.
- Osteocytes begin the sequence by signaling quiescent cells lining the bone surface to become activated *osteoblasts* (bone forming cells).
- Osteoblasts secrete a substance known as receptor-activator-nuclear-factor kappa beta (better known as RANK-ligand).
- RANK-L acts upon *osteoclast* (bone removing cells) precursors in the marrow/hematopoietic tissue to stimulate the differentiation into mature osteoclasts which then attach to the locally targeted area of bone to start resorption.
- Over a usual period of 2 weeks the osteoclasts secrete both acids and proteases to dissolve both the mineral and proteinaceous matrix of bone, creating a "resorption pit." At the completion of the resorption phase the osteoclasts undergo apoptosis and resorption ceases.
- The osteoblast then follows behind the osteoclast and over a 3-month period, deposits new bone matrix which is then mineralized in the presence of an osteoblast-derived matrix vesicle where the enzyme alkaline phosphatase (ALP) assists in the very local production of a super-saturated calcium-phosphate product.
- In some cases the osteoblasts become trapped in the newly formed bone, becoming osteocytes while other osteoblasts return to a quiescent bone-lining phase.

KEY POINTS

Bone tissue is normally in a constant state of self-remodeling, a process that is critical for growth, maintenance, and repair. Bone remodeling is initiated by osteocytes and follows an orderly sequence of osteoclastic bone removal and then osteoblastic bone formation. Disorders in the initiation, timing, rate, extent, and coupling of these resorption/formation processes is the basis of most metabolic bone diseases. Clinical bone medicine requires a detailed understanding of these processes and an ability to measure various individual aspects.

4.1.3 FUNCTIONAL CATEGORIZATION OF BONE DISEASES

Broadly speaking, metabolic bone disorders may be classified into five common groups:

- disorders of mineralization of bone (also called "rickets" in children and "osteomalacia" in adults)

- disorders of proteinaceous bone matrix (often congenital mutations in collagen structure)
- low bone mass without abnormality in mineralization or matrix ("osteoporosis")
- low bone mass/abnormal metabolic bone function as a consequence of other chronic diseases or drugs
- rare disorders causing abnormally high bone mass (also called "sclerosing bone diseases"), not further discussed here

This functional and clinical classification is only possible through an understanding of normal bone remodeling; it is largely disorders of the remodeling process that lead to the various metabolic bone diseases. The clinical chemistry laboratory can play a very useful role in helping the clinician to understand both normal and abnormal bone remodeling which is necessary for both diagnosis and therapy monitoring of the various metabolic bone diseases.

4.2 GENERAL MEDICAL ASPECTS OF METABOLIC BONE DISEASES

4.2.1 DISORDERS OF MINERALIZATION

Bone tissue depends upon correctly oriented depositions of calcium hydroxyapatite crystals in order to provide the "hardness" of bone. Mineralization, in turn, depends upon several prerequisites:

- adequate calcium
- adequate phosphate
- appropriate pH in the mineralization compartment to permit crystal formation
- ability to generate a super-saturated local concentration of calcium and phosphate to permit crystal propagation
- absence of local mineralization inhibitors

Therefore any medical disorder that alters one of the above conditions may result in improper or insufficient mineralization of bone, resulting in "soft" bone that is prone to deform (while growing) or break.

4.2.1.1 Inadequate calcium

This is more fully covered in the chapter that addresses calcium disorders. But in brief, a lack of calcium is a common cause of rickets (in children) or osteomalacia (in adults). Common causes of hypocalcemia include:

- hypoparathyroidism—either congenital or acquired lack of parathyroid hormone
- nutritional deficiency—typically severe and seen in societies with major malnutrition problems

- severe vitamin D deficiency—the classic historical cause of rickets, cured by sun exposure and vitamin D therapy. Once thought to be part of the history of medicine, this condition is now being seen with increasing frequency among certain ethnic groups even in the developed world [4]
- bowel disease causing malabsorption—typically celiac disease or severe inflammatory bowel disease. In recent years, this has become increasingly recognized as a consequence of certain bariatric surgery procedures as well [5]
- inability to convert vitamin D precursors into the active form (calcitriol)—most commonly seen as a consequence of end stage renal disease or extremely rarely from congenital absence of the necessary converting enzyme (1-alpha-hydroxylase) or dysfunctional vitamin D receptor
- note that all the above diseases are chronic—short-term hypocalcemic disorders such as accompanying acute pancreatitis are not typically linked to metabolic bone disease
- serum calcium measurement is the first step in screening for these diseases in the metabolic bone patient

4.2.1.2 Inadequate phosphate

Chronic hypophosphatemia has a lengthy differential diagnosis which is largely a collection of either inherited or acquired renal phosphate wasting diseases. A simplified categorization is therefore:

- Inherited phosphate wasting disorders such as X-linked hypophosphatemic rickets, autosomal dominant hypophosphatemic rickets, and other very rare syndromes. The majority of these disorders involve over-production or decreased catabolism of the phosphaturic hormone fibroblast growth factor 23 (FGF-23).
- Acquired phosphate wasting disorders typically due to tumoral over-production of FGF-23.
- Serum phosphate is the screening test for these diseases in the metabolic bone patient.
- Chronic, severe vitamin D deficiency may also lead to some degree of hypophosphatemia accompanied by hypocalcemia (see earlier).

4.2.1.3 Inappropriate pH for mineralization

Virtually any chronic disease-associated systemic acidosis may result in bone loss due to the inability to initiate crystal formation in an acid milieu. Examples where this may play a role in defective mineralization include:

- chronic renal failure with azotemia
- chronic respiratory acidosis in severe hypoventilation syndromes
- inherited disorders of the renal tubules leading to type-2 renal tubular acidosis (proximal bicarbonate wasting)
- acquired disorders of renal tubular function such as Fanconi syndrome (which often includes calcium, phosphate, glucose, and amino acid urinary wasting)

- serum electrolytes (bicarbonate) is the screening test for these diseases in the metabolic bone patient

4.2.1.4 Absence of mineralization promotors (alkaline phosphatase enzyme)

ALP is critically important for the generation of a local super-saturated calcium-phosphate solution to promote crystal formation and propagation. Congenital absence or inactivity of ALP (due to mutation in the gene coding tissue-non-specific ALP) results in the accumulation of extracellular inorganic pyrophosphate which inhibits mineralization, as well as insufficient calcium and phosphate for normal mineralization. This condition is known as hypophosphatasia. There are various degrees of severity of disease, ranging from lethal prenatal forms to milder forms often undiagnosed until adulthood or affecting only the teeth.

- Serum ALP is the screening test for this disease in the metabolic bone patient; subnormal levels should be viewed as suspicious for hypophosphatasia.
- Further investigation of low ALP includes measurement of pyridoxal-5′-phosphate (PLP), the active metabolite of vitamin B6. Since PLP is a substrate for ALP activity, hypophosphatasia results in PLP accumulation and subsequent high level of vitamin B6/PLP [6].
- Sequencing of the ALPL gene is confirmatory.

4.2.1.5 Presence of mineralization inhibitors

Chronic ingestion of any compound known to inhibit mineralization may also lead to metabolic bone disease. The classic example is chronic excess aluminum toxicity, historically acquired from use of aluminum-based phosphate binders in treatment of renal failure—related hyperphosphatemia.

> **KEY POINTS**
>
> Age-associated bone loss is not a disorder of bone mineralization (osteomalacia) and therefore the diagnosis of age-associated osteoporosis requires that other metabolic bone diseases be excluded from consideration. There are five basic requirements for normal bone mineralization and abnormalities in any one of them may lead to a mineralization disorder. It is therefore prudent to measure serum calcium, phosphate, ALP, and serum electrolytes (bicarbonate) at the time of presentation. Abnormalities in any one of these analytes should prompt further investigation for an alternate cause of the bone problem.

4.2.2 DISORDERS OF BONE MATRIX

Any disease affecting the production or structure of type-1 (bone) collagen may result in low bone mass and propensity to fracture. The classic example is osteogenesis imperfecta, the common subtypes of which result from a mutation in the COL1A1 or COL1A2 genes coding for type-1 collagen. This collection of diseases has a wide range of severity ranging from a lethal peri-natal form (type 2) to severe, infantile form with multiple fractures and limb deformities (type 3) to a milder form that may go completely undiagnosed (type 1). The clinical hallmark is the presence of a strong family history of young age-onset fractures along with a classic blue-colored sclera of the eyes. This is not a disorder of mineralization and so there are no clinical chemistry biomarkers for diagnosis. Owing to wide phenotypic variability, the diagnosis is usually made on a clinical basis or through genetic testing.

4.3 LOW BONE MASS WITHOUT DISORDERS OF COLLAGEN STRUCTURE OR MINERALIZATION (OSTEOPOROSIS)

4.3.1 OSTEOPOROSIS OVERVIEW

Age-associated osteoporosis is by far the most common metabolic bone "disease" with population prevalence figures high enough to qualify as a public health issue. The vast majority of fractures after the age of 50 are strongly associated with age-related low bone mass and it is these fractures that are targeted for prevention by medical therapies.

4.3.1.1 Osteoporosis diagnosis: What exactly is being "diagnosed"?

If osteoporosis was strictly and completely defined as low bone density, it could be defined according to the radiologic test, dual-absorptiometry bone density scanning (DXA-BMD). For many years the DXA-BMD was indeed the standard "test" for diagnosing osteoporosis. The World Health Organization (WHO), in concert with researchers and bone densitometry developers, developed a way of assigning an individual "T-score" to a given patient result, according to the distribution of bone density seen in an ethnically homogeneous population. The T-score was the number of standard deviations above or below a young sex-matched mean for the bone density as measured in g/cm^2. The WHO has decided that a T-score of " -2.5" or below (i.e., lower than 2.5 SD below the mean) would be called "osteoporosis" and that T-score results >-2.5 but <-1.0 would be termed "osteopenia" in order to convey the idea of increased fracture risk at bone densities that are not yet termed osteoporosis.

This approach was not without merit but has gradually lost favor in the clinical community because of several factors:

- It turns natural bone loss into a concept of disease state which ultimately affects nearly every single aged person who lives long enough.
- Many people with "osteoporosis" on DXA-BMD do not fracture.
- Many fragility fractures occur in people who do not have "osteoporosis" by DXA-BMD.

- There are multiple other fracture risk factors that are independent of DXA-BMD and which may be equally strong predictors of fracture. Reliance purely upon DXA-BMD may over-simplify fracture risk estimates.
- "Osteoporosis" defined solely by DXA-BMD is an asymptomatic state which may have no clinical ramifications if it never leads to fracture.
- Pharmaceutical implications for prevention of DXA-BMD defined "osteoporosis" necessarily become applicable to 100% of the population since bone loss is expected with ageing. Pharmaceutical prevention of all bone loss may not bring clinical benefit for many patients who are very low risk for fracture.

The clinical definition of osteoporosis has always been independent of DXA-BMD and is defined as the occurrence of a fragility fracture in a menopausal woman or elderly man. This definition has maintained the importance of focus upon fracture risk, fracture morbidity, and fracture treatment costs.

In recent years the focus has shifted to become much less focused on diagnosing "osteoporosis" according to DXA-BMD but rather toward finding people who are at high fracture risk on account of the combination of low bone density, age, frailty, and other fracture risk factors. As such, the use of 10-year fracture risk estimation calculators such as FRAX (C) and others have become the standard of care. Rather than diagnosing "osteoporosis," clinicians are now encouraged to determine 10-year fracture risks in elderly persons (many of whom will have osteoporosis) and to use such data as the starting point for discussions about possible fracture risk reduction interventions. A DXA-BMD determination may form a part of this assessment but therapeutic decisions are no longer made according to the T-score classification in isolation.

KEY POINTS

Osteoporosis is extremely common with natural ageing. Its presence may be defined according to radiologic tests (bone density) but the modern paradigm views osteoporosis as a component of age, frailty, and low bone mass. All of these, in combination with other fall and fracture risk factors, serve to increase the risk of disabling fragility fractures. Estimation of 10-year fracture risk is the preferred starting point for clinical management.

4.3.1.2 Population screening

For the reasons outlined earlier, population screening with DXA-BMD has never been shown to be cost-effective in a broad sense. Most osteoporosis guidelines [7] now recommend targeted screening in high-risk populations including those patients:

- older than the age of 65 (possibly older than the age of 85 for men)
- chronic high-dose glucocorticoid use

- personal history of prior fragility fracture
- family history of hip fracture

There are many other minor risk factors known to either increase fracture risk or increase the risk of bone loss [8] such factors may also be considered in the decision to order DXA-BMD.

4.3.1.3 Other systemic diseases causing low bone density

There is a lengthy list of chronic medical disorders and drugs that are known to have effects upon bone mass, bone quality, and bone resorption. This includes:

- gastrointestinal diseases like celiac disease, inflammatory bowel disease, and chronic liver failure
- rheumatologic disorders such as rheumatoid arthritis, lupus
- hematologic diseases including myeloma, leukemia, and iron overload syndromes
- pulmonary disorders including cystic fibrosis and chronic obstructive lung disease
- chronic renal failure is associated with multiple specific bone disorders collectively known as renal osteodystrophy
- endocrinologic problems like hyperthyroidism, hyperparathyroidism, and Cushing's disease and hypogonadism in both men and women
- drugs including glucocorticoids, proton pump inhibitors, heparin, SSRIs
- in many of these conditions, there is very little known about optimal assessment and specific therapy. The metabolic bone clinician will often need to extrapolate clinical trial information from general osteoporosis studies but biochemical markers of bone metabolism may help to elucidate the underlying pathophysiology of the bone disorder.

4.4 OVERVIEW OF CLINICAL USE OF BIOCHEMICAL MARKERS OF BONE METABOLISM

The clinical chemistry laboratory plays an important role in the assessment and monitoring of metabolic bone disease. Owing to their rapid changes in response to therapy, they provide early and dynamic information about the bone metabolic state. Increasingly, biomarkers of bone metabolism are also used in research settings to help understand novel bone disorders and gauge response to therapies.

Understanding of the normal "coupled" and sequential processes of bone resorption and formation is critical to intelligent use of what are collectively referred to as bone turnover markers (BTMs).

4.4.1 POTENTIAL CLINICAL USES OF BONE TURNOVER MARKERS

- *Document the presence of active bone loss*—Excessive bone resorption accompanies many bone disorders when osteoclast activity is prolonged or "uncoupled" from osteoblast activity. Detection of this state may help the clinician predict the presence or rate of expected bone loss. Additionally, use of drugs that inhibit osteoclast function may be highly effective in treating bone diseases characterized by such increased net bone resorption.
- *Document the presence of physiologic increased remodeling*—A transient but physiologic increase in bone resorption is part of a growing skeleton as well as fracture healing. The difference between this scenario and that above is that physiologic increased bone resorption is followed by a matching increase bone formation and thus there may be an increase of both resorption and formation markers.
- *Ascertain low bone remodeling states*—Some bone diseases are characterized by a "low turnover" state (e.g., long-term immobilization, diabetic bone disease, renal failure) wherein there may be a marked decrease in osteclast number, lifespan, or activity. Bone resorptive markers may be very low and may be accompanied by resultant low bone formation and a decrease in formation markers. Use of drugs that stimulate osteoblast activity (with or without osteoclast coupling) may be of special interest in such patients.
- *Biochemically measure a drug effect*—Use of "anti-resorptive" medications such as bisphosphonates or denosumab are expected to rapidly and markedly decrease osteoclast activity. In such settings the bone resorption markers may be used by the clinician to prove a clinical effect of such drugs, to investigate suspected drug malabsorption (markers may be high) or even to monitor drug adherence. BTMs change quickly in response to initiation of antiresorptive therapy with effects seen in as little as 3 months. This may allow for rapid measure of efficacy since bone density changes typically require at least one or more years [9].
- *Prediction of fracture risk*—Limited studies in postmenopausal osteoporosis suggest that an elevation of bone resorption markers at baseline may be an additional independent predictor of future fracture [10].
- *Anabolic therapy response measurement*—Most current osteoporosis medications are targeted toward inhibition of osteoclast function and result in a decrease in bone resorption followed by a decrease in bone formation [11]. However, anabolic therapy such as with teriparatide (1−34 parathyroid hormone) produces the reverse effect by causing an initial pure bone formation effect followed by an increase in bone resorption. In this setting, one may see an initial rise in bone formation markers, demonstrating new bone formation long before a bone density change is seen [12].

4.4.2 CLINICAL CONTROVERSIES REGARDING USE OF BONE TURNOVER MARKERS

- *Unknown value in drug selection decisions*—As outlined earlier, it is theoretically attractive to choose antiresorptive or anabolic (osteoblast stimulating) drugs based upon the underlying bone metabolic state for which the BTM may help define. However, there are no clinical trials proving the superiority of this approach compared to a more general treatment of patients without knowledge of their relative resorption/formation status [13].

- *Uncertain relevance as a measure of therapy efficacy*—Although bone resorption markers are expected to decrease with antiresorptive therapy, it is still unproven that antifracture efficacy is dependent upon the demonstration of such a decrease or that a proportional or target level necessarily predicts any clinical outcome. The corollary to this is that there is no evidence to support improved therapeutic efficacy after making a therapy change based upon bone resorption marker levels.

- *Uncertain value as a patient-specific measure*—The relationship between bone loss, fracture risk, and drug responses are thus far described at the population level. With very wide inter- and intrasubject variability in bone resorption markers, it is still uncertain as to the exact clinical interpretation of such markers measured in individual patients. Some have argued that BTMs are best suited to population studies and not to clinical patient management [14].

- *Uncertain correlation with formal histomorphometry*—Bone biopsy histomorphometry studies represent the gold standard in defining the parameters of bone remodeling. Some studies comparing histomorphometry measures have noted a significant discrepancy between tissue-level remodeling and circulating biomarkers, raising concern about their reliability in the clinic. At this time, bone resorption markers cannot be considered a full substitute for bone biopsy and histomorphometry when needed, especially in situations like bone disease complicating chronic renal failure [15].

- *Limited clinical scenarios for diagnostic use*—Bone resorption markers are generally not valid for interpretation in the setting of renal impairment. Chronic kidney disease may cause a false increase in blood measures of BTM and false decrease in urine BTM measures. This is a particular weakness since renal bone disease may represent the most common set of bone diseases where remodeling parameters would be useful for management.

- *Intrasubject variability*—Bone resorption markers have wide intrasubject variability even when standardized to fasting, first morning collection [16]. Thus a change of >40% may be necessary before being able to conclude that it represents a clinical difference compared to a prior measure. This very wide coefficient of variation means than many reported changes in resorption markers will be of little clinical interpretive value.

KEY POINTS

BTMs are very attractive in concept since they permit estimations of both bone resorption and bone formation. In contrast to DXA-BMD, bone markers are easily measured and change rapidly in response to clinical intervention. However, the routine adoption of BTM in clinical practice has not yet received wide approval due to many uncertainties in their actual relevance to osteoporosis management. They are likely a very useful research tool for studying drugs and diseases in a population but further clinical studies are needed to confirm their appropriate use in individual patients.

4.5 LABORATORY CONSIDERATIONS TO ASSESS BONE METABOLISM (MARKERS OF BONE METABOLISM)

4.5.1 MARKERS OF BONE FORMATION

Several tests of bone formation have been used although they vary in terms of specificity to bone and clinical availability.

4.5.1.1 Alkaline phosphatase

As discussed earlier, ALP is an integral component in bone formation as it is required to generate the basic environment to promote deposition of hydroxyapatite crystals at the bone surface. Therefore ALP concentrations will be elevated in cases of increased bone formation. However, ALP is also synthesized in various tissues including the intestines, kidney, liver, placenta, and bone. Further, ALP exists in circulation as multiple isoenzymes (products of different genes) and isoforms (products of the same gene with different posttranslational modifications). ALP isoenzymes and isoforms contribute to the total activity, which is the most commonly measured parameter in the clinical laboratory, thus decreasing the specificity of ALP as a marker of bone formation. Despite its limited specificity, ALP may be used with the clinical findings to assess bone formation [17].

In the laboratory, ALP is generally measured in terms of its activity. An important preanalytical consideration for measuring ALP is that it requires magnesium as a cofactor for activity. Thus blood should be collected in tubes without chelating agents such as EDTA (purple top tube) or citrate (navy blue top tube) that can remove magnesium and result in artificially low ALP activity. Serum (red top tube) or plasma (heparin, green top tube) can be used to measure ALP. The most common method measures total ALP activity in serum or plasma through addition of an uncolored substrate that is converted to a colored end product in the presence of an alkaline pH. Improved specificity for the bone form of ALP may be accomplished by the use of gel-based separations with detection by addition of an uncolored substrate overlaid on the gel. The substrate is then converted to a colored product by the ALP that migrated to the position in the gel matrix. Additionally the contribution of the bone isoform to the total activity may be inferred following a heat denaturation step. This method takes advantage of the heat-labile nature of the bone isoform of ALP, while the liver form retains its activity. The total ALP in the sample is measured first, followed by incubation at 56°C and a second measurement of ALP. Because the bone form is heat-labile the heating step deactivates the enzyme, thus the contribution to total ALP may be calculated by taking the difference in the values. This method gave rise to the common laboratory phrase used to describe ALP isoforms as "bone burns, liver lives." Although isoform analysis remains available, the test is highly labor intensive and is therefore rarely ordered.

Age-specific reference ranges are important when interpreting ALP activity, as adolescents exhibit higher bone turnover and formation versus adults whose baseline levels of ALP will be lower, which is consistent with bone maintenance rather than growth. Elderly patients generally exhibit lower bone turnover, therefore ALP is expected to be somewhat lower in these patients. Other causes of elevated bone ALP include fracture as osteoblasts are activated to repair the bone by osteocytes residing within the matrix which sense the fracture. It is of utmost importance that the other sources of ALP be considered when interpreting elevated ALP results. Total ALP activity increases in liver (e.g., biliary obstruction due to metastatic tumor or stone) and kidney disease, as well as pregnancy, as the placental form of ALP becomes predominant. Thus ALP levels should be interpreted in the context of the clinical picture [17].

4.5.1.2 Osteocalcin (bone gla-protein)

Osteocalcin (OC), also known as bone gla-protein due to the presence of three gamma-carboxyglutamic acid residues in its amino acid structure, is the most abundant noncollagenous protein in bone. While the physiological relevance of OC is not well described, its synthesis is known to be stimulated by 1,25-hydroxyvitamin D and is elevated in states of increased bone formation. Alternatively, OC is also synthesized upon osteoblast activation, thus it is generally regarded as a marker of bone turnover, rather than a specific marker for bone formation [18]. OC exhibits a very short half-life ($\sim$5 min) in serum and is secreted through the kidney. OC is generally measured in serum (red top tube) by ELISA-based technologies and thus is reported in mass units. The reference intervals for OC are dependent on the type of immunoassay employed, as differences in the antigen preparation will yield significantly different results due to the presence of intact and fragments in circulation. Therefore clinicians should consult clinical chemists when OC results do not agree with other clinical findings. Also, OC may not be useful as a marker of bone formation in patients undergoing 1,25-hydroxyvitamin D supplementation as vitamin D is a direct promotor of OC synthesis. Despite this, OC may be useful in cases where ALP levels may be misleading [17].

4.5.1.3 C-terminal propeptide procollagen type-1 N/C propeptide

Approximately 90% of bone is made up of type-1 collagen, which lends it its tensile strength. As mentioned before, type-1 collagen is synthesized by the *COL1A1* gene and is deposited within the bone matrix with a helical trimeric structure made up of three collagen α subunits. The helical structure results in long strands of organic matrix which becomes highly cross-linked during maturation to provide the structure. During its synthesis the C- and N-terminal ends of the collagen fiber contain bulky propeptide domains that are cleaved off during maturation. The C-terminal propeptide (P1CP) is larger ($\sim$100 kDa) and is cleaved from the maturing collagen molecule first, while the procollagen type 1 N/C propeptide (P1NP) end is smaller ($\sim$35 kDa), and tends to be cleaved following trimer formation. For this reason, P1CP is generally found in circulation as a monomer, while P1NP is found in both monomeric and trimeric form. Once these propeptides are cleaved from the maturing collagen, they are released into circulation where they may be used as indicators for bone formation. These propeptides are present in a 1:1 molar ratio to collagen and may therefore be used as markers of collagen synthesis. Both propeptides are cleared from circulation by the liver, thus their concentration is not dependent on renal function. Although type-1 collagen is also synthesized in other tissue types, making P1NP nonspecific for bone formation, P1NP has been singled out by the International Osteoporosis Foundation-International Federation of Clinical Chemistry (IOF-IFCC) Bone Marker Standards Working Group for standardization. Specifically, reference standards are being developed and diagnostic cut-points are being validated [19]. Propeptides can thus be useful bone formation markers where ALP and OC levels are unclear such as in patients with kidney disease or those receiving calcitriol supplements and may become the analyte of choice for assessing bone formation in the near future.

4.5.2 MARKERS OF BONE RESORPTION

Bone resorption markers are largely measures of cross-linking molecules which are released into the circulation (serum then urine) during the process of collagen breakdown/osteoclastic resorption.

4.5.2.1 Amino/carboxy terminal cross-linking telopeptide of bone collagen

During bone resorption, osteoclast activity promotes collagen solubilization by proteolytic cleavage of the mature collagen contained within the bone matrix. As mentioned previously, mature collagen contains numerous cross-links at both the N- and C-terminal ends, giving collagen its tensile strength. Proteolytic cleavage of collagen releases both N- and C-terminal fragments of the mature collagen into circulation. These peptides are covalently bound to other fragments of collagen molecules via cross-links and are referred to as N- and C-terminal telopeptide cross-links (NTX/CTX). Telopeptides are small peptide fragments and thus are readily excreted by the kidney, thus allowing measurement in both urine and serum. These telopeptides (NTX-CTX) can be measured by immunoassays using turbidometry or nephelometry. Increased concentrations of telopeptides have been described in diseases causing elevated bone resorption such as osteoporosis, Paget's disease, metastatic bone disease, as well as both hyperparathyroidism and hyperthyroidism. However, significant inter- and intraindividual variations of 5%−60% have been noted, even when measuring 24 h or first void morning urine. This wide variation necessitates changes of a minimum of up to 40% to be considered biologically significant [16]. In addition, CTX is the bone resorption marker that has been singled out by the IOF-IFCC Bone Marker Standards Working Group for standardization, similar to P1NP as discussed earlier [19].

4.5.2.2 Pyridinoline/deoxypyridinoline

Pyridinoline (PYD) and deoxypyridinoline (DPD) are formed through reaction of the side chain of residues of collagen during trimerization in bone maturation. DPD and PYD act as the cross-links between collagen molecules and are formed during collagen maturation; thus their presence in circulation is the result of collagen degradation. Both DPD and PYD are found in both serum and urine; however, DPD is a more sensitive and specific marker of bone resorption than PYD due to its higher concentration in bone. Additionally, PYD is more widespread in tissues than DPD. High-performance liquid chromatography (HPLC) and immunoassays methods can be used to measure PYD and DPD, and reference intervals are highly dependent on the type of method used. Additionally, DPD concentrations are highly dependent on age and sex, being significantly higher in infancy and adolescence as well as in postmenopausal women. Owing to the high biological variability and nonspecific tissue distribution of DPD and PYD, these tests are not routinely used as markers of bone resorption [17].

4.6 SPECIAL SITUATIONS FOR INTERPRETATION OF BONE TURNOVER MARKER

4.6.1 PEDIATRICS

BTMs, both resorption and formation, reflect bone remodeling in both health and disease. The growing skeleton is a time of markedly increased remodeling, a process necessary for enlargement of bone size. Thus due to this physiologic increase in remodeling, BTM reference ranges are significantly higher in a pediatric population. Sex-specific reference ranges are needed that account for different ages. Ideally, pubertal status, weight, and height should also be considered but this is more difficult to standardize across age groups. Generally speaking, most BTMs follow a similar distribution when standardized by sex and age, indicating that they are all derived from the same metabolic processes [20]. Also, reference ranges differ for different methods (HPLC vs. immunoassay). Thus each laboratory should generate their own local sex- and age-specific reference range based on their patient populations for reporting BTM in patients younger than 18 years old.

KEY POINTS

BTMs in the preadult stages vary according to sex and age. Reference ranges have been published although locally generated data are always preferable. BTMs in patients younger than the age of 18 must always be reported according to an age-specific reference range to avoid misinterpretation since the measured levels will usually be substantially higher than the adult reference range.

4.6.2 POSTFRACTURE

Following a fracture, there is a physiologic increase in bone remodeling during the course of fracture healing. BTMs measured immediately after fracture seem to be similar to prefracture status but rise significantly by 4 months and remain elevated until 12 months postfracture. This physiologic BTM rise must be considered when choosing to measure or interpret BTM after having sustained a fracture [21].

4.6.3 RENAL IMPAIRMENT

Significant renal impairment will alter BTM dramatically both from a bone disease perspective as well as with respect to renal clearance issues.

- With increasing severity of renal (secondary) hyperparathyroidism, there is a marked increase in bone turnover. If left untreated, this may lead to a condition known as osteitis fibrosa—marked bone resorption with fibrosis in the marrow space. Patients may become hypercalcemic and fractures are common. Serum ALP often rises and is used as a marker of disease severity [22].
- As mentioned before, collagen cross-links both serum and urine are unreliable in the setting of impaired renal function [23].

KEY POINTS

With the exception of ALP, BTMs are unreliable in chronic kidney disease. There is some evidence to support the idea that ALP may be a reasonable marker of high turnover bone disease in advanced renal hyperparathyroidism.

4.7 EMERGING BIOMARKERS

Several other biomarkers have been identified that may have relevance to understanding of bone formation or resorption. Some of these novel markers reflect very different aspects of osteoclast/osteoblast numbers and function and/or are derived from different sources than the type-1 bone collagen. Thus these tests (described later) may be complementary to the current common tests; in some cases, further study may prove these newer tests to be superior by virtue of having fewer confounding preanalytical variables for consideration. It is unlikely that these markers will gain widespread use until their value is shown in large population studies and clinical trials.

4.7.1 SCLEROSTIN (SOST)

Sclerostin is secreted by the bone resident cell (osteocytes) and inhibits bone formation by inhibiting osteoblast differentiation. Sclerostin deficiency is a cause of sclerostosis (very high bone mineral density) [24]. Sclerostin is different than other markers of bone resorption or formation because it is not just a marker but thought to be part of the actual pathologic process in some cases. It is the only marker that may help define the presence of low bone formation.

- High sclerostin levels may play a part in disease processes characterized by decreased new bone formation [25].
- High sclerostin levels may predict fractures [26] and even mortality [27].
- There is interest in the role of high sclerostin in chronic kidney disease as a potential marker of renal osteodystrophy [28].

The role of serum sclerostin measurements is not yet clear in clinical care. There are intriguing population studies suggesting that sclerostin levels may help define particular bone disorders (such as diabetes-associated bone disease [25]). At this time however, there are wide variations in results obtained with various sclerostin assays [29] and debate as to whether serum levels have clear relevance to bone sclerostin production and activity [30].

The development of therapeutic antisclerostin monoclonal antibodies [31] will lead to increasing interest as to the relationships of serum sclerostin levels, therapy choices, and outcomes.

4.7.2 DICKKOPF-1

Dickkopf-1 (DKK-1) is another inhibitor of osteoblast signaling, similar to sclerostin. Serum levels tend to be higher in established osteoporosis [32] and may explain some of the "uncoupling" of bone formation from bone resorption that results in the postmenopausal bone loss. Serum levels have been found to correlate with depressed bone formation in hematological malignancies [33] or bone metastases but has not yet found a role in metabolic bone clinics.

4.7.3 **URINARY MIDOSTEOCALCIN**

This is a noncollagenous bone matrix protein which reflects bone resorption, bone loss, and predicts fracture risk [34].

4.7.4 **BONE SIALOPROTEIN**

Bone sialoprotein is a phosphorylated glycoprotein (noncollagen based) that has been shown to be a marker of bone resorption [35] but clinical applications and assays are still in development.

4.7.5 **TARTRATE-RESISTANT ACID PHOSPHATASE 5B ISOENZYME**

Early work shows that tartrate-resistant acid phosphatase 5b isoenzyme is an osteoclastic enzyme that reflects both the number and activity of osteoclasts. It may be a superior clinical resorption marker because it has much less diurnal and food-related variation compared with the traditional collagen cross-link markers [36].

4.8 **COMMON MISUSE OF BONE TURNOVER MARKER**

4.8.1 **FAILURE TO MEASURE ACCORDING TO STANDARDIZED CONDITIONS**

Bone resorption markers in particular have a daily oscillation with a change in magnitude that is similar to that seen as true changes in response to therapy. In order to avoid misinterpreting biologic variation as true change, sample acquisition must be standardized to the fasting, morning collection. Urinary markers of resorption may be cumbersome if measured through 24 h urine collection although it does take care of the diurnal variation problem. Spot urine second morning void samples after emptying the bladder upon wakening are faster to collect but the process may be difficult to explain to patients. As such, fasting serum markers are now preferred.

4.8.2 **OVER-INTERPRETATION OF CHANGES IN SERIAL MEASURES**

As mentioned earlier, it must be remembered that there can be wide intrasubject variation in bone resorption markers even at steady state and in the absence of any bone therapy. Therefore small to moderate changes in resorption markers cannot usually be considered to necessarily have clinical meaning. Large changes are more believable. It has been suggested that laboratories should report the 95% confidence interval or "least significant change" around serial turnover markers in order to better alert clinicians as to the presence or absence of a believable or true change between serial measures.

4.8.3 FAILURE TO RECOGNIZE THAT FORMATION AND RESORPTION ARE COORDINATED PROCESSES; HIGH RESORPTION DOES NOT MEAN BONE LOSS IN ALL SITUATIONS

In postmenopausal osteoporosis where there is a natural increase in uncoupled bone resorption and relative decrease in bone formation, it makes sense to see high markers of bone resorption as indicative of expected net bone losses. However, outside of menopausal osteoporosis, such resorption—formation uncoupling, may not be present. For example, a young patient who has just received a kidney transplant may have a high resorption marker as they transition from low-turnover bone disease to normal bone remodeling. If the resorption marker is viewed in isolation, one might assume that bone losses are occurring however, if a bone formation marker is measured, it may be high. This would indicate that the patient has normally coupled bone remodeling which might even lead to increased bone mass. Similar findings may be seen during therapy with teriparatide as increased bone remodeling (both resorption and formation) leads to net bone gains.

4.8.4 OVER-RELIANCE UPON RESULTS IN CLINICAL MANAGEMENT

The theory behind potential uses of bone remodeling markers is very attractive and the clinical applications of immediate biomarkers would seem to be sensible. However, it must always be remembered that there is still no definitive evidence showing that an individual patient's BTM necessarily aids in selection of therapy, prediction of fracture, or prediction of therapy outcome. There has been discussion in the osteoporosis field of a possible "target" BTM response to ensure antiresorptive efficacy but this is not yet defined. At best, BTM should be remembered to be a physiology-based research tool with potential for individual clinical application. Appropriate interpretation will require incorporation of the history and likely serial bone densitometry studies.

4.9 EXAMPLE CASE WITH DISCUSSION

A 60-year-old woman presents for discussion about bone health. She is otherwise well and takes no medications. She experienced menopause at the age of 52 and has not had any fragility fractures. Her primary care doctor decides to order a bone density which is surprisingly low with a hip T-score of -3.7. The doctor plans to do additional biochemical investigations to determine whether there is an alternate cause of bone loss.

Question 1: Should the doctor order a marker of bone resorption?

Response: There may or may not be a secondary bone disease here such as hyperparathyroidism. However, a marker of bone resorption will not help to make a diagnosis. If the turnover marker level is measured and is high, it only informs the doctor that bone loss is ongoing, something that could have been correctly assumed anyways on account of the fact that the patient is menopausal.

There is some evidence that high markers of bone turnover are independent predictors of future fragility fracture. However, this factor is not accounted for in standard risk estimation models and there is no guidance yet as to how to incorporate the two. At this time, routine BTM measures for fracture risk estimation is not recommended.

The BTM is ordered anyway and is found to be at the upper end of the reference range.

Question 2: Given that the BTM is not overtly high, does this mean the patient will not benefit from antiresorptive bisphosphonate therapy?

Response: This is an example of a clinician probably putting too much weight on a single BTM measure. It is worth remembering that there is fairly large intraindividual coefficient of variation in any one measure and upon repeat, the BTM in this patient might actually be overtly high. Postmeal BTM measures are falsely low and this might be another reason as to why the result was not overtly high.

Either way, there is no convincing evidence at this time to suggest that baseline BTM measures should influence choice of antifracture drug therapy. Fracture benefits have been shown in osteoporotic patients with both high and "normal" BTM at baseline.

The patient starts taking a bisphosphonate to reduce her risk of fracture. One year later, her practitioner repeats a bone density measurement and finds that it is surprisingly lower than at baseline.

Question 3: How could a BTM help at this point?

Response: decline in bone density while on drug therapy is a complex situation with many possible explanations and interpretations that cannot be fully discussed here. However, once underlying secondary bone disease is excluded, the over-riding question is whether the bisphosphonate is failing to exert its expected effect.

If the patient is taking the bisphosphonate regularly and is taking it according to the prescribed method (empty stomach with water only) and is absorbing the medication, it is expected that the drug will suppress osteoclastic bone resorption. Note that all three conditions must be met for this to be true. If the doctor orders a B-CTX cross laps, it should be found to be in the lower half of the reference range (which is derived from a premenopausal adult population). If the B-CTX level is high, then it will be up to the doctor to determine which of the three preanalytical variables is at fault: nonadherence, incorrect use, or malabsorption.

In this case the CTX was found to be high despite proper and regular adherence to therapy. The patient was investigated for malabsorption and discovered to have celiac disease which also likely explains why her bone density was so surprisingly low at baseline. She was switched to an intravenous formulation of bisphosphonate and 3 months later, repeat CTX measures were very suppressed. Bone density 1 year later showed a 3% increase as expected.

KEY POINTS

- Markers of bone resorption and bone formation reflect two separate yet linked processes of osteoclast and osteoblast function.
- Most metabolic bone diseases and therapies pertain to excessive bone resorption and so markers of osteoclast activity are the most commonly ordered BTM.
- BTMs are inherently attractive as a supplemental means of assessing bone physiology and pathology; they may predict fracture risk and may be used to determine therapy adherence or response.
- Routine use of BTM is still controversial due to limited clinical trial and individual patient data, showing that measures can inform therapy in a way that changes outcomes.

REFERENCES

[1] Cooper C. Epidemiology of osteoporosis. Osteoporos Int 1999;9(8):S2−8.

[2] Kanis JA, Odén A, McCloskey EV, Johansson H, Wahl DA, Cooper C. A systematic review of hip fracture incidence and probability of fracture worldwide. Osteoporos Int 2012;23(9):2239−56.

[3] Johnell O, Kanis JA. An estimate of the worldwide prevalence and disability associated with osteoporotic fractures. Osteoporos Int 2006;17(12):1726−33.

[4] Mithal A, Wahl DA, Bonjour JP, Burckhardt P, Dawson-Hughes B, Eisman JA, et al. IOF Committee of Scientific Advisors (CSA) Nutrition Working Group. Global vitamin D status and determinants of hypovitaminosis D. Osteoporos Int 2009;20 (11):1807−20.

[5] Madan AK, Orth WS, Tichansky DS, Ternovits CA. Vitamin and trace mineral levels after laparoscopic gastric bypass. Obes Surg 2006;16(5):603−6.

[6] Whyte MP, Mahuren JD, Vrabel LA, Coburn SP. Markedly increased circulating pyridoxal-5'-phosphate levels in hypophosphatasia. Alkaline phosphatase acts in vitamin B6 metabolism. J Clin Invest 1985;76(2):752.

[7] Kanis JA, McCloskey EV, Johansson H, Cooper C, Rizzoli R, Reginster JY. European guidance for the diagnosis and management of osteoporosis in postmenopausal women. Osteoporos Int 2013;24(1):23−57.

[8] Hannan MT, Felson DT, Dawson-Hughes B, Tucker KL, Cupples LA, Wilson PW, et al. Risk factors for longitudinal bone loss in elderly men and women: the Framingham Osteoporosis Study. J Bone Miner Res 2000;15(4):710−20.

[9] Garnero P, Sornay-Rendu E, Duboeuf F, Delmas PD. Markers of bone turnover predict postmenopausal forearm bone loss over 4 years: the OFELY study. J Bone Miner Res 1999;14(9):1614−21.

[10] Garnero P, Hausherr E, Chapuy MC, Marcelli C, Grandjean H, Muller C, et al. Markers of bone resorption predict hip fracture in elderly women: the EPIDOS Prospective Study. J Bone Miner Res 1996;11(10):1531−8.

[11] Russell RG, Watts NB, Ebetino FH, Rogers MJ. Mechanisms of action of bisphosphonates: similarities and differences and their potential influence on clinical efficacy. Osteoporos Int 2008;19(6):733−59.

[12] Chen P, Satterwhite JH, Licata AA, Lewiecki EM, Sipos AA, Misurski DM, et al. Early changes in biochemical markers of bone formation predict BMD response to teriparatide in postmenopausal women with osteoporosis. J Bone Miner Res 2005;20 (6):962−70.

[13] Delmas PD, Licata AA, Reginster JY, Crans GG, Chen P, Misurski DA, et al. Fracture risk reduction during treatment with teriparatide is independent of pretreatment bone turnover. Bone 2006;39(2):237−43.

[14] Civitelli R, Armamento-Villareal R, Napoli N. Bone turnover markers: understanding their value in clinical trials and clinical practice. Osteoporos Int 2009;20(6):843−51.

[15] Lehmann G, Ott U, Kaemmerer D, Schuetze J, Wolf G. Bone histomorphometry and biochemical markers of bone turnover in patients with chronic kidney disease Stages 3−5. Clin Nephrol 2008;70(4):296−305.

[16] Rosen HN, Moses AC, Garber J, Iloputaife ID, Ross DS, Lee SL, et al. Serum CTX: a new marker of bone resorption that shows treatment effect more often than other markers because of low coefficient of variability and large changes with bisphosphonate therapy. Calcified Tissue Int 2000;66(2):100−3.

[17] Brown JP, Albert C, Nassar BA, Adachi JD, Cole D, Davison KS, et al. Bone turnover markers in the management of postmenopausal osteoporosis. Clin Biochem 2009;42(10−11):929−42.

[18] Clarke B. Normal bone anatomy and physiology. Clin J Am Soc Nephrol 2008 (Suppl. 3):S131−9.

[19] Morris HA, Eastell R, Jorgesen NR, Cavalier E, Vasikaran S, Chubb SA, et al. Clinical usefulness of bone turnover marker concentrations in osteoporosis. Clin Chim Acta 2017;467(4):34−41.

[20] Rauchenzauner M, Schmid A, Heinz-Erian P, Kapelari K, Falkensammer G, Griesmacher A, et al. Sex- and age-specific reference curves for serum markers of bone turnover in healthy children from 2 months to 18 years. J Clin Endocrinol Metab 2007;92(2):443−9.

[21] Ivaska KK, Gerdhem P, Åkesson K, Garnero P, Obrant KJ. Effect of fracture on bone turnover markers: a longitudinal study comparing marker levels before and after injury in 113 elderly women. J Bone Miner Res 2007;22(8):1155−64.

[22] Sardiwal S, Magnusson P, Goldsmith DJ, Lamb EJ. Bone alkaline phosphatase in CKD−mineral bone disorder. Am J Kidney Dis 2013;62(4):810−22.

[23] Alvarez L, Torregrosa JV, Peris P, Monegal A, Bedini JL, De Osaba MJ, et al. Effect of hemodialysis and renal failure on serum biochemical markers of bone turnover. J Bone Miner Metab 2004;22(3):254−9.

[24] Balemans W, Ebeling M, Patel N, Van Hul E, Olson P, Dioszegi M, et al. Increased bone density in sclerosteosis is due to the deficiency of a novel secreted protein (SOST). Hum Mol Genet 2001;10(5):537−43.

[25] Morse LR, Sudhakar S, Lazzari AA, Tun C, Garshick E, Zafonte R, et al. Sclerostin: a candidate biomarker of SCI-induced osteoporosis. Osteoporos Int 2013;24 (3):961−8.

[26] Ardawi MS, Rouzi AA, Al-Sibiani SA, Al-Senani NS, Qari MH, Mousa SA. High serum sclerostin predicts the occurrence of osteoporotic fractures in postmenopausal women: the Center of Excellence for Osteoporosis Research Study. J Bone Miner Res 2012;27(12):2592−602.

[27] Gonçalves FL, Elias RM, Dos Reis LM, Graciolli FG, Zampieri FG, Oliveira RB, et al. Serum sclerostin is an independent predictor of mortality in hemodialysis patients. BMC Nephrol 2014;15(1):1.

[28] Pelletier S, Dubourg L, Carlier MC, Hadj-Aissa A, Fouque D. The relation between renal function and serum sclerostin in adult patients with CKD. Clin J Am Soc Nephrol 2013;8(5):819−23.

[29] Moysés RM, Jamal SA, Graciolli FG, dos Reis LM, Elias RM. Can we compare serum sclerostin results obtained with different assays in hemodialysis patients? Int Urol Nephrol 2015;47(5):847−50.

[30] Clarke BL, Drake MT. Clinical utility of serum sclerostin measurements. Bonekey Rep 2013;2:361.

[31] McClung MR, Grauer A, Boonen S, Bolognese MA, Brown JP, Diez-Perez A, et al. Romosozumab in postmenopausal women with low bone mineral density. N Engl J Med 2014;370(5):412−20.

[32] Butler JS, Murray DW, Hurson CJ, O'Brien J, Doran PP, O'Byrne JM. The role of Dkk1 in bone mass regulation: correlating serum Dkk1 expression with bone mineral density. J Orthop Res 2011;29(3):414−18.

[33] Voorzanger-Rousselot N, Goehrig D, Journé F, Doriath V, Body JJ, Clezardin P, et al. Increased Dickkopf-1 expression in breast cancer bone metastases. Br J Cancer 2007;97(7):964−70.

[34] Gerdhem P, Ivaska KK, Alatalo SL, Halleen JM, Hellman J, Isaksson A, et al. Biochemical markers of bone metabolism and prediction of fracture in elderly women. J Bone Miner Res 2004;19(3):386−93.

[35] Seibel MJ, Woitge HW, Pecherstorfer MA, Karmatschek MA, Horn E, Ludwig HE, et al. Serum immunoreactive bone sialoprotein as a new marker of bone turnover in metabolic and malignant bone disease. J Clin Endocrinol Metab 1996;81 (9):3289−94.

[36] Halleen JM, Alatalo SL, Janckila AJ, Woitge HW, Seibel MJ, Väänänen HK. Serum tartrate-resistant acid phosphatase 5b is a specific and sensitive marker of bone resorption. Clin Chem 2001;47(3):597−600.

Adrenal disorders

5

Gregory Kline, MD[1] and Alex C. Chin, PhD[2]

[1]*Clinical Professor, Department of Medicine, Division of Endocrinology,
University of Calgary, Calgary, Alberta, Canada*
[2]*Assistant Professor, Department of Pathology and Laboratory Medicine, University of Calgary
and, Clinical Biochemist, Calgary Laboratory Services, Calgary, Alberta, Canada*

CHAPTER OUTLINE

Endocrine Biomarkers. DOI: http://dx.doi.org/10.1016/B978-0-12-803412-5.00005-7

5.1 OVERVIEW OF ADRENAL DISEASES

Although they are very small in size, the adrenal glands which sit atop the kidneys produce a wide range of endocrine products which have potent effects on numerous body systems. Diseases characterized by either excessive or insufficient adrenal hormones may lead to serious chronic illness and even fatality. Clinical diagnosis may be simple in the most advanced forms of disease but diagnosis at such a late stage usually means the patient will have suffered significant and potentially irreversible consequences by that time. The situation is further complicated by the increasing recognition that some adrenal diseases may manifest in ways that are similar to other common nonadrenal health problems such as obesity and hypertension. It can thus be very challenging to discern the presence of actual adrenal disease in the patient with obesity and hypertension and subtle, potentially abnormal adrenal biochemical test results.

5.1.1 STRUCTURAL–FUNCTIONAL CLASSIFICATION OF ADRENAL DISORDERS

The adrenal gland is embryologically, histologically, and functionally divided into two main types of tissue: steroid-producing adrenocortical tissue and catecholamine-producing adrenomedullary tissue. The adrenal cortex is further subdivided into functional subsets that produce different types of steroid hormones according to the array of steroidogenic enzymes that are particular to each tissue subtype. These functional subdivisions are known as the zona glomerulosa which produces aldosterone as its end-product, the zona fasiculata which produces cortisol, and the zona reticulosum which produces sex-type steroids.

In clinical medicine the various adrenal diseases are thus characterized according to the excessive or deficient production of one or more of the adrenocortical or adrenomedullary hormones. However, because of their embryological and functional distinction, it is highly unusual to see clinical disorders that simultaneously affect both major hormone classes. Owing to the complexity of some adrenal disease definitions, the specific epidemiologic data will be presented in each relevant section.

5.1.2 BASIC ADRENAL PHYSIOLOGY

5.1.2.1 Steroidogenesis

This is a highly complex procedure in which a common precursor is transformed into a specific hormone, designed to have highly specific actions on different target organs. It may briefly be summarized as follows:

- All types of steroids produced in the adrenal cortex are derivatives of cholesterol which is taken up by LDL receptors on adrenal cortical cell membranes.
- A series of complex "shuttles" carry the cholesterol to the inner mitochondrial membrane where steroidogenesis occurs.
- The presence or absence of specific steroidogenic enzymes in each type of steroid-producing tissue determines the direction and final end-product of the particular cell type's steroid production.
- Steroidogenesis is a multistep pathway that produces many steroid intermediates en route to a final product.
- The intermediate products are also specific to the enzymes present in the various adrenocortical cell types.
- Some intermediate products, if produced to excess and released into the circulation, may have biological activities that are similar to the steroid end-product.
- Fig. 5.1 shows the classical steroidogenic pathways. It may be viewed as a "bent fork" with each tong representing a final product: cortisol, aldosterone, and sex steroids (estrogen/testosterone). Each pathway may share several early enzymatic steps in common.

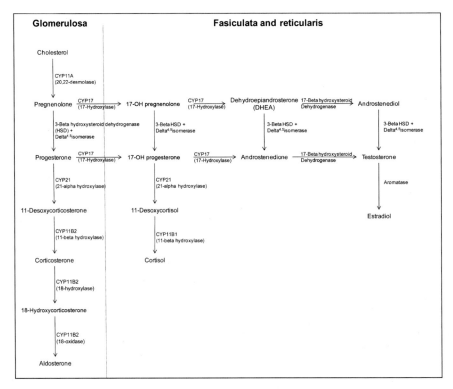

FIGURE 5.1

Biosynthesis of adrenocortical hormones.

5.1.2.2 Steroid action

Each steroid product has its own tissue-specific action:

- Cortisol (called a "glucocorticoid") plays important roles in glucose and fat metabolism, blood pressure and vascular tone, body water balance, and immune regulation.
- Aldosterone (called a "mineralocorticoid") has a predominant role in maintenance of blood pressure, prevention of dehydration, and regulation of kidney sodium salt and potassium handling. Under the action of aldosterone the kidney retains sodium (and water) in exchange for potassium lost in the urine.
- Sex steroids (estrogen/testosterone/other androgens) play major roles in determination of secondary sexual characteristics, reproductive functions, and bone metabolism.

5.1.2.3 Control of steroid synthesis

Each specific steroid product is synthesized and secreted both in a tonic fashion and upon demand by the body.

- Cortisol is produced when cells of the adrenal fasiculata are stimulated by the pituitary adrenocorticotropin hormone (ACTH).
- ACTH is produced by the pituitary in response to the hypothalamic hormone corticotrophin-releasing hormone (CRH).
- Circulating cortisol is sensed by the hypothalamus and pituitary which then modulate their secretion of CRH and ACTH accordingly in what is termed a "negative feedback loop."
- High cortisol suppresses CRH and ACTH in order to prevent further unnecessary production of cortisol. Low cortisol leads to an increase in CRH and ACTH in order to increase further cortisol production.
- Aldosterone is produced when cells of the zona glomerulosa are stimulated by the hormone angiotensin II.
- Angiotensin II is produced in response to the renal hormone "renin."
- Renin is released by the kidney in response to perceived hypotension.
- Aldosterone increases sodium and water resorption by the kidney which raises blood pressure and thus decreases renin production in its own "negative feedback loop."
- Adrenal androgens (and subsequent metabolic products like estrogen and testosterone) may be produced in small amounts in response to physiologic ACTH action due to the shared initial enzymatic steps with cortisol production.
- Adrenal androgens (in normal physiology) are minor contributors to the whole body sex steroid pool and are thus largely not involved in any kind of negative feedback loop.
- Other steroidogenic tissues such as ovaries or testes produce sex steroids in much larger amounts and under specific feedback control loops with the pituitary.

5.1.2.4 Adrenomedullary hormone (catecholamine) synthesis and action

- The adrenal medulla produces both epinephrine and norepinephrine as the major final catecholamine products.
- Other neural tissue in the body is capable of producing catecholamines such as norepinephrine and dopamine.
- Adrenal medulla is the only major source of the enzyme required to convert norepinephrine to epinephrine and thus epinephrine is not significantly produced outside the adrenal gland.
- Catecholamines are released from synapses in the central nervous system and the adrenal gland in response to sympathetic nervous system stimulation, classically called the "flight or fight" response.
- Catecholamines exert their effect via the alpha and beta-adrenergic receptors; different receptor affinities for the different catecholamines may explain some minor differences in biologic actions.

- In general, catecholamines increase the cardiac pulse rate and contractility and produce marked peripheral vasoconstriction (although epinephrine alone may produce pure vasodilation).
- The sum of the cardiovascular responses is a marked increase in blood pressure, heart rate, and cardiac output.
- Adrenomedullary hormones do not have classical negative feedback loops like adrenocortical hormones.

KEY POINTS

The adrenals make two kinds of hormones—steroids and catecholamines. The major steroids are cortisol, aldosterone, and to a lesser extent, sex steroids/androgens. The steroids are tightly regulated in production and subtype. The clinical features of adrenal hormone over-production and under-production may be predicted by knowledge of their individual physiologic effects.

5.2 GENERAL CLINICAL PRESENTATIONS

5.2.1 OVERVIEW OF CATECHOLAMINE EXCESS SYNDROMES

The catecholamine excess syndromes are best understood as the clinical extensions of their basic physiology. Catecholamines may be produced in excess by adrenal medullary tumors known as pheochromocytomas or by extraadrenal tumors of the sympathetic or parasympathetic nervous systems, known as paragangliomas. Together they may be referred to as "PPGL".

The "classic" clinical features are hypertension (which may be paroxysmal), headache, and palpitations although this triad is only found in a minority of cases. Other catecholamine actions may be present including constipation, hypotension (if purely epinephrine secreting), anxiety, tachycardia, and abdominal pain. PPGL may present with one or none of the above features and up to 50% of "incidentally" discovered adrenal pheochromocytomas may be normotensive [1]. PPGL may be benign or malignant and may occur individually or together as part of an inherited syndrome. Numerous germ line gene mutations have now been described as causing familial PPGL, some of which, such as SDHB mutations, are associated with a much higher rate of malignancy [2]. Although these are rare tumors, with a population prevalence of 0.2%−2% among hypertensives [3], the consequences of undiagnosed catecholamine excess may be fatal and therefore a general rule would be to always do a diagnostic screening test if the diagnosis is even possibly considered. Once there is clear biochemical evidence of catecholamine excess the clinician must next localize the tumor, using CT, MRI, or PET-CT scans in preparation for surgical removal. At present there are no clinically relevant syndromes of adrenal catecholamine deficiency.

> **KEY POINTS**
>
> Catecholamine excess states are usually caused by tumors of the adrenal medulla or peripheral sympathetic nervous system. The classical presentation is paroxysmal hypertension which may be severe and/or fatal. They are rare tumors but the most common error in clinical management is failure to consider the diagnosis at all.

5.2.2 OVERVIEW OF CORTISOL EXCESS SYNDROMES

As is common in all adrenal diseases, the clinical syndromes of specific hormone excess states simply reflect the physiology of the hormone action but taken to an extreme level. Cortisol excess may have a wide spectrum in clinical presentation, ranging from very mild to very severe (see Section 5.3). There is a lengthy list of possible clinical manifestations but few patients present with the full complement. The rate of onset of the clinical features is also highly variable, ranging from several years to just several weeks in the most severe (and often malignant) causes.

The most common clinical features are weight gain, hypertension, type-2 diabetes, osteoporosis, depression, frequent infections along with skin thinning, easy bruising, and poor wound healing. Affected patients often develop premature cardiovascular disease, which is a common cause of death. Less common but highly specific [4] clinical features include proximal muscle weakness, purple abdominal striae, supraclavicular fat pads, and a classic physical appearance known as a "moon facies." The constellation of the above clinical features is known as "Cushing's Syndrome," which may be used to describe the phenotype of such patients without regard to the specific etiology. In contrast, "Cushing's Disease" specifically pertains to an ACTH-producing pituitary tumor.

"Cushing's Syndrome" may be classified as "ACTH dependent" or "ACTH independent." "ACTH dependent" forms pertain to either an ACTH-producing (and usually benign) pituitary tumor (Cushing's disease (CD)) or, more rarely, an extrapituitary tumor (often malignant) with the ability to make ACTH. The latter situation is termed "Ectopic ACTH-dependent Cushing's Syndrome" and is classically caused by small cell lung carcinomas although many other tumors have also been shown to be rare causes of ectopic ACTH production. "ACTH-independent" forms of Cushing's syndrome usually pertain to the primary adrenal overproduction of cortisol with appropriate secondary suppression of pituitary ACTH secretion. This may be seen in the setting of benign or malignant adrenal tumors, along with rare genetic or endogenous disorders of adrenal function such as "AIMAH" (ACTH-independent macronodular adrenal hyperplasia). Long-term exogenous glucocorticoid use can produce an identical appearance and is termed "Exogenous" or "Iatrogenic" Cushing's syndrome. After the diagnosis of Cushing's syndrome the clinician must determine whether it is ACTH dependent or ACTH independent. Once that is known the clinician may then order the appropriate imaging study (e.g., MRI pituitary or adrenal) in order to find the tumor and plan for surgical resection.

> **KEY POINTS**
>
> Cortisol-producing tumors are uncommon but have many clinical features that may range from mild to severe with milder cases being more difficult to recognize. Infections and vascular complications are the usual serious consequences if not diagnosed and removed. Cortisol excess (Cushing's syndrome) may be broadly classified as ACTH dependent (usually from ACTH-producing pituitary tumors) or ACTH independent (usually from cortisol-producing adrenal tumors).

5.2.3 OVERVIEW OF CORTISOL DEFICIENCY SYNDROMES

Glucocorticoid deficiency syndromes may present at any age and the clinical features are often defined according to the age of the patient and the rate of development of the deficiency. More rapid and complete deficiencies tend to present acutely and may be life threatening. Slower onset and partial deficiencies may be insidious and take several years before a clear diagnosis is made. If the glucocorticoid deficiency occurs as a consequence of destruction of the adrenal glands, it is termed "primary" adrenal insufficiency or failure. If the glucocorticoid deficiency occurs as a consequence of pituitary tumor/disease and lack of ACTH-driven cortisol production, it is termed "secondary" adrenal insufficiency.

In primary adrenal insufficiency (PAI), both adrenals are usually affected by the disease process. Surgical removal of one adrenal does not typically result in endocrine deficiency syndromes. When both adrenals are destroyed or removed, the patient typically suffers deficiency of both glucocorticoids and mineralocorticoids and the clinical presentation is characteristic of both hormones being lost. The most common cause of PAI is due to autoimmune destruction of the adrenals ("Addison's Disease" (AD)) but may also result from bilateral adrenal hemorrhage, infarction, or infection such as tuberculosis or HIV. Neonatal forms of PAI are usually a result of congenital absence of one or more steroidogenic enzymes. The loss of glucocorticoids results in fatigue, anorexia, nausea, weight loss, abdominal pain, hypotension, and hypoglycemia. Additional loss of mineralocorticoids results in hypotension (sometimes severe), salt-wasting, salt craving, and hyperkalemia. As the cortisol levels decline, the pituitary increases production of ACTH in a usually futile attempt to normalize the cortisol levels. ACTH in turn is produced by cleavage of the precursor molecule pro-opio-melano-corticotropin (POMC). The "M" in POMC pertains to another cleavage product "melanocyte stimulating hormone (MSH)" which drives cutaneous melanin synthesis. Thus when the hypothalamic-pituitary unit wishes to make more ACTH, it does so via POMC and increased MSH is a byproduct. The clinical relevance is that MSH causes marked hyper pigmentation, a classic sign of complete primary adrenal failure. PAI may be fatal if untreated. Both cortisol and aldosterone must be given as replacement hormones and usually produce near-instant clinical improvement for affected patients.

In secondary adrenal insufficiency the major problem is the lack of hypothalamic CRH or pituitary ACTH. This, in turn, usually results from a pituitary tumor or surgery but may also be caused by stroke, brain injury, or as a consequence of pituitary radiation. Rare genetic diseases of the pituitary may also be seen particularly in neonates or young children. Mineralocorticoid production is intact, being under the control of renal renin–angiotensin stimulation. The ACTH deficiency thus causes exclusive loss of glucocorticoid function. The normal mineralocorticoid function and lack of ACTH-mediated hyper pigmentation help to differentiate secondary adrenal insufficiency from PAI. With normal serum potassium and blood pressure the clinical presentation of pure glucocorticoid deficiency may be more subtle and typically presents as fatigue, poor appetite, nausea, and weight loss. In more severe and complete ACTH deficiency, hypotension and hyponatremia may be detected and the presence of these two features may be associated with high risk of morbidity and mortality should the diagnosis not be made. As with any pituitary disease or tumor, more than one pituitary hormone may be affected and thus each pituitary hormone–target organ pairing must be evaluated.

KEY POINTS

Glucocorticoid deficiency syndromes may be difficult to diagnose by virtue of nonspecific features such as fatigue and nausea. However, such deficiency may be fatal if undiagnosed and clinicians must maintain a low threshold for considering the diagnosis. When the disease is due to bilateral adrenal destruction, concomitant mineralocorticoid deficiency also presents as hyperkalemia and hypotension with high ACTH-related hyper pigmentation. When due to pituitary disease such as tumors causing low ACTH, pure glucocorticoid deficiency is called "secondary" adrenal insufficiency.

5.2.4 OVERVIEW OF ALDOSTERONE/MINERALOCORTICOID EXCESS SYNDROMES

With its major role in sodium/potassium homeostasis, aldosterone excess typically presents as hypertension with hypokalemia. Intravascular volume is increased and blood pressure rises to maintain natriuresis without producing an edema state from the excess sodium retention. As sodium is resorbed in the kidney under the effect of aldosterone, potassium is the ion for which it is exchanged and thus the potassium is lost in the urine, causing hypokalemia. Approximately 30% of primary aldosterone excess (termed "primary aldosteronism" or "PA") are a result of a benign adrenocortical tumor secreting aldosterone. The remaining 70% are thought to be made up of various etiologies that have a net effect of causing inappropriate adrenal aldosterone production from both adrenal glands (sometimes generally called "bilateral adrenal hyperplasia"). The original and classic clinical presentation may be severe and presents as drug resistant hypertension with hypokalemia but it is estimated that up to 70% of PA cases actually do not have hypokalemia. A low potassium level may simply be a marker of more severe disease with higher aldosterone levels.

PA appears to be associated with a higher risk of heart disease, arrhythmia, and renal injury compared to "essential" or "primary" hypertension alone [5]. Once a diagnosis of PA is made the clinician must decide whether to pursue strictly medical therapy with a mineralocorticoid receptor antagonist or whether to perform additional investigations to try and discover whether the patient has a unilateral adenoma versus bilateral adrenal aldosterone secretion. Those with unilateral adrenal disease may be offered surgical resection which often cures or dramatically improves the hypertension. For those patients in whom additional, localizing studies are going to be done, the clinician typically will request a CT adrenals and adrenal vein sampling to biochemically prove which of the two adrenals harbors the tumor. Because incidental, nonfunctional adrenal masses are common in an older population, and because many aldosteronomas are <1 cm in size, most CT adrenal findings cannot be taken as definitive proof of aldosteronoma localization alone. Without adrenal vein sampling, incorrect surgical decisions may be made in 15% or more of cases [6]. Although aldosterone is by far the most common mineralocorticoid to cause clinical disease, there are a number of other rare situations where defects in mineralocorticoid production or metabolism may lead to abnormally high levels of nonaldosterone mineralocorticoids that can produce a similar disease picture.

KEY POINTS

Primary aldosterone excess typically presents with hypertension and possible hypokalmia. In this setting a benign aldosterone-producing tumor may be present but requires careful biochemical studies and lab-based localization studies to detect it prior to consideration of removal.

5.2.5 OVERVIEW OF ALDOSTERONE DEFICIENCY SYNDROMES

Because of its key role in retaining sodium for maintenance of intravascular volume, the deficiency of aldosterone predictably manifests as salt-wasting hypotension and high blood potassium levels. As mentioned in Section 5.2.3, aldosterone deficiency as part of complete adrenocortical destruction (AD) often accompanies cortisol deficiency which also causes nausea, weight loss, anorexia, and hyponatremia. In cases of adrenal loss, plasma renin levels are typically high (much like ACTH rises with cortisol deficiency). Selective loss of aldosterone with normal cortisol is much less common but may be seen in a disorder known as a "type 4 renal tubular acidosis" and relates to a deficiency of renal renin production. This problem may be associated with diabetic kidney disease among others.

> **KEY POINTS**
>
> Aldosterone deficiency is rarely seen as an isolated entity; combined aldosterone and cortisol deficiency are the classic endocrine problems of primary adrenal failure (AD). The aldosterone deficiency is recognized by the presence of hyperkalemia.

5.2.6 OVERVIEW OF ADRENAL SEX STEROID DISORDERS

In both men and women the adrenal gland is a minor contributor to physiologic sex steroid production with testes and ovaries being the primary sex steroid producer. The majority of adrenal sex steroids are categorized as androgens which may interact with the androgen receptor or be further transformed into the classical sex steroids testosterone and estrogen. Massive over-production of adrenal androgens may cause features of androgen excess in women including masculine hair growth on face and body, voice deepening, and acne. Over-production of adrenal androgens in a male is less easily perceptible unless it occurs during childhood when it may present with precocious puberty. Very rare adrenal production of estrogen may cause breast growth and hypogonadism in a man. When it occurs, excessive adrenal sex steroid production is typically caused by benign or malignant adrenal tumors. Rare congenital disorders of steroidogenesis also present with adrenal sex steroid excess during childhood. Adrenal sex steroid deficiency is not known to cause a clinically meaningful problem, provided the testes and ovaries are functional.

> **KEY POINTS**
>
> Rare adrenal tumors or disorders of steroidogenesis may cause marked over-production of adrenal sex steroids which are primarily androgens. This is a rare cause of virilization in women or precocious puberty in both sexes.

5.3 LABORATORY CONSIDERATIONS IN ADRENAL DISEASE DEFINITION AND DIAGNOSIS

5.3.1 THE PROBLEM OF "SPECTRUM BIAS"

Unlike many other endocrinopathies, disorders of adrenal hormone excess are not just hard to diagnose in milder cases but also hard to even define from a clinical and biochemical standpoint. Aldosterone excess may present as "hypertension" with no other features thus making it clinically indistinct from common essential hypertension. Cortisol excess may present as obesity with diabetes and depression, lacking other specific features and thus mimicking a very common population finding. Both problems may be present across a spectrum of severity that overlaps with the general population. Case definition therefore often is completely dependent upon biochemical test results to differentiate from "non-endocrine" causes of the same features.

Thus biochemical tests that define disease must be developed and validated; such validation studies depend upon the ability to show different results in a "non-adrenal disease" population compared with a gold-standard "adrenal disease" population. Until the age of all-encompassing genetic test—based definitions of disease, a gold-standard diagnosis in adrenal disease likely includes pathological and functional examination of the adrenals retrieved by adrenalectomy. Adrenalectomy cannot be performed on patients who do not already have a confirmed adrenal diagnosis. The problem can thus be seen to be circular: definitive diagnoses require confirmation by adrenalectomy and adrenalectomy cannot be done without definitive diagnosis. Biochemical test validation requires access to a definitively "diseased" population and a definitively "diseased" population cannot be defined without biochemical tests.

As a consequence of the above situation, researchers have largely had to be content with the development of biochemical tests, which are validated in cases where there are gross clinical abnormalities, often compared to obviously healthy patients. This dichotomous classification of disease is probably only valid for interpreting test results in either clearly healthy or clearly diseased patients. Patients whose disease manifestations (either clinical or biochemical) fall somewhere along the spectrum may be consequently less easily identified. Furthermore, "diagnostic" biochemical definitions of adrenal hormone excess syndromes may be less accurate for the identification of all true forms of endocrine disease that may be present in a population [7].

In order to properly interpret test results, reference ranges or clinical cut-offs need to be derived using appropriate data from the population specific to the clinic and the laboratory. Clinical laboratory tests can be either developed commercially by a manufacturer or within the laboratory (laboratory developed tests), but derivation of the reference ranges may be incomplete and is associated with data that may not be appropriate for the local patient population, particularly for commercially available tests. Indeed the probability that a positive or abnormal test result according to the prevalence of disease (predictive value) will be different between centers. While most laboratories will adopt the reference ranges from manufacturer data according to clinical laboratory standards, these data may not be complete especially if the study population is not similar to the local population. Therefore laboratories should validate their own reference ranges according to the disease prevalence in their local population. However, since this is rarely possible for most clinical laboratories, the most prudent approach to adrenal laboratory test interpretation should be to consult an endocrinologist who can assess the patient and put any laboratory result into an accurate clinical context before deciding if it will be called "true positive" or "false positive." Once the test is put into practice the laboratory can also work with the endocrinologist to fill in the gaps associated with test interpretation through retrospective review of laboratory data and linkage to clinical presentation to come up with acceptable clinical cut-offs for disease conditions. The clinical importance is that the clinician and clinical chemist need to work together using all available clinical and biochemical data—sometimes observed over long periods of time—to come to the best clinical diagnosis, keeping in mind that no single test may be definitive for all cases.

5.3.2 THE "GOLD STANDARD" DIAGNOSIS

It is now recognized that adrenal excess syndromes may have a variety of etiologies, ranging from tumor hormone production to steroid enzyme defects to functional disorders of the control of adrenal function. Some tumors of the adrenal may be nonfunctional and the mere presence of an adrenal mass by imaging does not confirm endocrine relevance. Traditional adrenocortical histopathology in ex vivo adrenal tissue cannot confirm the prior presence of a functional adrenal syndrome. Therefore for both research and clinical purposes, "confirmation" of a definite adrenal functional disease likely requires a combination of documented resolution of preoperative features, normalization of documented preoperative adrenal biochemical tests, and supportive adrenal pathology that may include immunohistochemistry and/or studies of adrenal enzyme expression [8].

5.3.3 LIQUID CHROMATOGRAPHY—TANDEM MASS SPECTROMETRY VERSUS IMMUNOASSAY IN ADRENAL DISEASE DIAGNOSIS

For the past 40 years, most adrenal hormones have been measured by chromatography or (radio)immunoassay, both of which have known limitations, particularly regarding disease specificity while being amenable to high throughput and less complexity. Notwithstanding the considerations of Section 5.3.1, most modern adrenal disease definitions are based upon studies performed in an age of immunoassay. However, the high specificity attributed to antibody—antigen interactions is not always infallible. Cross-reactive substances in patient specimens can lead to false positive results. Conversely, potential polymorphisms or alterations in epitope availability may lead to false negative results. The need for accurate and sensitive methods such as mass spectrometry to detect biomarkers of interest is important for both patient care and research studies. For more information on immunoassays, please see Chapter 1, Variables affecting endocrine tests results, errors prevention and mitigation.

There is increasing evidence that adoption of liquid chromatography—tandem mass spectrometry (LC—MS/MS) methodology in routine clinical laboratories will produce results that are sometimes dramatically different than what has been seen with older methods [9]. Indeed, clinical societies are taking note and a recent position statement from the Endocrine Society and announcement by the Journal of Clinical Endocrinology and Metabolism referred to a strong recommendation for LC—MS/MS methods for reporting sex steroids for clinical use and research studies, respectively. It is likely that similar statements will be employed for other biochemical markers [10]. While manufacturers work toward developing ready-to-use reagents and LC−MS/MS instruments for clinical use, it is becoming increasingly important for both laboratory professionals and clinicians to become familiar with this technology for method development and test interpretation. Improved detection methods by LC−MS/MS or development of new biomarkers employing this technology will translate to better patient care in addition to higher quality and highly reproducible research studies. "Diagnostic" biochemical cut-offs should therefore not be assumed to be the same between newer and older methods. It is likely that entirely new and validated biochemical definitions of disease will be required in order to ensure diagnostic accuracy and take advantage of the improved LC−MS/MS methods for steroid and catecholamine measurements. Nevertheless LC−MS/MS will be the preferred, if not standard, method to detect endocrine biomarkers in the near future.

KEY POINTS

The biochemical definitions of many adrenal hormone excess syndromes are a work in progress; many validation studies lack an adequate gold-standard comparison and thus are often best suited to help diagnose the most obvious and clinically recognizable disease forms. Many adrenal endocrine disorders may have both clinical and biochemical features that overlap with the general population. A firm endocrine diagnosis may require several lines of investigation, integrated with careful clinical observation over time. Clinicians should be cautious about prematurely "ruling in" or "ruling out" suspected endocrine disorders based on limited investigations particularly when laboratory methods are not consistent with the methods used to derive clinical practice guidelines if available. LC−MS/MS will be the technology of choice to assess endocrine markers in the near future and require reevaluation of clinical decision points.

5.4 SPECIFIC LABORATORY APPROACHES TO ADRENAL DISEASE: PHEOCHROMOCYTOMA/PARAGANGLIOMA

5.4.1 CATECHOLAMINE AND METABOLITE MEASUREMENT

5.4.1.1 Urinary measures—clinical considerations

The 24 h outpatient urine collection for urine metanephrine/normetanephrine (UMN/UNM) is easily accomplished by most patients. It requires very little patient preparation and is ideally suited to use as a "first-line" screening test for PPGL in most routine cases, as recommended by recent guidelines [11]. Plasma metanephrines (PMNs) and plasma normetanephrines (PNMs) are also recommended as first-line screening tests and will be discussed further later [11].

Clinical causes of false "negative" UMN/UNM are relatively rare; hence, the reported overall sensitivity of at least 90% [12]. Incomplete 24 h collections can be a cause of a false negative result although the laboratory retains processes to determine the completeness of collection by confirming with the patient and measuring urine creatinine. Nevertheless some experts have recommended that two separate determinations of UMN/UNM be considered to ensure adequate sensitivity, especially in cases of high clinical suspicion. Some types of paragangliomas (typically those in the head and neck region) are nonsecretory and may have normal UMN/UNM. Very rarely, an adrenal pheochromocytoma may have normal UMN/UNM, especially if large, malignant or poorly differentiated and lacking the necessary enzymes for such terminal catecholamine synthesis. In such cases, it may be useful to measure 24 h urine dopamine (produced earlier in the catecholamine synthesis pathway) or its main metabolite, 3-methoxytyramine (3-MT) [13]. Having said that, isolated findings of slightly elevated urine 3-MT are very rarely of any clinical significance but can often lead to an unnecessary series of imaging and other investigations to find the "occult" dopamine secreting pheochromocytoma which is usually nonexistent. We recommend against the routine reporting of 3-MT levels in a 24 h urine catecholamine metabolite result unless there is a careful discussion between the clinical chemist and endocrinologist.

From a clinical perspective, it is far more common to face elevated levels of UMN/UNM in the absence of known PPGL, the so-called "false positive." As mentioned above the test specificity is highly dependent upon the reference range in use. For laboratories using a nonhypertensive reference range, it is to be expected that many patients with essential hypertension will have an abnormal result. With the very high prevalence of hypertension in the population, this translates to a high number of "false positives" which may require further investigation or specialist input. If a "hypertensive" reference range is used by the local laboratory, "false positive" results will be much less frequent albeit with possible overall reduction in sensitivity for true disease. Common clinical conditions causing non-PPGL elevations in UMN/UNM include untreated sleep apnea [14] and generalized anxiety. A number of medications may cause physiological increases in catecholamines and their metabolites including tricyclic antidepressants, selective serotonin reuptake inhibitors (SSRIs), serotonin-norepinephrine reuptake inhibitors (SNRIs), inhaled beta-agonists as found in nasal decongestants, and antiparkinsonian agents such as levodopa/carbidopa. Adrenergic drugs of abuse such as cocaine must also be considered. It is important to recognize that standardized reference ranges for UMN/UNM are those defined in the outpatient setting; acute illness or in-patient settings may be associated with marked elevations in catecholamine production [15]. Elevated levels of UMN/UNM collected during acute illness must be interpreted with extreme caution to avoid incorrect attribution to an adrenal tumor. When faced with an elevated UMN/UNM result, the clinician should take a careful history and physical exam as most "false-positive" causes can usually be easily detected. Levels of UMN/UNM that are more than 300% above the upper reference limit are almost always due to PPGL and the clinician is justified in proceeding straight to imaging studies. However, given the potential life-threatening nature of true PPGL, we recommend that some form of reassessment or additional testing be considered for cases that are deemed to be "false positive" after the clinical assessment. This may involve removal of the offending medication/and repeating the 24 h UMN/UNM or might involve measurement of PMN/PNM under carefully controlled conditions as outlined later. With careful patient preparation, PMN/PNM may have nearly 99% sensitivity (see Section 5.4.1.4) and thus a negative/normal result is useful to exclude further PPGL directed tests. Imaging tests of the adrenals are not a good way to investigate a suspected false positive UMN/UNM, rather it is preferable that the abnormal UMN/UNM result be either confirmed or biochemically refuted before getting to the stage of imaging. Nonfunctional adrenal masses are common and thus their mere presence does not diagnose pheochromocytoma. Paragangliomas may be extraadrenal and easily missed on imaging thus normal adrenals do not necessarily exclude PPGL either.

5.4.1.2 Plasma catecholamine/metabolite measures—clinical considerations

Many of the same clinical considerations pertaining to UMN/UNM interpretation also apply to plasma samples and should be considered. Unlike urine measurements, plasma measurements require a high degree of careful patient preparation and positioning prior to collection although collection and handling of a 24 h urine specimen may be equally cumbersome. Nevertheless PMN/PNM testing requires special considerations for general use in routine community clinical laboratory sites. Incorrect positioning and sample collection is an important source of false-positive results [16], which has potential to lead to many unnecessary additional investigations. When assessing a high level of PMN/PNM, the first questions should be directed to an evaluation of how the sample was collected and repeat measurement with careful positioning and preparation may be the most appropriate next step if there are any doubts regarding the results.

In large clinical series the sensitivity of properly collected PMN/PNM has been reported to be close to 96% [17]. Thus the PMN/PNM is also considered to be a good initial test for screening for PPGLs [11] and probably the ideal test to "rule out" PPGL in cases of suspected false positive UMN/UNM. If PMN/PNM is performed and is nonelevated, it is likely that PPGL can be confidently excluded from further consideration. The very high sensitivity of PMN/PNM has also led some guidelines to suggest that it be the test of choice for diagnosis or annual screening of individuals at very high risk for PPGL such as those with hereditary PPGL syndromes [18]. Even with such high sensitivity, specificity has been reported to be excellent with PMN/PNM when properly collected. However, it is still <100% specific and so clinical interpretation is still required especially for cases of mild elevation in the levels. Nonetheless with the appropriate clinical history or risk factors, abnormal high PMN/PNM (especially if reproducibly abnormal) should almost always lead to consideration of an imaging study. Guidelines for the most appropriate initial imaging study are available [11].

Normal PMN/PNM may be seen in some paragangliomas (particularly of the head and neck) that are nonsecretory. Therefore in a patient with known carotid masses or thought to have a possible germ line mutation associated with head and neck paragangliomas (such as succinate dehydrogenase complex subunit D (SDHD)), the PMN/PNM results must be understood to be unhelpful diagnostically.

5.4.1.3 Catecholamines/metabolite measures—laboratory considerations

The hallmark of PPGL is tumoral over-production of catecholamines. Much of the secreted catecholamines undergo metabolism to other compounds that are stable and easily detected. As such the catechol-O-methyl metabolites metanephrine and normetanephrine have become the most commonly ordered tests and are now recommended for initial investigation.

Collection of plasma and urine specimens is an important factor in testing for catecholamines and their metabolites. Catecholamines in particular are sensitive to changes to sympathetic nerve and adrenal stimulation due to changes to posture as well as mental and physiological status. Conversely the measurement of the catecholamine metabolites metanephrines and normetanephrines is advantageous since they are less susceptible to changes to sympathetic and adrenal activity. Although there is less stringency in obtaining samples, studies have shown that there are significant differences according to the posture of the subject whereby plasma concentrations for metanephrines and normetanephrines are higher in seated as compared to supine positions. In particular, metanephrines and normetanephrines are rapidly cleared upon assumption of the supine position and may reduce the false positive rate. Therefore it is recommended that individuals should be resting in the supine position for 30 min before collection [16] and it would be informative for the clinician to know how the reference ranges were derived.

While urine specimen collection may be considered to be more convenient as compared with supine, resting plasma specimens, 24 h urine collection is not always simple. Although it is less invasive, it is not always convenient to collect urine over a course of 24 h, nor is this always done in a reliable fashion by the patient. Furthermore handling of 24 h urine specimens by the laboratory requires additional considerations. Proper collection and storage prior to delivery is required by the patient. After delivery, laboratory specimen accessioning staff needs to ensure that the specimen is properly mixed and aliquotted since it is impractical to deliver such large containers to the testing laboratory. Alternatively, spot urine collections can be done to avoid errors associated with 24 h urine collection such as physical activity. Random urinary catecholamines or associated metabolites can be normalized against urinary creatinine excretion. However, creatinine excretion can also be influenced by muscle mass, kidney function, diet, physical activity, and diurnal variation. Sample handling should also be considered whereby appropriate preservatives such as hydrochloric acid are needed to maintain urine pH since catecholamines are particularly prone to oxidation under alkaline conditions and deconjugation under acidic conditions. If plasma or urine samples cannot be analyzed immediately, they should be stored frozen.

Influences from diet and drugs can cause either direct analytical interference or physiological changes that affect plasma or urine catecholamine and metabolite levels. Dietary contributions by fruits, vegetables, and nuts can cause increases in sulfated catecholamines and catecholamine metabolites. Traditional high-performance liquid chromatography (HPLC) electrochemical detection methods are prone to direct analytical interference where certain medications coelute with normetanephrine or metanephrine and result in false elevations of these analytes. With the advent of more specific detection methods using tandem mass spectrometry, these direct analytical interferences can be avoided. However, there are confounding interferences by medications which affect normetanephrine and metanephrine levels through direct physiological effects and will affect all measurement methods. Table 5.1 lists several interfering medications which should be discontinued, if possible, for at least 3 days before testing. It is recommended that patients undergo an overnight fast before collection of blood for measurement of plasma catecholamine and catecholamine metabolites.

In addition to the role of posture in determining reference ranges, it is useful to know how the results can be interpreted in the context of clinical conditions such as hypertension. When compared to normotensive patients, hypertensive patients generally have higher levels of plasma and urinary catecholamine and catecholamine metabolites. Knowledge of appropriate cut-offs in the hypertensive patient population may especially be useful to reduce potential false positives since these patients are more likely to be tested for pheochromocytoma according to the clinical presentation.

5.4.1.4 Plasma catecholamines: special situations

5.4.1.4.1 Metabolite pattern and prediction of genetic pheochromocytoma and paraganglioma syndromes/tumor type

With the increasing recognition that up to 33% of PPGL may be linked to an underlying germ line mutation, there has been a major focus on the clinician's role in identifying such PPGL syndromes both for prognostic information as well as individual and family screening programs. Some familial PPGL syndromes have a predilection for adrenal pheochromocytomas much more than paragangliomas, including those with mutations in RET (multiple endocrine neoplasia 2), TMEM, and MAX genes. Therefore marked elevations in PMN or UMN may indicate the presence of significant epinephrine release which is indicative of adrenal tissue as the only major site or source of the enzymes required to produce epinephrine from norepinephrine. This biochemical finding may thus spur the clinician to look more carefully for the other features of RET mutations particularly and to guide subsequent selection of genetic testing. Germ line syndromes characterized by paragangliomas such as the SHDx, NF1, VHL, and others usually present with high PNM only, thus guiding both the imaging studies, clinical and genetic investigations [19].

Table 5.1 Medications that Cause Falsely Elevated Increases in Catecholamine Metabolites

Drug	Plasma		Urine	
	Normetanephrines	Metanephrines	Normetanephrines	Metanephrines
Analytical interference on some HPLC electrochemical detection methods				
Acetaminophen	++	–	++	–
Labetalol	–	–	++	++
Sotalol	–	–	++	++
α-Methyldopa	++	–	++	–
Buspirone	–	++	–	++
Sulfasalazine	++	–	++	–
Pharmacodynamic interference affecting all measurement methods				
Tricyclic antidepressants	++	–	++	–
Phenoxybenzamine	++	–	++	–
Monoamine oxidase-inhibitors	++	++	++	++
Sympathomimetics	+	+	+	+
Cocaine	++	+	++	+
Analytical interference on some HPLC electrochemical detection methods and also increasing 3-methoxytyramine dopamine metabolite				
Levodopa	+	+	++	+

(–) no increase; (+) mild increase; (+ +) clear increase.
Adapted from Lenders JW, Duh QY, Eisenhofer G, Gimenez-Roqueplo AP, Grebe SK, Murad MH, et al. Phecchromocytoma and paraganglioma: an endocrine society clinical practice guideline. J Clin Endocrinol Metab 2014;99:1915—42.

Patients with known germ line mutations in PPGL-related genes are recommended to undergo frequent biochemical screening for PPGL and PMN/PNM is the test of choice. Patients with malignant PPGL also require long-term serial PMN/PNM measures as part of their surveillance for active or onset of new disease.

5.4.1.4.2 Plasma normetanephrine in "confirmatory tests"

Rarely, patients may present with high PNM or UNM which are clinically believed to be "false positive" after assessment and yet continue to be persistently abnormal. In such cases the clinician may feel obligated to perform additional testing to prove the absence of PPGL, especially if the patient reports concerning symptoms or has equivocal imaging results. A "clonidine suppression test" has been described for use in such a setting. This is built on the premise that non-PPGL elevations in norepinephrine are of central nervous system origin and usually linked to anxiety or medication use. The full protocol is described elsewhere [20] but basically consists of serial measurements of PNM before and after an oral dose of clonidine, which activates alpha-2 adrenergic receptors in the brain and sympathetic nerve endings to suppress norepinephrine release without affecting catecholamine release from pheochromocytomas. Failure to suppress PNM into the normal range following clonidine is reliable evidence for an actual diagnosis of PPGL. The clonidine suppression test is much more accurate when PNM is measured, rather than plasma norepinephrine. Accuracy is best in those with the most marked elevations in PNM; however, such patients are already highly likely to harbor true disease. In patients with mild elevations in PNM the suppression test sensitivity may drop to <90% and must therefore be interpreted in the broader clinical context. This test should not be performed in those with hypotension or those who are concomitantly using alpha or beta-adrenergic receptor blocking drugs which may precipitate severe hypotension. It is recommended that such testing be done only with continuous experienced nursing supervision and not in a community laboratory setting.

5.4.1.4.3 Plasma normetanephrine in renal insufficiency

Many patients with chronic kidney disease have hypertension and thus it is not unusual for some to be investigated for PPGL. However, UMN/UNM cannot be collected and/or interpreted in individuals with advanced kidney disease. PMN/PNM clearance is decreased in kidney disease and thus the standard reference ranges should not apply. However, if PMN/PNM levels are normal according to the usual reference range, the test would be highly useful for excluding PPGL consideration. Limited studies and case reports in kidney disease patients have suggested an informal reference range of 100% above the usual upper limit of normal (ULN) for the definition of high PMN/PNM [21]. Owing to the relative lack of validation studies in this population, any abnormal PMN/PNM in a kidney failure patient should probably be reviewed by a clinician with extensive experience in PPGL presentations.

KEY POINTS

To screen for PPGL, either urine or plasma metanephrine/normetanephrines should be measured. There are numerous reasons for false positive elevations in both tests and careful patient preparation, especially with plasma measurements, is critical. Both tests have very high sensitivity and persistently normal results can confidently exclude almost all PPGL. However, due to the potential seriousness of a true PPGL diagnosis, all abnormal screening tests must be carefully reviewed by both the chemist and clinician; in many cases, additional testing will be necessary to avoid a missed diagnosis.

5.4.2 OTHER AND HISTORICAL TESTS FOR PHEOCHROMOCYTOMA AND PARAGANGLIOMA

5.4.2.1 Chromogranin A

Discussed in more detail in the neuroendocrinology Chapter 9, Neuroendocrine tumors, chromogranin A is a biologically inert and nonspecific marker of neuroendocrine tumors. It has its own list of reasons for nontumor-related high levels and cannot always be assumed to be indicative of a neuroendocrine tumor syndrome. In the context of suspected PPGL with equivocal elevations in PMN/PNM, consideration could be given to measuring it as an additional (albeit nondiagnostic) test that may help to decide upon whether there is sufficient concern to embark upon detailed imaging studies for PPGL. Added to abnormal PMN/PNM, chromogranin A measurements may perform at least as well as the traditional clonidine suppression test [22] but ongoing caution is advised with interpretation, to avoid missing any non-PPGL confounding situations for both PMN/PNM and chromogranin A.

5.4.2.2 Other urinary catecholamine metabolites

Vanillylmandelic acid (VMA) is produced in the liver and is a major product of norepinephrine and epinephrine metabolism excreted in the urine. Therefore VMA is a poor diagnostic marker for pheochromocytoma given that very little is derived from catecholamine and metanephrines. Homovanillic acid (HVA), in contrast, is not significantly produced in the liver. Instead a significant portion is derived from dopamine in the gastrointestinal tract and the brain. Although VMA and HVA have fallen out of favor for pheochromocytoma diagnosis, they have been found to be useful in combination for detection of neuroblastoma and other neuroendocrine tumors. Urinary HVA and VMA are measured from either 24 h or random urine collections standardized to urinary output for creatinine. Given the challenges of collecting 24 h urine specimens from some patient populations such as pediatrics, random specimens are collected more often. Methods for VMA and HVA detection have been developed for gas chromatography, HPLC, and LC−MS/MS. LC−MS/MS is now the favored method given the minimal sample preparation and high specificity especially when multiplexed for other useful analytes.

5.4.2.3 Urine or serum catecholamines

Largely replaced by measurements of metabolites in most laboratories, older studies suggest poorer sensitivity and specificity compared to UMN/UNM or PMN/PNM. Plasma catecholamines are exquisitely sensitive to patient preparation and may rise with anxiety or simple needlestick. If measured, they should be drawn from an indwelling line with the patient in a supine, resting position for at least 30 min prior to collection. Some authors have suggested possible use of plasma catecholamines for patients with kidney failure undergoing PPGL investigation where marked elevations may be considered suspicious but it is unknown as to whether this approach has any advantage compared to PMN/PNM.

5.4.3 COMMON MISUSE OF THE LABORATORY IN INVESTIGATION OF SUSPECTED PHEOCHROMOCYTOMA AND PARAGANGLIOMA

5.4.3.1 Failure to consider possible clinical causes of false positive elevations in urine metanephrine/normetanephrine or plasma metanephrine/normetanephrine prior to ordering the test

Many of the causes of false positive elevations can be foreseen: medications, untreated sleep apnea, severe acute illness, and acute anxiety are all extremely common and usually known to the clinician. In these settings a high result for UNM/PNM is to be expected and yet true PPGL cannot be either diagnosed or excluded. If anything, the clinician will now create the additional dilemma of having to definitively "explain" the high results either by unnecessary imaging studies or additional biochemical tests. It would be preferable, if possible, to wait for better clinical conditions and patient preparation prior to ordering the test.

5.4.3.2 Failure to recognize that pheochromocytoma and paraganglioma are rare diseases and that most urine normetanephrine/plasma metanephrine elevations will actually not represent true pheochromocytoma and paraganglioma

It is worth remembering that unless the PNM/UNM result is more than 300% above the ULN, a single slightly elevated UNM/PNM is not diagnostic of PPGL. Patients should not be told they have a tumor and adrenal imaging tests should not be routinely performed without careful consideration and/or repeat biochemical testing.

5.4.3.3 Failure to follow up on an abnormal plasma metanephrine/urine normetanephrine result

While most slight elevations in PNM/UNM are indeed not due to PPGL, one cannot assume that the results are purely false positive. Some PPGL produce very minor amounts of excess catecholamines and up to 15% [13] of true PPGL could be missed if clinicians only acted upon gross elevations in UNM/PNM. At the very least the presence of abnormal UNM/PNM that is not definitively explained may cause confusion and mismanagement in the patient's future, in the event that other healthcare providers see the result and make decisions or assumptions about its meaning. We recommend that every abnormal UNM/PNM result should have some kind of formally documented assessment to ensure it is either explained or the PPGL are found.

5.4.3.4 Use of a diagnostic approach that aims to "catch" the pheochromocytoma and paraganglioma during a clinically symptomatic episode

A long standing misunderstanding is that PPGL are biochemically quiescent until they suddenly release a "surge" of catecholamines to produce patient symptoms. It is now known that PPGL continuously secrete catecholamines and that catecholamine levels do not have any apparent correlation with clinical symptomatology. Thus an approach of waiting to test until symptoms potentially delays a diagnosis. Additionally, continued biochemical testing after several normal results is unnecessary given the very high sensitivity of both UNM/PNM done at any point and irrespective of symptoms.

5.4.3.5 Continued testing and imaging for pheochromocytoma and paraganglioma in the face of a normal set of urine metanephrine/normetanephrine and/or plasma metanephrine/normetanephrine

In the absence of a known or suspected familial PPGL syndrome, PPGL are rare tumors that in turn are not usually occult. The majority of adrenal pheochromocytomas are large and obvious as are many paragangliomas (if full body imaging is done). The sensitivity of both UMN/UNM and PMN/PNM approaches 100%. Clinical symptom complexes that simulate classical pheochromocytoma are extremely common and usually not indicative of underlying PPGL. Thus while it is very wise to consider and screen for PPGL at least once in the right setting, if the biochemical screening is normal (and especially if imaging is done and is normal), PPGL should truly be excluded and further testing is not indicated without consultation from an expert in PPGL diseases.

KEY POINTS

PPGL are very rare diseases yet critically important when present; this fact often leads to frequent screening tests done for patients with hypertension. By virtue of being a very rare disease, most slightly high PNM/UNM results will be false positive but this must not be assumed. Expert consultation is recommended and/or repeat testing with special attention to patient preparation and sample collection to minimize false positive results.

5.4.4 EXAMPLE WITH DISCUSSION

A 55-year-old man presents with severe, resistant hypertension requiring four drugs for control, including an angiotensin-converting enzyme (ACE)-inhibitor, thiazide diuretic, and diltiazem. He denies any paroxysms of headache or tachycardia and there is no family history of endocrine disorders. Other medical conditions include severe obesity (BMI 44 kg/m^2), type-2 diabetes, hyperlipidemia for which he takes metformin, and a statin.

Question 1: *What biochemical investigations should be made regarding his resistant hypertension?*

Answer: The differential diagnosis of secondary hypertension is beyond the scope of this chapter however, does include several endocrine diagnoses including PA, CD (discussed later), and pheochromocytoma. Although pheochromocytoma would be a very rare diagnosis (especially without a classic history or familial form), it is a potentially serious diagnosis with high mortality if missed. Therefore it would be reasonable to perform a screening test for pheochromocytoma in this case and a 24 h UMN/UNM is ordered.

Question 2: *The 24 h UNM result is 40% above the upper limit of the reference range. Does this mean he is likely to have a pheochromocytoma?*

Answer: At this point, these results should be viewed as being probable false positives. PPGL is an extremely rare condition, therefore, the majority of slightly abnormal high UMN/UNM will end up being false positives (i.e., poor positive predictive value). Most true PPGL will have UMN/UNM levels well over 100% above the upper reference limit. As well, even essential (primary) hypertension patients may have mild elevations in 24 h UMN/UNM without having PPGL and so it is useful to consider the reference ranges reported—are they for normotensive persons or those with primary hypertension?

Question 3: *If you think this is a false positive, other than "primary hypertension," what other conditions/situations should be considered by the team as an explanation?*

Answer: Many medications can interfere with UMN/UNM measurements both from a direct analytical aspect or from a true but physiologic impact upon catecholamine secretion from the central nervous system. A clue to CNS-derived catecholamines is the fact that the UMNs are normal. Epinephrine (and its subsequent metabolite metanephrine) is secreted almost exclusively by the adrenal medulla and so pure CNS-mediated catecholamine increases are typically of norepinephrine/normetanephrine alone. Thus an elevation in UMN should be viewed as highly suspicious for PPGL but pure elevations in normetanephrine could be either adrenal/paraganglioma or CNS derived.
If possible, medications interfering with catecholamine synthesis or measurement should be stopped for at least 3 days in advance of measurement (see Table 5.1). In this case, none of his medications could explain a false positive.

The other possibility here is undiagnosed obstructive sleep apnea (OSA). This condition of recurrent nocturnal hypoxia can produce significant nocturnal increases in catecholamines (basically a night-time flight-or-fight response). With BMI of 44, OSA is highly likely.

Question 4: If OSA is the suspected cause of a false positive, what could be done to prove this point (so as to avoid further, unnecessary radiologic investigations for PPGL)?

Answer: Measurement of supine, resting, plasma free normetanephrines/metanephrines (PNMs/PMNs) should be independent of the collective 8 h recurrent hypoxia-induced catecholamine rise. As well the PNM/PMN measures are thought to have near 100% sensitivity and so, if normal, are very useful for ruling out PPGL. In this case, this is what was done and the results were indeed normal. The patient was referred to overnight sleep testing which confirmed OSA. After starting continuous positive airway pressure (CPAP) therapy a repeat 24 h UNM/UMN was repeated and was normal.

5.5 SPECIFIC LABORATORY APPROACHES TO ADRENAL DISEASE: CUSHING'S SYNDROME

5.5.1 GENERAL CONSIDERATIONS

5.5.1.1 Pseudo-Cushing's versus true Cushing's versus iatrogenic Cushing's

As outlined in Section 5.2.2 the approach to a hypercortisolism (CS) diagnosis usually begins with a patient having at least some of the typical clinical features which may be mild or severe. Iatrogenic CS from exogenous steroid use should be excluded prior to any biochemical testing—it is a diagnosis made by history, not chemistry. Careful consideration of all possible exogenous glucocorticoid sources must include topical, inhaled, oral, and injected forms, along with unregulated substances consumed without physician supervision. Some patients may even not be aware of their exogenous glucocorticoid use and so questioning must be very broad—the patient who is getting "injections" on a repeated basis for their sore hip may well be getting high-dose glucocorticoid without their knowledge.

Non-iatrogenic CS is a very rare disease. Therefore like pheochromocytoma, most slightly abnormal biochemical tests for CS will represent false positives. As such the results of all biochemical testing for CS must be interpreted not only numerically but also with a full clinical picture in mind. Section 5.3.1 details the difficulty of setting reference ranges and interpretations for tests of rare diseases which have a wide spectrum of severity and symptomatic overlap with extremely common features of a normal population. Therefore many patients undergo assessment for possible CS and some will have occasional "abnormal" results. However, the astute clinical chemist−clinician team will be very cautious about assigning a diagnosis of endogenous CS to the patient with nonspecific clinical findings and mild biochemical abnormalities. True CS from endogenous endocrine tumors is a very serious health problem and once a diagnosis is assigned, the patient will be obligated to undergo a surgery (either adrenal or pituitary) or commence long-term and potentially dangerous medications for control of CS. With these high stakes, we recommend that all presumed diagnoses of CS be reviewed with an expert to ensure diagnostic accuracy. Patients who, after long-term clinical observation, do not convincingly display progressive classic features of CS are often deemed to have "pseudo-cushings," which is most commonly a setting of alcoholism, obesity, diabetes, depression, and hypertension. Several caveats apply to this designation and if in doubt, expert consultation is justified.

5.5.1.2 The three phases of diagnosing Cushing's syndrome

An orderly approach is necessary and many endocrinologists consider a CS diagnosis as analogous to that of a prosecuting attorney who must gather all possible lines of evidence to synthesize a correct interpretation of events.

The first step is a screening test. The word "screening" does not pertain to population screening but rather the selection of a fairly simple initial test that should be widely available and used to determine whether any additional testing will be warranted when CS is suspected. The three most commonly used screening tests are the 24 h urine cortisol, the midnight salivary cortisol, or the 1 mg overnight dexamethasone suppression test (DST), all reviewed in detail later.

Following an abnormal result of a screening test, patients should undergo a "confirmatory" test which is intended to ensure the reliability and reproducibility of the abnormal result. This may not be necessary if the patient is severely cushingoid with a grossly abnormal first-step test. However, the confirmatory step is exceedingly important in almost all other cases for the results of the confirmatory test will ultimately define the patient as having or not having CS. The typical confirmatory tests include a 2-day DST, a 4-h intravenous DST, or some endocrinologists simply choose to perform the other two tests from the screening test panel with the understanding that if abnormal results are seen in all tests, the likelihood of true disease is probably greater. Very rarely, even the common "confirmatory tests" may not give adequate diagnostic confidence in either direction and an endocrinologist may choose to do a CRH stimulation test with or without additional dexamethasone suppression.

Once a formal diagnosis of endogenous CS is confirmed, the third step is localization of the source of the problem. This typically involves measurement of serum ACTH as outlined in Section 5.2.2. For patients with ACTH-dependent CS, pituitary imaging and chest X-ray is next and if a tumor is not found, the final step involves measurement of ACTH during direct sampling of the pituitary venous effluent in the inferior petrosal sinus.

5.5.1.3 Disease variants that may affect all laboratory tests for cortisol excess

5.5.1.3.1 "Subclinical Cushing's syndrome"

It is now widely recognized that some forms of endogenous glucocorticoid excess may have extremely subtle, if any, clinical manifestations. This is particularly so in those with a cortisol-producing adrenal mass. In such cases the degree of abnormality in various tests of cortisol excess may be very slight or even upper-normal. Some patients for example may have slightly abnormal responses to a DST but completely normal 24 h urine free cortisol (UFC). The existence of several slight abnormalities in tests of cortisol excess in the setting of nonspecific or no clinical features is termed "sub-clinical Cushing's." The laboratory definition of "sub-clinical Cushing's" is still debated [23] and the overall clinical relevance not clarified with no consistent evidence of benefit to detection and treatment of such subtle cortisol abnormalities.

5.5.1.3.2 "Cyclic Cushing's syndrome"

This exceedingly rare variant of CS describes patients whose disease may "cycle" through stages of complete clinical and laboratory inactivity with other phases of overt CS both by clinical and laboratory measures. Thus depending on the phase of disease activity the testing for CS may range from complete normality to grossly high cortisol levels by any test. This form of CS may require several years of clinical observation and testing to clarify the diagnosis but should be considered in those who appear cushingoid yet have highly variable results of tests of cortisol excess.

KEY POINTS

Cushing's syndrome/disease is a rare diagnosis with clinical features that have substantial overlap with common disorders like hypertension, diabetes, and obesity. A true diagnosis of Cushing's requires that a "case" be constructed, with multiple different tests of cortisol production, nonsuppressibility or diurnal dysregulation. Assay specific "normal ranges" for serum cortisol immunoassays is not particularly relevant since the diagnosis rests upon abnormal cortisol dynamics in response to suppression testing.

5.5.2 SPECIFIC TESTS FOR GLUCOCORTICOID EXCESS ("SCREENING TESTS")

5.5.2.1 Cortisol measurement

There is no investigation of glucocorticoid excess that does not involve measures of cortisol in one way or another. It is therefore critical to understand the laboratory aspects of cortisol measurement even before trying to interpret the meaning of various numerical results obtained during dynamic tests. Cortisol exhibits diurnal variation and is directly associated with ACTH levels which peak in early morning and reach a nadir at midnight. More specifically, these levels are directly related to the sleeping patterns where cortisol levels are highest before awakening and lowest at sleep, so this pattern can be significantly altered for night shift workers. Approximately 90% of cortisol circulates in the blood mainly bound to corticosteroid-binding globulin (CBG; transcortin) and other plasma proteins such as albumin to a lesser extent, while the remaining 10% circulates as free or biologically active hormone. Total cortisol is directly proportional to CBG levels which can be increased in states of estrogen excess such as pregnancy or oral contraceptive use while free cortisol remains normal. Additional preanalytical considerations that are associated with increased cortisol include stress, depression, glucocorticoid therapy, hypoglycemia, and hyperthyroidism. Conversely, cortisol can be decreased in protein-losing states or during decreased protein production.

Measurement of total cortisol in the plasma is mainly done by immunoassays and displacement of cortisol from binding proteins is an essential step to avoid steric hindrance and ensure proper presentation of the epitope for effective antibody—antigen interactions. Displacement of cortisol can be achieved through use of typical protein displacement agents such as 8-anilino-1-napthalene-sulfonic acid (ANS) and salicylates, or the alteration of reaction conditions such as pH or heat. Most immunoassays have enough binding protein displacing agents adequate for normal protein states. However, this poses a problem when measuring cortisol in patients with enhanced protein levels such as pregnancy. Furthermore immunoassays are limited by the quality and specificity of the antibody. Indeed, many immunoassays may cross-react with similar steroid-like molecules such as prednisolone, prednisone, and 11-deoxycortisol [24,25]. Although specificity can be improved with more specific antibodies, a growing number of laboratories are circumventing this problem through adoption of mass spectrometric techniques which are more specific and are not affected by the problem of protein displacement inherent in the method design.

5.5.2.2 The 24 h urine free cortisol measurement

While blood cortisol is associated with binding proteins, urinary cortisol is found in free form. Free cortisol is the biologically active form and exists in equilibrium through binding with transport proteins CBG and albumin. Several methods exist in measuring free cortisol in serum such as ultrafiltration, dialysis, and gel filtration, but these are technically demanding and not amenable for routine use. Instead, determination of free cortisol is done by taking advantage of the fact that cortisol exists in free form through normal filtration in the urine. Measurement of 24 h urinary free cortisol is still the most commonly performed screening test for CS. However, collection of a 24 h urine specimen requires particular considerations. Patients need to be instructed to refrain from any glucocorticoid-containing medications (inhaled, topical, ingested, or injected) for as long as possible according to the elimination kinetics prior to doing the 24 h urine collection. However, the most common source of error is the collection of urine specimens over a 24 h time period where both incomplete and excessive collections can lead to under- or over-estimation of cortisol levels, but measurement of urine creatinine can indicate the integrity of the timed specimen. Random urine samples are not particularly useful due to the episodic and diurnal release of cortisol. As mentioned in the previous section, laboratory considerations for collecting 24 h urine specimens include proper acidification and refrigeration for adequate sample stability, aliquotting of a thoroughly mixed specimen for frozen storage particularly if the test is not done in close proximity to the collection time, and extraction for eliminating cross-reactive substances and improving specificity.

If 24 h urine cortisol levels are above the reference range (especially if on a repeated basis), glucocorticoid excess must be suspected. Multiple studies in differing populations have yielded highly variable reports of sensitivity and specificity of UFC, depending upon the reference range in use and the population studied. At 100% specificity the sensitivity for CS may be only 71% and at 100% sensitivity the specificity drops to 73% [26]. Some evidence suggests that an elevation in UFC represents a "late" event in the development of Cushing's syndrome, becoming abnormal only once the disease has progressed to a state of high amounts of cortisol production and may therefore have poor sensitivity for earlier or more mild forms of Cushing's syndrome.

False positive results are very common, especially with 24 h urine cortisol levels that are less than twice the upper reference range. Common explanations include:

Physiologic increases in cortisol production:
- acute illness/severe stress
- alcohol withdrawal
- prolonged strenuous physical activity
- pregnancy

Physiologic increases in urinary cortisol clearance
- polyuric states (often primary polydipsia) [27]
- excessive urine collection—beyond 24 h

Disease-associated states/Pseudo-Cushing's states
- untreated sleep apnea [28]
- obesity
- polycystic ovary syndrome (often associated with obesity)
- alcoholism
- depression
- diabetes, especially uncontrolled
- anorexia nervosa/starvation
- care must be taken with the above "causes" of hypercortisoluria since conditions like diabetes, obesity, depression, anovulation/polycystic ovaries are also potential manifestations of true Cushing's syndrome

Drugs
- carbamazepine
- fenofibrate
- synthetic glucocorticoids (unless cortisol measured by LC−MS/MS)
- black licorice (inhibits 11-BHSD2 conversion of cortisone to cortisol)

False negative results may also occur in patients with true Cushing's syndrome. This may be seen in situations such as incomplete 24 h urine collections leading to underestimation of free cortisol levels, mild cases of Cushing's syndrome where cortisol release is not significantly increased, and accelerated endogenous metabolism of cortisol. In addition, there may be a high amount of within-subject variability in UFC levels, even with established Cushing's syndrome [29], thus it is important to recognize a particular patient's normal cortisol ranges with respect to lifestyle and individual physiology.

False positive results may occur in chronic kidney disease where inability to excrete cortisol metabolites that may cross-react with less-specific immunoassays. Indeed, older immunoassays for cortisol frequently have cross-reactivity with cortisol metabolites in urine, something that is reflected in the determination of the reference range. Therefore adoption of extraction techniques to remove possible interfering substances and employment of highly specific LC−MS/MS methods for urine cortisol measures may require significant lowering of the upper reference limit for 24 h urine cortisol [30].

5.5.2.3 Late night salivary cortisol

More recently, very sensitive methods to measure salivary cortisol have been developed and this type of specimen collection has been touted to be convenient for both the patient and the laboratory in terms of specimen collection and measurement without extraction, respectively. In a similar manner to urinary free cortisol, late night salivary cortisol (LNSC) has the advantage of being free from factors that interfere with serum cortisol−binding proteins since saliva is essentially an ultra filtrate of plasma. Multiple studies suggest that this test has a sensitivity of >92% and specificity of >95% for true CS compared to a population with simple obesity [31]. This method is increasingly preferred by many endocrinologists because of its ease of collection and high sensitivity for CS. Given this convenience, saliva can be collected multiple times without the constraints of 24 h urine collection to either confirm successive biologically active cortisol measurements over a time period or to demonstrate diurnal variation between AM and PM measurements. However, there are a number of potential problems associated with saliva collection by the patient who may not be aware of preanalytical factors. Therefore it is important for the laboratory and the clinician to provide the necessary information for specimen collection and storage before transport to the laboratory for analysis. The timing of specimen collection is especially important given the diurnal variation of cortisol release. Some clinicians prefer to have the specimens collected on two consecutive nights to address repeatability of the results. Others may prefer a morning and midnight collection to assess diurnal variation. To reduce potential interfering substances the patient should refrain from drinking alcohol, eating, or brushing teeth prior to collection periods. There are different collection devices that may require chewing of a cotton swab or drooling into a tube and will require instructions for use. Furthermore once the specimens have been collected, they will need to be stored appropriately at 4°C for up to a week or frozen if transport to the laboratory is delayed. Once the specimens have reached the laboratory, it is important for both the patient and laboratory staff to confirm the patient identification details and collection time of the samples for appropriate interpretation of the results.

False "positive" elevations in LNSC may be seen in

Clinical conditions or physiologic variation of normal circadian rhythm
* depression
* shift workers
* critical illness
* convalescence post-myocardial infarction (MI)
* diabetes
* smokers
* stress

Measurement interference
* patients using chewing tobacco or licorice due to salivary 11-BHSD2 inhibited conversion of cortisol to cortisone
* contamination with oral/inhaled glucocorticoids

5.5.2.4 Overnight dexamethasone suppression test

Considered by many experts to be the single most sensitive test for cortisol excess, the premise is based upon the fact that physiologic elevations in cortisol (i.e., related to stress, etc.) should be fully suppressible by a potent exogenous glucocorticoid. Tumors secreting ACTH or cortisol usually lack the exquisite feedback control and therefore cannot be "shut off" by even supraphysiologic glucocorticoids. Dexamethasone is given because it is not detected in standard cortisol immunoassays thus will not interfere with the serum cortisol measurement.

The standard overnight low-dose DST consists of 1 mg dexamethasone administered at 2300 h with serum cortisol measured the next day at 0900. Many different interpretive cut-offs have been suggested over many years of clinical reporting of this test. Older approaches to interpretation have suggested that the 0900 cortisol is deemed abnormally high if >140 nmol/L (5 μg/dL) but this cut-off has been shown to have particularly poor sensitivity. More recently, endocrinologists use a 0900 cortisol of <50 nmol/L (2 μg/dL) to indicate normality although the specificity for actual CS may be as low as 80% at this level [32]. Using the paradigm of "screening test followed by confirmatory test," guidelines for CS diagnosis have deliberately recommended use of the lower cut-off, accepting the poor specificity, in order to avoid missing a case of true CS. The confirmatory test is intended to have a high specificity to permit correct identification of false positives on the screening test. The DST is more cumbersome for patients since it requires filling a prescription for dexamethasone and then correct timing of the dosing and next day blood collection. However, performed properly and in the absence of confounders (see later), many endocrinologists use the DST as the final arbiter for interpretation of otherwise uncertain results from other test modalities. In almost all clinical cases a DST cortisol of <50 nmol/L (2 μg/dL) can be considered sufficiently accurate to rule out active CS at the time of the test.

False positive DST depends in part upon the level of cortisol deemed to be abnormal. However, other causes of a false "non-suppressed" cortisol include:

- drugs that accelerate dexamethasone metabolism (such as carbamazepine, phenytoin, etc.)
- alcoholism
- incorrect or non-ingestion of dexamethasone by the patient (intentional or non-intentional)
- drugs that increase CBG and thus falsely elevate total cortisol measures (estrogens, mitotane)
- oral contraceptives (it is critically important that female patients avoid use of an estrogen containing oral contraceptives for 1 month prior to any serum cortisol measurements)
- stress
- obesity
- infection
- acute or chronic illness
- severe depression
- pregnancy

False negative DST may be seen in patients that experience decreased metabolism of dexamethasone caused by:

- drugs such as itraconazole, ritonavir, fluoxetine, and diltiazem which effectively equate to a high-dose dexamethsone effect
- liver failure
- kidney failure

In order to mitigate the chance of false positive or false negative results, the clinician may consider concomitant measurement of a serum dexamethasone level which should be higher than 5.0 nmol/L (196.2 ng/dL).

KEY POINTS

Three screening tests for Cushing's syndrome are recognized: the 24 h urine cortisol, the LNSC, and the 1 mg overnight DST. A more careful diagnosis often uses combinations of the above tests. There are numerous preanalytical variables that affect each test; thus the appropriate screening test must be selected with care. Owing to the seriousness of true Cushing's syndrome, expert endocrinology consultation is recommended in order to assess the clinical features necessary for interpretation of borderline or unexpected chemistry results.

5.5.3 SPECIFIC TESTS FOR GLUCOCORTICOID EXCESS ("CONFIRMATORY TESTS")

5.5.3.1 Combinations of first-line tests

Current CS guidelines recommend that positive screening tests be followed by additional testing using one of the other first-line screening tests [33]. Consistent abnormalities on 2 or more tests listed above are frequently sufficient to be considered as "confirmatory" in the appropriate clinical setting. More complex tests of possible glucocorticoid excess should ideally be performed in a specialized endocrine testing unit under the supervision of an experienced endocrinologist. More complex testing may be indicated if there is disagreement among the results of multiple first-line tests and/or incongruence with the clinical setting.

5.5.3.2 The 4 h IV dexamethasone suppression test

This is a commonly used confirmatory test although not formally validated on large population samples. This test is subject to same limitations as any other DST with serum cortisol measures. Intravenous dexamethasone infusion ensures the drug is actually administered and not subject to variable absorption. The full protocol is described elsewhere [34] but basically consists of 4 mg IV dexamethasone administered over 4 h with hourly serum cortisol measures along with serum cortisol measured next day at 0900 h. "Normal" controls and "pseudo-CS" show suppression of serum cortisol by more than 80% to <200 nmol/L (7.2 µg/dL) at 5 h and <100 nmol/L (3.6 µg/dL) at 0900. Patients with CS typically have some degree of suppression at 5 h but next day serum cortisol will be >130 nmol/L (4.7 µg/dL) [35]. Other small studies have suggested a next day 0900 cortisol of >276 nmol/L (10.0 µg/dL) to be indicative of CS [34].

5.5.3.3 The 2-day, 2 mg dexamethasone suppression test (Liddle's test)

This test is based on a similar principle to other DST protocols. It requires 0.5 mg dexamethasone given at 6 h intervals starting at 0900 on Day 1 for a total of eight doses. Serum cortisol (and possibly dexamethasone) is measured at 0900 on Day 3. Abnormal final cortisol is defined as >50 nmol/L (1.8 µg/dL) although some have advocated >38 nmol/L (1.4 µg/dL). Overall sensitivity and specificity are not clearly superior to shorter DSTs. The test can be cumbersome for patients and difficult to explain; errors in dexamethasone administration could seriously compromise the test validity. It is generally not performed by most endocrinologists given the lack of apparent advantage to other tests.

5.5.3.4 Dexamethasone-suppressed corticotropin-releasing hormone stimulation test

The dexamethasone-suppressed corticotropin-releasing hormone (CRH) stimulation test is the most complex dynamic test for CS, requiring an experienced endocrine testing unit that can administer CRH according to a strict protocol with serial timed cortisol and ACTH measures thereafter. The premise of the test is that ACTH-producing adenomas of CD have enhanced sensitivity to CRH infusion with a marked rise in both ACTH and cortisol. Dexamethasone administered prior to CRH follows the standard 2-day, 2 mg protocol and is intended to increase the specificity of the disease by fully suppressing ACTH in normal subjects. However, some data suggest that the addition of dexamethasone to the protocol does not actually increase the specificity in diagnosis. If dexamethasone is used, consideration should be given to measuring pre-CRH dexamethasone levels with an expected level >1600 ng/dL [36]. The interpretation of a dex-CRH stimulation is still not widely standardized, but an increase of cortisol above 38 nmol/L (1.4 μg/dL) by 15 min post-CRH injection was found to have a sensitivity of 100% in a small series [37]. An increase of ACTH to >5.9 pmol/L (27 pg/mL) after CRH injection was shown to have a sensitivity of 95% and specificity of 97% [38]. However, other interpretive criteria are reported [39]. The clear lack of agreement between these reported interpretation criteria raises questions about test reproducibility and accuracy, especially when such tests are often viewed as "the final word" in difficult CS investigations. Because ectopic ACTH-producing tumors do not retain CRH sensitivity, some have used the dexamethasone-CRH stimulation test as a stage 3 "localizing test" to differentiate Cushing's disease from ectopic ACTH syndromes although this is almost never necessary.

5.5.3.5 Desmopressin (DDAVP) stimulation tests

Although it has been described several times, the desmopressin (DDAVP) stimulation test has not yet received wide acceptance although it would be substantially less costly than a CRH stimulation test. The test principle is based upon the fact that desmopressin (DDAVP) stimulates ACTH secretion much the same way as CRH. Some reports indicate that a post-desmopressin (DDAVP) ACTH level of >15.8 pmol/L (71.8 pg/mL) had a sensitivity and specificity of 91% and 95% for the differentiation of CD from Pseudo-Cushings [40]. The same report indicated that Pseudo-Cushings almost never had a post-desmopressin (DDAVP) rise of more than 8.1 pmol/L over baseline. Other authors report a peak ACTH level >6 pmol/L (27 pg/mL) to have a sensitivity of 87% and specificity of 91% [41]. Combination criteria such as basal cortisol > 331 nmol/L (12.0 μg/dL) with a change in ACTH > 4 pmol/L (18 pg/mL) gave a sensitivity of 97% and specificity of 100% [42]. All the above show that there is still no uniformly agreed criteria for interpretation of these protocols; local validation is absolutely necessary before general adoption into practice.

KEY POINTS

In borderline cases (either clinically or biochemically), additional testing is recommended to confirm or refute a diagnosis of Cushing's syndrome. These tests often involve dedicated nursing staff and are best performed under the controlled conditions of an Endocrine Testing Unit.

5.5.4 SPECIFIC TESTS FOR GLUCOCORTICOID EXCESS ("LOCALIZING TESTS")

5.5.4.1 Serum adrenocorticotropin hormone

Once it has been firmly established that a patient has endogenous glucocorticoid excess, the third step is to localize the source of the disease. The primary test at this stage will be a serum ACTH level, which may be measured at any time of day once CS is confirmed. If ACTH is suppressed, it indicates a primary adrenal source for the cortisol excess, usually an adrenal mass. Computed tomography or MRI can then be ordered to visualize the adrenals. If the ACTH level is high or "inappropriately normal," it indicates a diagnosis of pituitary CD or nonpituitary "ectopic" ACTH-producing tumor. The easiest next step is usually MRI of the pituitary—a pituitary tumor of >1 cm in size is convincing evidence of pituitary Cushing's tumor. If the MRI pituitary shows no tumor or possible but small tumor <1 cm in size, further testing will be needed to confirm the source. A simple chest X-ray may help detect an obvious lung tumor if ectopic ACTH syndrome is considered. Ectopic ACTH syndrome is often (but not always) associated with a highly malignant and rapidly progressive Cushing's appearance, often with extremely high cortisol and ACTH levels along with rapid pigmentation.

5.5.4.2 Special tests to differentiate Cushing's disease and ectopic adrenocorticotropin hormone

5.5.4.2.1 Dexamethasone-suppressed corticotrophin-releasing hormone stim test

As mentioned earlier, there is some evidence that the dexamethasone-suppressed CRH stimulation test may be of use at this stage.

5.5.4.2.2 High-dose dexamethasone suppression test

The high-dose DST involves administration of 8 mg dexamethasone at 2300 h and measurement of a next morning cortisol. The premise is that even CD retains some degree of dexamethasone responsiveness whereas ectopic ACTH should be completely nonresponsive. A next-AM cortisol that is more than 50%−80% lower than baseline suggests CD with variable reported specificity. However, at this stage of diagnosis, it is critical to try and get as close to perfect specificity as possible in order to avoid having a patient undergo unnecessary pituitary surgery. Since the high-dose dexamethasone test is unlikely to be sufficiently diagnostic, it is rarely employed by endocrinologists who generally prefer inferior petrosal sinus sampling (IPSS) as the more diagnostic tool.

5.5.4.2.3 Pro-opio-melano-corticotropin and agouti-related protein levels

Plasma POMC and agouti-related protein (AgRP)—recent observations suggest that ectopic ACTH-producing tumors produce markedly elevated levels of POMC

(>26 fmol/mL) and AgRP (>280 pg/mL) and that if one or both peptides are found to be high, it appears to be an accurate means to identify ectopic ACTH as the cause [43].

5.5.4.2.4 Inferior petrosal sinus sampling

IPSS: currently considered the gold-standard test to confirm a diagnosis of pituitary source primary ACTH excess in the setting of established hypercortisolemia, this test measures ACTH levels drawn from both the left and right petrosal sinuses during selective venous sampling by an interventional radiologist. The ACTH levels are compared to simultaneously drawn peripheral venous samples at baseline and following injection of either desmopressin (DDAVP) or CRH. A petrosal sinus ACTH: peripheral ACTH level > 3:1 is considered to be diagnostic of pituitary CD. Some authors recommend simultaneous measurement of serum prolactin with each ACTH level drawn during IPSS in order to confirm proximity to the pituitary venous effluent, especially in cases complicated by irregular venous anatomy [44]. IPSS can only be considered diagnostic if performed during a time of active CS; measurement of 24 h urine cortisol both before and after IPSS is recommended to ensure that hypercortisolemia was present during the time of IPSS. Paradoxical decreases in IPSS-collected ACTH levels have been reported in cases where synthetic ACTH is mistakenly used instead of CRH during IPSS [45]. In dynamic endocrine tests requiring pre- and poststimulation timed sample collection, the clinical team and laboratory have a major responsibility to ensure that all samples are correctly identified and labeled according to timing and specific location (i.e., right vs. left petrosal sinus). Given the complexity of the procedure, it is helpful to notify the laboratory so that all samples are accounted for and that there are no issues with instrument performance. Reporting of results in tabular format according to time, test name, and anatomical site can facilitate appropriate interpretation.

KEY POINTS

Only after hypercortisolism is clinically and biochemically confirmed should "localizing" tests be done to discover the source of the ACTH- or cortisol-producing tumor. Simple biochemical tests beyond serum ACTH are increasingly of historical value; most experienced centers will use IPSS as the most reliable localizing step.

5.5.4.3 Special tests pertaining to primary adrenal hypercortisolism

Although an adrenal cortisol−producing mass is the most common etiology, another known entity is that of "AIMAH"—ACTH-independent macronodular adrenal hyperplasia. This entity is due to the adrenocortical expression of aberrant receptors which drive cortisol synthesis in response to ligands such as TSH, catecholamines, LH, etc. A 3-day testing protocol is available to help identify the putative ligand which may permit simple endocrine therapy to suppress the ligand-driven cortisol production [46].

5.5.5 SPECIAL SITUATIONS IN THE INVESTIGATION OF CUSHING'S SYNDROME

5.5.5.1 Pregnancy

Urinary free cortisol levels increase up to 300% in pregnancy; however, it is unclear whether this is due to physiologically increased production or increased clearance [47]. Therefore pathological CS can only be diagnosed in situations where more severe levels of cortisol are present. In addition, dexamethasone suppressibility appears to be less in pregnancy and thus dexamethasone-based tests are not recommended [48]. LNSC testing may still be useful to detect abnormal circadian rhythm, but has not been specifically validated as a diagnostic test during pregnancy.

5.5.5.2 Chronic kidney disease

UFC measures are thought to be unreliable in the setting of impaired renal function. It is debatable whether dexamethasone-based tests are reliable in kidney failure patients [49]. Validation studies of CS tests in chronic kidney disease are urgently needed.

5.5.5.3 Cyclic forms of hypercortisolism

Cyclical CS is a rare finding among CS patients but may confound investigations by the appearance of intermittently normal studies. Periodic repeat testing may be required to make the diagnosis. Some authors recommend frequent LNSC measures although such long-term testing is not indicated in patients who lack clinical features to support a Cushing's consideration.

5.5.5.4 Surreptitious use of glucocorticoids

Use of exogenous glucocorticoids is rampant both for medical and nonmedical reasons. Some "alternative" medications may contain significant amounts of glucocorticoids which may or may not be listed as ingredients. As well, even low-dose inhaled glucocorticoids may be sufficiently absorbed so as to be supraphysiologic, and combination with a drug that inhibits metabolism may actually lead to a severe cushingoid appearance [50]. In many such cases the patient will look cushingoid but serum, salivary, and urine measures of cortisol will actually be low or even undetectable and typically associated with undetectable serum ACTH. If necessary, chromatographic measurement of various synthetic and fluorinated steroids and associated metabolites to confirm ingestion may be measured to confirm the diagnosis but a careful history and investigation into all ingested (or injected and inhaled) compounds will make the diagnosis clear.

5.5.5.5 Ectopic corticotrophin-releasing hormone–producing tumors

Very rarely, neuroendocrine tumors (often malignant and of pancreatic origin) may secrete CRH, leading to ACTH-dependent CS but lacking pituitary abnormalities. If suspected, serum CRH may be measured and should be suppressed in the setting of CS; elevated levels should prompt additional investigations with expert endocrinology input.

5.5.5.6 Glucocorticoid resistance syndromes

Thought to be exceedingly rare, partial resistance to glucocorticoids due to a mutation in the gene for the nuclear glucocorticoid receptor may present with persistently abnormally high measures of urine or serum cortisol, often with incomplete dexamethasone suppression. Patients are classically noncushingoid (due to glucocorticoid resistance) and may have elevated ACTH with secondary elevations in aldosterone and adrenal androgens. Partial glucocorticoid resistance has also been described in cases of steroid-resistant asthma and multiple myeloma [51].

KEY POINTS

Besides patient, laboratory, and medication factors that may confound tests for CD, there are several rare disease factors that may result in confusing tests of cortisol excess. This includes cyclical secretion of cortisol as well as surreptitious administration of synthetic glucocorticoids. If the patient has a clinically cushingoid appearance but inconclusive laboratory tests for Cushing's, expert assessment is strongly recommended to consider these rare but important factors.

5.5.6 EXAMPLE WITH DISCUSSION

A 24-year-old woman presents with recent amenorrhea, hypertension, and mild acne, along with a 14-pound weight gain over 6 months.

Question 1: If Cushing's syndrome is suspected, what is the first test that should be ordered?

Answer: After ensuring that no exogenous glucocorticoids are in use (from any source), the clinician may choose to do a 24 h urine cortisol or an overnight 1 mg DST. Some centers will order a late night salivary cortisol as well.

Question 2: The 24 h urine cortisol is 150% above the upper reference limit and the dexamethasone-suppressed cortisol level is 388 nmol/L (14 μg/dL). What tests should be performed next?

Answer: With the classic presentation earlier the pretest probability for true CS is relatively high and with two separate abnormal measures of cortisol dynamics, the diagnosis of CS is secure. No additional biochemical testing is needed to prove the high cortisol levels. At this point the investigation should move to phase 2 where it will be determined whether the CS is ACTH dependent or independent. Thus an ACTH level should be measured.

Question 3: *The ACTH level is normal. Does this rule out CD?*

Answer: Absolutely not! In the setting of sustained hypercortisolism the "normal" ACTH level is in fact very inappropriate and thus abnormal. This is a case therefore of ACTH-dependent CD. The differential diagnosis at this point is pituitary tumor versus ectopic (usually malignant) tumor and an MRI pituitary is the next step.

Question 4: *The MRI pituitary shows a 2−3 mm pituitary nodule. Does this confirm CD?*

Answer: Not quite. In the absence of known malignancy (and a routine chest X-ray here is not a bad idea), pituitary CD is indeed the most likely but a 2−3 mm "adenoma" is still pretty nonspecific. IPSS for ACTH should be performed prior to any surgery. This test was done and confirmed a high level of ACTH (compared to inferior vena cava (IVC)) coming from the right side of the pituitary (petrosal sinus). Thus ectopic ACTH is excluded confidently. The patient went to surgery where a 3 mm ACTH secreting pituitary adenoma was resected and the patient cured.

5.5.7 COMMON MISUSE OF THE LABORATORY IN THE INVESTIGATION OF PRIMARY CORTISOL EXCESS

5.5.7.1 *Premature diagnosis of "confirmed hypercortisolism"*

A single abnormal screening test for cortisol excess is rarely sufficient to make a firm diagnosis, especially in the setting of nonspecific clinical features. False positive results in the screening tests are common and extensive wasting of resources may result by moving too quickly to tests of localization. Such localizing tests are often "negative" or nondiagnostic when a true diagnosis of Cushing's syndrome/disease has not been properly established.

5.5.7.2 *Premature exclusion of possible hypercortisolism*

True CD is a serious diagnosis that must not be missed. Milder forms may be more subtle both clinically and biochemically. A single screening test may have insufficient sensitivity to truly rule out CD in all cases. Any patients for whom a diagnosis of cortisol excess is a legitimate consideration should probably have more than one screening test done before concluding that the disease is absent.

5.5.7.3 *Failure to consider common confounders prior to testing*

Many of the screening tests for CS have common reasons for both false positive and false negative test results, notwithstanding the pretest probability of disease. Measurement of cortisol while actually taking glucocorticoids therapeutically is a particularly common error, including those administered topically or by injection. Tests of cortisol excess should not be performed in patients with a confounding factor until measures have been taken to eliminate the confounder. Appropriate consultation with the clinical chemist will help determine assay cross-reactivities and the course of action when such agents are currently being administered.

5.5.7.4 Measurement of AM/PM or random blood cortisol

The availability of antemeridian/postmeridian (AM/PM) reference ranges does not absolutely provide useful clinical information in regard to disease. They simply provide a piece of the puzzle for further diagnostic workup according to the clinical presentation of the patient. Indeed, CS is only diagnosed through the specific methods outlined earlier. Furthermore random serum cortisol levels are almost without meaning for appropriate clinical interpretation since additional knowledge is needed in terms of the diurnal variation individual to the patient and status of disease.

5.5.7.5 Failure to interpret adrenocorticotropin hormone in the context of the cortisol test results

Most blood ACTH assays report a first morning reference range that is derived from healthy controls. However, the reference range cannot be applied to a population of hypercortisolemic patients. If the patient is confirmed to be hypercortisolemic, any blood ACTH measures must be interpreted in that light. So a suppressed ACTH is actually "normal/appropriate" and a "normal" ACTH is actually "inappropriate" and indicative of ACTH-dependent CS. In the hypercortisolemic patient a normal ACTH level according to the reference range should not be an indication of the absence of disease.

5.6 SPECIFIC LABORATORY APPROACHES TO ADRENAL DISEASE: ADRENAL INSUFFICIENCY

5.6.1 GENERAL CONSIDERATIONS

5.6.1.1 Serum cortisol measurements

The reader is referred to Sections 5.5.2 and 5.5.5 for a discussion of the general limitations and considerations to any measures of cortisol including overt or surreptitious exogenous glucocorticoid use which may suppress the endogenous hypothalamic-pituitary–adrenal axis and must be considered whenever any type of cortisol measurements is found to be "low." The remainder of the following discussion presumes that falsely low cortisol levels as a result of exogenous glucocorticoid use has been excluded.

Adrenal insufficiency is often considered in patients with acute or critical illness. One must remember that blood cortisol assays measure both free and protein-bound cortisol. Thus in any hypoproteinemic state such as critical illness, total cortisol may appear to be "low" and yet may not reflect a true indication of adrenal function.

5.6.1.2 Adrenal disease versus "Adrenal Fatigue"

A full discussion of this issue is beyond the scope of this textbook; however, the medical practitioner should be aware of the growing phenomenon whereby patients who are often stricken with chronic fatigue, are given a label of "adrenal fatigue" with or without laboratory testing. When testing is performed, it is often with random salivary cortisol measures and nontraditional interpretations of the results. If called upon to verify such a diagnosis, it is critical that the patient discontinue all exogenous glucocorticoids of any type prior to testing. At this time, there is no evidence that adrenal functional disease can explain symptoms in patients with normal ACTH levels and normal tests of adrenal reserve.

5.6.1.3 Time course of primary adrenal insufficiency

PAI often has a very acute and sometimes life-threatening presentation but other cases may be more insidious and ideally diagnosed prior to such acute decompensation. The diagnosis of PAI may be easily made when presenting with acute, end-stage disease but earlier stages in disease may be associated with normal cortisol levels and thus depend on more careful endocrine testing when a high degree of suspicion is present. The first detectable biochemical abnormality in PAI is an elevation in basal ACTH levels, followed by loss, but not absolute, of aldosterone and finally by loss of cortisol production [52]. Thus the phase of disease must be considered when selecting a test to investigate the possible diagnosis in a patient who presents with less dramatic symptoms. For patients with suspected PAI a basal serum ACTH should be considered as the initial screening test. If ACTH is high, careful additional investigation and/or clinical observation should ensue and early consultation with endocrinology is advised. It is critical to remember that in pituitary disease the serum ACTH may be normal or low and thus cannot be used as a screening test.

5.6.1.4 The "two-step" investigation of adrenal failure: static versus dynamic tests of adrenal reserve, then localization

As discussed in Section 5.2.4, adrenal failure may be due to primary destruction of the adrenals or due to lack of ACTH-stimulated adrenal function from pituitary disease. However, for all possible diagnoses, the first step must always be the test of whether the adrenal glands can make cortisol normally. Unlike endocrine organs like thyroid, normal adrenal function is not always determined by basal levels of cortisol. Because of the critical role of increased physiologic cortisol production in acute illness, normal adrenal function is clinically and biochemically defined as the ability to secrete a normal amount of cortisol when subjected to a stress, either real or manufactured. Once a dynamic test of adrenal reserve has been performed and found to be abnormal, the second step is to identify the level of injury as either primary adrenal or secondary to pituitary disease. At this point an ACTH level should differentiate the two where high levels occur in PAI and normal or low levels occur in pituitary dysfunction.

KEY POINTS

Adrenal insufficiency is a serious disease of either the adrenal or pituitary. It must not be confused with ill defined and unproven concepts such as "adrenal fatigue." The hallmark of true adrenal insufficiency is low cortisol, often requiring a dynamic test for definitive diagnosis.

5.6.2 SPECIFIC TESTS FOR ADRENAL INSUFFICIENCY

5.6.2.1 Random cortisol

As outlined earlier, random cortisol levels are rarely sufficient alone to diagnose the presence or absence of adrenal dysfunction and a dynamic test must be done. There are two exceptions to this principle:

1. When a patient presents with the classic clinical features of acute adrenal insufficiency, the disease is often at a far-advanced state and adrenal destruction is often nearly complete. A random cortisol, if measured, will often be <50 nmol/L (1.8 μg/dL) and such low levels, in concert with the clinical presentation can immediately confirm the diagnosis. In such a setting a dynamic test of adrenal function is not needed and will only delay the appropriate patient diagnosis and therapy. However, an ACTH level should still be drawn *prior to initiation of any glucocorticoid replacement*, in order to determine whether the diagnosis is PAI or pituitary disease.
2. When a patient presents with signs or symptoms of suspected hypocortisolism and a random serum cortisol is found to be >500 nmol/L (18.1 μg/dL), adrenal insufficiency should be almost entirely excluded and routine use of glucocorticoid therapy is not indicated.

5.6.2.2 The (1–24) adrenocorticotropin hormone stimulation test

This test is based on the idea that a bolus of synthetic ACTH can simulate a physiologic stress and thus test the "adrenal reserve" as a means of defining a normal functioning gland. It is especially useful for diagnosing PAI and has the advantage of being safe, easy and, quick to perform in almost all healthcare systems. Rapid turnaround time for laboratory cortisol reporting is especially important to ensure that critical adrenal failure is diagnosed promptly. The classic test involves the administration of 250 μg of (1–24)ACTH (Cosyntropin or Synacthen) by IV push. Blood cortisol is measured at time "0" (baseline) and again at 30 and 60 min postinjection. Interpretation criteria are slightly variable according to the reference but generally, a cortisol level of >500 nmol/L (18.1 μg/dL) has long been considered the cut-off that defines normal adrenal reserve [53]. Newer immunoassays with improved specificity for cortisol may generate substantially lower peak cortisol levels and thus the definition of a normal cortisol response to stimulation is assay-specific. The change in cortisol levels during the test is not as important since patients with adrenal insufficiency may already be secreting cortisol at the maximal rate.

5.6.2.3 Variations on the (1−24) adrenocorticotropin hormone stimulation test

5.6.2.3.1 The 1 μg test

It has been argued that a 250 μg bolus of [1−24] ACTH far exceeds any reasonable level of ACTH that might be secreted by the pituitary gland under maximal stress therefore perhaps producing artificially high adrenal responses. An alternate dose of 1 μg [1−24] ACTH has been proposed as a more sensitive "physiologic" stress test for the adrenals, especially for use in patients with suspected pituitary disease. Acceptance of this alternate dose test has been variable in the endocrinology literature with some arguing its superiority on the basis of physiology whereas others argued against its use on the basis of lack of demonstrated superiority in diagnosis. Not generally available in a 1 μg vial, the dose must be manually prepared by dilution of a 250 μg dose. The alternate protocol requires a measurement of a 30 min cortisol alone, which is the time of typical maximal response, and a level >500 nmol/L (18.1 μg/dL) is still generally used to define normality [54] with the same caveats as listed above for newer cortisol immunoassays.

5.6.2.3.2 Serum aldosterone levels

Since aldosterone secretion is also stimulated by ACTH, it may sometimes be useful to measure aldosterone along with cortisol during a stimulation test. A normal aldosterone response to 250 μg ACTH is defined as >500 pmol/L (18.1 μg/dL) [55] although this is derived from a very small number of human samples. A normal aldosterone response suggests that the etiology of the cortisol deficiency is likely to be due to pituitary ACTH deficiency [56]. Patients with PAI (AD) had aldosterone levels <140 pmol/L (5 ng/dL) following stimulation.

5.6.2.3.3 Stimulated salivary cortisol levels

The use of salivary cortisol measurements instead of blood cortisol levels has been reported in ACTH stimulation tests. With salivary cortisol measured by radioimmunoassay (RIA), levels of >24.3 nmol/L (0.88 μg/dL) have a high degree of agreement with serum cortisol measures for the definition of normal adrenal function [57]. When used during insulin hypoglycemia, a salivary cortisol >20.1 nmol/L (0.73 μg/dL) is thought to indicate normal results [58]. Other than perhaps slightly greater ease of sample collection, there is no compelling reason to routinely use salivary cortisol measures in this or other dynamic adrenal testing. Salivary cortisol may be useful as a test of adrenal function in hypoproteinemic patients or in patients using oral estrogens which may lower or raise serum cortisol levels, respectively.

5.6.2.4 Other dynamic tests of the hypothalamus−pituitary−adrenal axis

While the ACTH stimulation test is ideal for PAI, other tests are often employed to check for adrenal insufficiency where hypothalamic/pituitary disease is suspected. Although it is still debated, the argument against the ACTH stimulation test is that it does not necessarily require a normal functioning pituitary in order to stimulate an adrenal cortisol response. The other tests include the insulin-induced hypoglycemia test (IIHT) and the glucagon stimulation test (GST) both of which require an intact hypothalamus and pituitary in order to stimulate cortisol production. Protocols for both are described in detail elsewhere [59] while older tests such as the CRH stimulation test and metyrapone test are largely historical. In both cases, serial cortisol measures are performed and interpretation based on the peak value. However, several studies have shown that the results and diagnostic interpretations taken from IIHT or GST or ACTH stimulation tests are poorly reproducible [59]. This may be related to the different mechanisms by which the hypothalamus−pituitary−adrenal (HPA) axis is stimulated in the different tests. In routine clinical practice a uniform diagnostic cut-off of 500 nmol/L is often taken to be indicative of a normal response but this may be a significant over-simplification that is not strongly supported by the medical literature.

For the IIHT a peak cortisol >500 nmol/L (18.1 µg/dL) has been the traditionally used definition of normal [58] but this value was derived at a time when older cortisol assays were in use. Newer studies of the IIHT with modern cortisol immunoassays have suggested a lower values of 300−400 nmol/L (10.9−14.5 µg/dL) for normality [60]. Because the IIHT is considered the gold standard for diagnosis of adrenal insufficiency, there are no figures pertaining to sensitivity and specificity.

The GST is a newer test of the HPA axis that does not carry the risks of hypoglycemia-related complications of the IIHT. A peak cortisol of <250−300 nmol/L (9.1−10.9 µg/dL) may be considered abnormal although comparisons to the IIHT results indicate that the GST as a whole may only have sensitivity and specificity of 92% and 100% [61].

> **KEY POINTS**
>
> In situations of very high clinical suspicion a very low random cortisol is often sufficient to confirm adrenal insufficiency. Less dramatic cases require dynamic testing, usually an "ACTH stimulation test" which measures adrenal reserve. Although a peak serum cortisol of >500 nmol/L (18.1 µg/dL) is widely known to be a definition of normal, this cut-off may not necessarily be accurate for all methods of cortisol measurement. Each lab should use a lab definition that is locally generated or at least validated for the assay in use.

5.6.3 SPECIAL SITUATIONS IN ADRENAL CORTISOL TESTING

5.6.3.1 Congenital adrenal hyperplasia

Congenital adrenal hyperplasia (CAH) is a generic term for a collection of disorders characterized by a congenital absence of a steroidogenic enzyme which may range from partial to complete deficiency. There are a variety of clinical phenotypes depending upon which enzyme is deficient. The most common form of CAH is due to a mutation in CYP21A2 resulting in a deficiency of the p450c21 (21-hydroxylase) enzyme which converts progesterone to 11-deoxycorticosterone (a precursor of aldosterone) and also converts 17-OH-progesterone to 11-deoxycortisol (a precursor of cortisol). Deficiency of p450c21 therefore results in aldosterone and cortisol deficiency of variable degrees. The cortisol deficiency results in a high level of ACTH which is unable to increase cortisol synthesis but at high levels can drive adrenal androgen synthesis which is a steroidogenic process that does not require the p450c21 enzyme. The aldosterone and cortisol deficiency may present in infancy or early childhood with salt-wasting, hypotension, hypoglycemia, hyperkalemia, and failure to thrive. The ACTH-induced androgen excess may cause a female (46XX) fetus to become virilized in utero, resulting in ambiguous genitalia at birth. In a 46XY male the excess androgens may lead to the appearance of precocious sexual maturity. A mild form of 21-hydroxylase deficiency may not present until adolescence where mild androgen excess may lead to acne and anovulation. Clinically, this is known as "non-classic" 21-hydroxylase deficiency.

In all forms of CAH the laboratory approach to diagnosis generally consists of measurement of the end-product hormones such as cortisol and aldosterone which are usually low, along with measurement of the steroid intermediate product that immediately precedes the expected enzyme deficiency. Elevated steroid precursors and deficiency of end-product hormones can confirm the clinical diagnosis which is then confirmed through genetic testing. Although random measurements of the respective hormones are often diagnostic in severely affected cases, an ACTH-stimulated set of measurements may be necessary for confirmation where affected 21-hydroxylase deficient persons typically have a marked elevation in 17-OH-progesterone to levels above 45 nmol/L (14.9 µg/L) [62]. In 21-hydroxylase deficiency, cortisol and aldosterone are typically low to very low and 17-OH-progesterone levels are very high, rising dramatically in response to ACTH infusion. With a population frequency of 1 in 500, neonatal screening for 21-hydroxylase deficiency is standard practice in many countries by measurement of a first morning 17-OH-progesterone in newborns by Day 3. CAH should be suspected in all cases of childhood adrenal insufficiency and expert endocrinology consultation is valuable to identify the exact enzyme defect and to initiate urgent adrenal replacement therapy. 17-OH-progesterone is commonly measured during treatment follow up of 21-hydroxylase deficiency patients, with higher levels indicating poorer overall metabolic control. Glucocorticoid and mineralocorticoid replacement should generally reduce the high ACTH levels and thus decrease the drive to produce 17-OH-progesterone. Nonadherence to therapy or gross under-treatment may see marked ACTH rises and subsequent high 17-OH-progesterone and androstenedione levels.

The majority of methods for 17-OH-progesterone measurement are either by immunoassay or LC−MS/MS. Given that newborn blood is high in steroid metabolites, efforts should be made to decrease potential cross-reactivities particularly with immunoassays. Therefore it is recommended that all newborn blood specimens should be subject to extraction protocols to reduce cross-reactive steroid metabolites and increase the specificity of the immunoassay. Conversely, LC−MS/MS methods typically begin with an extraction protocol to reduce cross-reactive substances and increase the concentration of the analyte of interest. Furthermore, coupled with more specific detection capabilities, LC−MS/MS methods have decreased false positive CAH screens [63].

KEY POINTS

CAH is a variety of clinical syndromes that are usually variations of glucocorticoid ± mineralocorticoid deficiency in neonates or young children. Depending on the steroidogenic enzyme defect, there can be subtypes that present with either deficiency or excessive adrenal sex steroids. Pediatric endocrinology consultation is strongly recommended after treatment of life-threatening glucocorticoid deficiency.

5.6.3.2 Adrenal function in critical illness

True adrenal insufficiency of any etiology may present with life-threatening shock, hypoglycemia and hyponatremia with altered level of consciousness. When the patient has overt hyper pigmentation with hyperkalemia, PAI can be an easy diagnosis but lacking these features, the diagnosis may be far more difficult especially if it coexists with other critical illness such as sepsis. The physiology of normal HPA function during critical illness is complex and varies according to the duration of time spent in a critically ill state. This complicates the interpretation of any ACTH or cortisol measures drawn for diagnosis. Critical illness is associated with a profound decrease in serum proteins including cortisol-binding globulin. Therefore measures of total serum cortisol may be greatly misleading in appearing very low or inappropriately nonelevated on account of the low amount of protein-bound cortisol. Free cortisol measurements from blood specimens may provide more accurate determinations of "bioavailable" cortisol [64]. Salivary cortisol, if obtainable, may also yield more accurate results although it has not been validated in this setting. Even with valid free cortisol determinations, the definition of adrenal insufficiency in critical illness is uncertain; some intensivists have considered a diagnosis of "relative adrenal insufficiency" in those whose cortisol levels seem "inappropriately low" for the acuity of illness. However, this diagnosis and the biochemical definition have not been standardized, and treatment responses based upon such testing in this setting appears to be independent of biochemical results. Therefore at this time it is recommended that the diagnosis of adrenal insufficiency be reserved for cases that are classically identifiable by the usual investigation and in the presence of a legitimate clinical etiology.

5.6.3.3 Hyperestrogenic states

The use of pharmacologic oral estrogens or pregnancy may increase cortisol-binding proteins thus giving a falsely high serum total cortisol measurement. This may obscure a diagnosis of adrenal insufficiency; discontinuation of oral estrogens or salivary cortisol measures may permit accurate diagnosis when suspected clinically.

5.6.3.4 Rare abnormalities in cortisol-binding globulin

Familial CBG deficiency is a rare cause of falsely low serum total cortisol levels, which may simulate the findings in adrenal insufficiency. Measurement of free cortisol or family screening may help clarify the situation [65]. Mitotane therapy for adrenal cortical carcinoma or severe Cushing's syndrome may falsely raise CBG levels making biochemical measures of adrenal function difficult to interpret [66].

5.6.3.5 Other tests for adrenal insufficiency

5.6.3.5.1 Antiadrenal antibodies

Autoimmune destruction of the adrenals is the most common cause of acquired adrenal insufficiency. The presence of autoantibodies such as anti-21-hydroxylase antibodies may be useful for confirmation of the autoimmune etiology. In patients with familial autoimmune disorders such as AIRE gene mutations, or in patients with other autoimmune endocrinopathies, a strongly positive antiadrenal antibody titer may be useful for predicting the eventual appearance of PAI [67]. Antiadrenal antibodies can be detected by immunofluorescence microscopy on primate adrenal gland substrate or by radiobinding assay. The immunofluorescence method detects anti-21-hydroxylase, anti-17-alpha hydroxylase, and other cytochrome p450 antigens [68,69]. Interpretation is done according to the detection of autoantigens that are localized to the microsomes of the adrenal cortical cells. Given the high probability of autoimmune disease in this patient population, samples are also tested for potential autoantibodies such as antimitochondrial antibodies, antinuclear antibodies, and antismooth muscle antibodies which will interfere with the immunofluorescence microscopy result. If such autoantibodies are present, then the immunofluorescence microscopy result is considered invalid. Acknowledging the lower specificity and potential interferences, the more specific radiobinding assay is the preferred method [70,71].

5.6.3.5.2 Adrenal androgens (dihydroepiandrosterone-sulfate)

Complete PAI or ACTH deficiency may also result in loss of adrenal androgens. There are several adrenal androgens that may be measured but dihydroepiandrosterone-sulfate (DHEA-S) is usually the measurable metabolite of choice due to its long half-life and stability in plasma. Low DHEA-S may be a feature of adrenocortical failure but has poor sensitivity and specificity as a screening test [72]. Indeed, low concentrations are normally observed in the pediatric and elderly populations.

5.6.4 EXAMPLE WITH DISCUSSION

A young man with a history of type-1 diabetes presents with recurrent hypoglycemia, not explained by errors in insulin use. AD is considered by the physician. He is otherwise stable and has no other specific symptoms.

Question 1: What tests should be ordered at this early stage?

Answer: The physician is wise to consider AD as an additional autoimmune endocrinopathy in this patient. In its early stages, AD may be hard to recognize without any hyper pigmentation. The simplest test would be a first morning cortisol and electrolyte panel looking for hyponatremia or hyperkalemia.

Question 2: The electrolytes show potassium of 5.1 mmol/L and morning cortisol of 130 nmol/L (4.7 µg/dL) (normal 180–480 nmol/L; 6.5–17.4µg/dL). Is Addison's ruled in or out at this stage?

Answer: Not yet. Many AD patients will maintain normal electrolytes until their cortisol and aldosterone deficiency is severe; the low-ish morning cortisol is suspicious but not necessarily diagnostic. It is also important that the clinician ensure that the patient has not taken any kind of exogenous glucocorticoid that might be affecting the result.

Question 3: What test should be done next?

Answer: A dynamic test of adrenal function is needed, usually a 250 µg ACTH stimulation test. In this case the test was done and showed a baseline cortisol of 75 nmol/L (2.7 µg/dL), 30′ cortisol 105 nmol/L (3.8 µg/dL), and 60′ cortisol 115 nmol/L (4.2 µg/dL). Although the peak cortisol to define normality is highly dependent upon the assay in use, there are no immunoassays or LC–MS/MS measures in which this particular response can be normal since most normal results will be a peak of at least 300–500 nmol/L (10.9–18.1 µg/dL). Thus this test confirms adrenal insufficiency in this case.

Question 4: Is this adrenal or pituitary disease?

Answer: With the history of autoimmune endocrine disease and the borderline high potassium level, this will almost certainly be primary adrenal failure (AD). However, to be sure, a serum ACTH level should be drawn prior to starting any cortisol replacement. In this case the ACTH level was 10X above the upper reference range, confirming primary adrenal failure.

Question 5: Any other tests to be considered?

Answer: In this case, autoimmune adrenal disease is most likely but should be confirmed by measuring antiadrenal antibodies. If an autoimmune nature cannot be confirmed, the clinician will have to consider other causes such as HIV or even X-linked adrenoleukodystrophy.

5.6.5 COMMON MISUSE OF THE LABORATORY IN THE INVESTIGATION OF ADRENAL INSUFFICIENCY

5.6.5.1 Testing serum cortisol in patients who are actively using exogenous glucocorticoids

Inhaled steroids may suppress the hypothalamic-pituitary—adrenal function. Although the inhaled glucocorticoid may not be detected by the cortisol assay, the otherwise suppressed adrenal gland may not be secreting much cortisol over and above what is being taken exogenously. Thus the serum cortisol, and often ACTH, will be low without necessarily indicating actual hypopituitarism or requiring glucocorticoid treatment in addition to the exogenous steroids already being taken.

5.6.5.2 Failure to consider the effect of hypoproteinemic states on total cortisol in the blood

Hypoalbuminemic patients such as those with critical illness may have low total serum cortisol levels despite normal HPA axis function and normal amounts of free cortisol which remains in equilibrium to maintain homeostasis. Assays that detect free cortisol in blood are not widely available but this may change with increased use of LC—MS/MS methods. As such, reference intervals that define the normal healthy population will be required to standardize existing tests to free cortisol measurements.

5.6.5.3 Excessive investigation when the diagnosis is already clear

Adrenal insufficiency is a potentially life-threatening diagnosis and once the diagnosis is made, treatment is urgently needed. Since dynamic testing of adrenal reserve can take time to arrange, treatment should not be delayed in cases where the diagnosis is already highly likely and the random cortisol level is <50 nmol/L (1.8 μg/dL).

5.6.5.4 Failure to consider the effects of surreptitious glucocorticoid or narcotic use

Many unregulated health products may contain unlisted amounts of glucocorticoids and related compounds. It is important to enquire about and stop any unrecognized health supplements before proceeding to testing or diagnosis. Similarly, chronic narcotic use may suppress the HPA axis [73] resulting in very low basal cortisol levels but often normal responses to an ACTH stimulation test. It is uncertain as to whether narcotic-related low basal cortisol levels are associated with clinical consequences; expert consultation is advised prior to initiation of long-term steroid replacement. In addition to physiological effects, interference by such glucocorticoid compounds may be detected by laboratory methods. Immunoassays may provide a false positive or negative result according to assay design, while LC—MS/MS methods have the capability to detect other unexpected compounds or metabolites in the patient specimen.

5.6.5.5 Failure to do any testing and making a diagnosis based on symptoms alone

Chronic adrenal insufficiency may have many features that overlap with chronic fatigue, which is a very common presentation. However, it is inappropriate and dangerous to initiate long-term steroid therapy without clear evidence of true adrenal insufficiency. Alternative tests such as salivary cortisol day-curves have not been standardized with respect to true adrenal disease and are no substitute for use of the currently validated tests of adrenal function.

5.7 SPECIFIC TESTS OF ADRENAL FUNCTION: PRIMARY ALDOSTERONE EXCESS

5.7.1 GENERAL CONSIDERATIONS

5.7.1.1 Primary aldosteronism: discrete disease or mechanistic spectrum

Many hyper-functional adrenocortical disorders like CD or pheochromocytoma produce classic and easily recognizable signs and symptoms, often due to a discrete adrenal tumor. Aldosterone excess states are also classically recognized as hypertension with hypokalemia and often due to a discrete adrenal tumor. However, it is now known that there are many other causes of autonomous aldosterone excess in the absence of an adrenal mass, including antiangiotensin II antibodies, ectopic adrenal receptors, and genetic mutations in ion channels involved in aldosterone secretion. As outlined below, primary aldosterone excess is typically defined according to the presence of an abnormal aldosterone–renin ratio (ARR). However, population studies have suggested that the ARR is normally distributed in hypertensive populations which argues against the current understanding of autonomous aldosteronism as being a discrete and unique disease [74]. It may be that mineralocorticoid hypertension blends seamlessly into the general hypertension population with classically identified PA representing just the more extreme end of the spectrum.

5.7.1.2 The value of hypokalemia as a screening test for PA in patients with hypertension

Although spontaneous or diuretic-induced hypokalemia is one of the traditional hallmarks of PA, multiple well-characterized PA cohorts have now shown that the majority of confirmed PA cases actually do not have overt hypokalemia. Therefore although electrolyte measurement to detect hypokalemia as a screening test for PA is widely recommended, the overall sensitivity is estimated at <30% which makes for a very poor screening strategy. As a consequence of this poor sensitivity, some hypertension societies recommend that all hypertensive patients be screened for PA by measurement of the ARR [75]. The feasibility and cost-effectiveness of this approach is not yet known.

KEY POINTS

Mineralocorticoid hypertension/PA is common and may have some biochemical overlap with what has traditionally been called "essential or idiopathic hypertension." Some subtypes of PA are surgically curable and should not be missed. If clinicians plan to do frequent biochemical screening for PA, it is crucial that the laboratory and clinician work together to validate a chemical definition of what will be flagged as "abnormal." The clinician will need a carefully considered approach to further investigate any abnormal cases in order to separate the surgically curable forms from those who will not benefit from invasive testing and surgery.

5.7.2 THE ALDOSTERONE-TO-RENIN RATIO

5.7.2.1 The concept of the aldosterone-to-renin ratio

The hallmark of primary (autonomous) aldosterone excess is the presence of inappropriately high serum aldosterone despite low renin. The actual serum aldosterone levels in confirmed cases of PA may be highly variable and are rarely useful or diagnostic by themselves. It is only in comparison to the level of renin that any aldosterone measurement may be deemed appropriate (i.e., due to renin-stimulated production) or inappropriate (i.e., autonomous PA). When expressed as a ratio of aldosterone over renin, the higher the ratio, the more convincing is the abnormality of the aldosterone level. As a calculated ratio, the ARR is inherently dependent upon the denominator. At very low levels of renin the ARR may become very high, even with relatively low-normal serum aldosterone levels. Whether all cases of high ARR, even those with low-normal aldosterone levels, represent true PA is still debatable. Some authors have proposed that an abnormal ARR must also include a certain level of high serum aldosterone in order to be considered indicative of actual autonomous aldosterone excess, as opposed to an artifact of extremely low renin levels alone [76]. While this seems to make theoretical sense, it is still unknown as to what extent this practice might actually decrease the sensitivity of the ARR for detecting true PA. Rather than insisting on a somewhat arbitrary serum aldosterone level minimum for defining high ARR, many experts now use a "minimum reporting level" for plasma renin according to the assay sensitivity. Therefore many investigators who report on the ARR describe a minimum renin level to avoid over-inflation of the ARR.

5.7.2.2 Patient preparation for aldosterone-to-renin ratio measures

It is critically important that the patient be prepared medically in advance of the ARR measurement. This includes

• Correction of any hypokalemia with oral supplements.

Table 5.2 Known Medications That Affect the ARR

Medication	Effect on Aldosterone	Effect on Renin	Effect on ARR	Effect on Laboratory Result
Beta-adrenergic blockers	↓	↓↓	↑	False positive
Central agonists (clonidine, α-methyldopa)	↓	↓↓	↑	False positive
Nonsteroidal antiinflammatory drugs (NSAIDs)	↓	↓↓	↑	False positive
Potassium wasting diuretics	→ ↑	↑↑	↓	False negative
Potassium sparing diuretics	↑	↑↑	↓	False negative
ACE (angiotensin-converting enzyme) inhibitors	↓	↑↑	↓	False negative
ARBs (angiotensin II type-I receptor blockers)	↓	↑↑	↓	False negative
Calcium blockers (dihydropyridines)	→ ↓	↑	↓	False negative
Renin inhibitors	↓	↓↑	↑↓	False positive or false negative

Adapted from Funder JW, Carey RM, Mantero F, Murad MH, Reincke M, Shibata H, et al. The management of primary aldosteronism: case detection, diagnosis, and treatment: an endocrine society clinical practice guideline. J Clin Endocrinol Metab 2016;101:1889–916.

- Cessation of drugs known to affect the ARR (Table 5.2) with use of nonconfounding medications for blood pressure control in the interim. It should be noted that renin inhibitors inhibit the renin activity but raise renin concentration. This will consequently raise or decrease the ARR depending on whether renin is measured by activity or by concentration, respectively. A detailed discussion regarding the differences in methods will be given later.
- Ad libitum sodium intake (no salt restriction) in the two days prior to testing.
- Blood collection should be performed prior to 1000h and collected with the patient in a seated, upright position after at least 1 h of ambulation. This is to maximally stimulate renin production and thus increase the specificity of interpretation of renin levels that are still found to be low or suppressed.

5.7.2.3 Aldosterone measurement

Given that aldosterone is present in picomolar (ng/dL) concentrations, it is very technically challenging to measure this analyte in the blood. Indeed, LC—MS/MS methods even require a more sensitive detector. As with most hormones, measurement methods for aldosterone have been developed using RIAs. With advances in detection methods and the complexities of using radioisotopes, aldosterone methods now employ nonradioactive principles. When collecting specimens for aldosterone measurement, it is important that the patient be in the upright position for at least 1 h. However, the laboratory may validate reference intervals for those who are either in the supine or upright positions, which may be important for certain patient populations.

5.7.2.4 Renin determination/measurement

The measurement and interpretation of renin requires some special considerations regarding its biochemistry and role in physiology. Renin is a proteolytic enzyme that needs to be proteolytically activated from its inert prorenin form which can be 10 to 100-fold higher concentration in the blood. This occurs in two steps where prorenin first undergoes a reversible conformational change to activated prorenin, and then second, it is proteolytically cleaved to produce active renin. There are two general methods for determining renin in the blood by measuring either activity of the activated renin or directly measuring the mass concentration of the renin. Both methods are susceptible to cryoactivation, acidification, or partial proteolysis of prorenin, which should all be minimized in vitro. In particular, specimens should not be chilled due to irreversible activation of prorenin with subsequent apparent increases in renin activity and concentration.

Plasma renin activity reflects the biological enzymatic conversion of angiotensinogen to angiotensin I. This method relies on measuring the rate of production of angiotensin I by either (radio)immunoassay or more recently, LC–MS/MS. There are a number of variables that affect plasma renin activity such as pH, incubation time, and endogenous angiotensinogen concentration. The pH of the reaction can affect the reversible conformation change to activated prorenin as described above and lead to overestimation of renin activity. To minimize the effects of prorenin activation, some assays deliberately proteolytically activate prorenin to renin. Given the effect of endogenous angiotensinogen, a variant of this assay, termed the plasma renin concentration assay, utilizes exogenously applied angiotensinogen at a high concentration to ensure that the rate limiting step is only due to renin activity. The ARR has traditionally been studied using plasma renin activity and is thus often still regarded as the gold standard. However, with increasing global use of direct renin concentration, novel validation studies of the ARR are now underway.

Direct renin concentration has now become a more convenient alternative for clinical laboratories. It is not subject to the variability of enzymatic bioactivity assays noted above and exhibits better reproducibility of results. However, there are a number of considerations when directly measuring renin mass concentration. Given that prorenin is 10 to 100-fold higher in the blood, potential antibody cross-reactivity between prorenin and renin need to be taken into account. Furthermore direct renin concentration is more susceptible to physiological changes in the body such as the role of estrogens in increasing angiotensinogen in the blood. When there is an increase in angiotensinogen substrate, this might lead to elevated levels of angiotensin I, angiotensin II, and aldosterone. However, homeostatic balances prevent hypertension by maintaining plasma renin activity but reducing direct renin concentration. A drop in direct renin concentration can consequently increase the ARR and lead to false diagnosis of PA [78].

As above, the hallmark of autonomous or inappropriate aldosterone secretion is the finding of suppressed renin in a hypertensive patient. Patient preparation for ARR measurement is not without some risk when stopping blood pressure lowering medications, most of which are stopped specifically because they may raise renin levels and thus give a falsely lower ARR. In the setting of true and clinically meaningful aldosterone excess however, the renin remains low even when taking drugs that would otherwise raise the renin level in a non-PA patient. Therefore it has been suggested that clinicians could order a plasma renin at first, without adjusting antihypertensive medications as a "first look" test for PA. Patients with overtly high renin are extremely unlikely to have true PA and further medication adjustment for a proper assessment of the ARR is expected to be of very low diagnostic value. Patients whose unadjusted plasma renin is low do not necessarily have PA but signal the clinician that there may be value in more stringent further investigations [79].

5.7.2.5 Interpretation of the aldosterone-to-renin ratio

When measured in a hypertensive person, an elevated ARR result likely signifies the presence of primary, inappropriate aldosterone excess. However, the definition of a high ARR has not yet reached uniform agreement. Many different authors have cited different cut-offs for what constitutes an abnormal ARR although each is usually based upon comparison to a small cohort of "confirmed" and obvious PA cases. Because the ARR is probably normally distributed in a hypertensive population which may have varying degrees of mineralocorticoid hypertension, it becomes problematic to state that any one ARR cut-off necessarily separates clear-cut PA from clear-cut essential hypertension [74].

Ideally, each laboratory should independently verify its own ARR cut-off in conjunction with a local hypertension clinic that is skilled in defining what will be called "PA" versus "non-PA." In the absence of such local validation the clinical chemistry laboratory will need to look at the literature reported cut-offs that pertain to the assay that will be locally used. Again, in consultation with the local hypertension clinic, the laboratory may want to choose a "lower" ARR cut-off in order to have maximum sensitivity in potential detection of PA cases; this might be most appropriate when expert hypertension consultation services are available to deal with "false positive" test results. In other places, where experienced endocrine hypertension services are not available, the laboratory may choose a "higher" ARR cut-off in order to have high specificity for readily identifiable PA cases. Published guidelines contain a useful summary of potential definitions of abnormal ARR, depending on the assays and reporting unit in use [77].

At the authors' institutions an abnormal ARR is defined as >550 when aldosterone is measured by immunoassay and expressed as pmol/L with plasma renin activity in ng/mL/h. If a direct renin concentration is used and reported in mIU/L, a ratio >60 is taken as evidence of possible primary aldosterone excess. Where direct renin concentration is reported in ng/L, a ratio of >100 is considered abnormal. It must not be assumed that all ordering physicians will know exactly how to interpret an ARR and unless the "ARR" is a specific test that may be requested, the clinical chemistry laboratory may only receive a request for individual aldosterone and renin results. Thus we recommend that each laboratory add an interpretive comment that is automatically reported with each aldosterone report to indicate the potential degree of abnormality along with a comment pertaining to the potential need for further investigation of less dramatic abnormalities. An example would be: "an aldosterone−renin ratio >550 (pmol/L/ng/mL/h;) or more is considered high when measured in the setting of hypertension. Particularly high ARR results are likely to be quite specific for PA whereas borderline high results may need further confirmatory testing." Furthermore, ordering physicians should be cognizant of potential preanalytical interferences that will affect interpretation of the ARR such as medications (Table 5.2) and require that patient be on salt intake ad libitum [80].

5.7.2.6 *Confounders in aldosterone-to-renin ratio interpretation*

Concomitant use of medications that raise plasma renin levels (diuretics, mineralocorticoid receptor antagonists, ACE-inhibitors, angiotensin receptor blockers) may all "falsely" decrease the ARR although this is less likely in severe PA with intense renin suppression (Table 5.2). SSRIs and NSAIDs are reported to cause false increases in the ARR, while estrogen- or drosperinone-containing OCPs are associated with false elevations in ARR only when a DRC assay is used. This is due to the estrogenic effect on plasma angiotensinogen production by the liver, leading to a higher amount of substrate available to renin which is then consumed in the production of angiotensin I, leading to a decrease in DRC [81]. In a well-controlled homeostatic mechanism, renin activity will remain relatively unchanged while renin concentration may fluctuate with availability of substrate.

Chronic kidney disease is particularly problematic with studies suggesting that cross-reactive polar metabolites cause overestimation of aldosterone and ARR measurements due to immunoassay interference [82,83]. Both high and low ARR have been associated with renal disease, however and it is unknown whether the conflicting findings reflect analytical confounders or whether the renal injury is actually just a consequence to prior aldosterone-mediated hypertension.

Age of the patient should also be considered as renin levels may decrease with increasing age and would not necessarily imply true primary aldosterone excess [84].

KEY POINTS

The ARR is the most common screening test for PA. There is no uniform agreement as to the definition of "abnormal," so clinical endocrinology and the laboratory should collaborate for proper validation of assay performance. A lower definition of abnormal ARR may have the highest sensitivity for PA and a higher ARR may have the highest specificity for true/surgical disease. Many medications can affect the ARR, usually by increasing renin levels. Thus a high ARR despite medications that increase renin should be viewed as highly abnormal and indicative of probable PA. The laboratory should be aware of assay interferences (either analytical or physiological) and be able to consult with the clinical team for proper interpretation of the results.

5.7.3 FURTHER TESTING OF THE PATIENT WITH A HIGH ALDOSTERONE-TO-RENIN RATIO

5.7.3.1 Management-directed laboratory testing

With the publication of the PATHWAY-2 study, it is now known that mineralocorticoid receptor antagonist therapy is highly useful in the management of resistant hypertension [85]. This is true even in the absence of a "formally" diagnosed case of PA. Therefore while multiple additional tests are available to confirm a diagnosis of PA, such tests should be limited to patients in whom ultimate surgical adrenalectomy would be considered. In patients with multiple medical comorbidities deemed to be high surgical risks, there is no value in doing extensive testing beyond an ARR alone.

5.7.3.2 Confirmatory testing

Several tests have been advocated to "confirm" a diagnosis of PA in subjects with a high ARR, assuming that such testing then excludes "false positive" ARR results from further PA consideration. This testing might only be needed in cases with borderline high ARR and/or low pretest probability. Some have argued that it does not add to the diagnostic investigation when the pretest probability is already high [86].

- The postcaptopril test measures aldosterone following an oral dose of captopril. Persistently high aldosterone is presumed to be indicative of aldosterone autonomy from the usual renin stimulation.
- The furosemide upright test measures renin after a dose of furosemide. This is based on the principle that plasma renin activity is dependent on an individual's hydration and sodium status. Administration of furosemide, a potent diuretic, will normally provide the stimulus to increase renin levels. Persistently suppressed renin is presumed to indicate the presence of intense suppression by aldosterone.
- The intravenous saline suppression test measures aldosterone following 2000 cc of 0.9% saline given over 4 h. This is based on the premise that rapid expansion of intravascular volume will normally lead to decreases in

aldosterone. Persistently elevated serum aldosterone is presumed to indicate autonomy from sodium-suppression of renin.

- The fludrocortisone suppression test measures serum aldosterone after 2 days of exogenous fludrocortisone administration. Flucrocortisone is a potent mineralocorticoid which will normally suppress aldosterone production. Persistently elevated aldosterone is presumed to indicate autonomy and nonsuppressibility by (unmeasured) mineralocorticoids.

These tests, if conducted, should be performed by experienced clinicians in a dedicated endocrine testing unit that can accurately document the timing of drug administration and sample collection. High volume salt or mineralocorticoid loading may provoke severe hypokalemia or hypertension in PA patients. Careful patient selection and monitoring during such testing is mandatory.

5.7.3.3 Adrenal vein sampling conduct

Given that a high ARR, with or without confirmatory testing, is common in hypertensive patients and incidental adrenal masses, which are often nonfunctioning, may have a prevalence of up to 8%, one cannot always assume that the coexistence of high ARR and adrenal mass necessarily proves that the mass is an aldosteronoma. Surgical adrenalectomy for PA is only offered to suitable patients with confirmed unilateral source of aldosterone excess, which is often due to an adenoma. Thus lateralization of PA must be proven prior to referral to surgery. Aldosterone measurement from the adrenal veins during selective venous sampling (AVS) is the current gold standard for proof of unilateral PA. AVS is technically difficult and should only be performed in highly experienced centers of excellence where adrenal venous catheterization rates can be proven to be achieved in more than 90% of attempts. Careful coordination with radiology, endocrinology, and the laboratory are essential to ensure proper patient preparation associated with ARR measurement, correct collection and labeling of the multiple samples, and accurate reporting of results in a fashion that is interpretable by the clinician. Consultation with endocrinology is recommended in order to develop a formal protocol for AVS including a decision as to whether samples will be collected with or without the use of cosyntropin stimulation. Near simultaneous measurement of cortisol and aldosterone from the IVC and each adrenal vein (pre- and postcosyntropin) is the standard approach.

5.7.3.4 Adrenal vein sampling interpretation

5.7.3.4.1 Confirmation of correct catheter placement

Although the radiologist may feel that correct adrenal vein catheterization was achieved by imaging, it is critical that such catheter placement be confirmed by biochemistry. In the setting of PA, adrenal venous cortisol is used as the "control" hormone that is assumed to be relatively equally secreted by both adrenals. For each adrenal sample the laboratory calculates the ratio of cortisol in the adrenal vein compared to the (simultaneously measured) IVC cortisol sample, denoted as

"$R_{a/ivc}$" and "$L_{a/ivc}$." This ratio is often called the "selectivity index." If the cortisol ratios are >3:1, it is considered biochemical evidence of accurate catheter placement. Some authors use different ratios, ranging from 2:1 to 5:1 as indicative of proper positioning, so local validation is recommended for each laboratory.

In our institution, if the selectivity index is <3:1, the adrenal vein sampling from that adrenal should be considered to have failed, and no further attempts to interpret the results should be made.

5.7.3.4.2 Determination of unilateral versus bilateral aldosterone excess

After determining that the AVS was successful according to selectivity indices, the next step is to determine whether the aldosterone excess derives from one or both adrenals. In order to account for differences in catheter positioning along the length of the adrenal vein and account for potential dilution effect, each aldosterone sample should be corrected by calculating the aldosterone/cortisol ratio. This generates three ratios $L_{a/c}$, $R_{a/c}$, and $IVC_{a/c}$ for each collection, pre- and postcosyntropin. These ratios may now be compared to determine asymmetry with the highest adrenal ratio compared with lowest adrenal ratio (the Lateralization Index, LI) and the lowest adrenal ratio compared to IVC ratio (the Suppression Index, SI). A Lateralization Index of >3:1 between the adrenals suggests lateralization to the dominant side with very high specificity [87]. A suppression index of 1.4 or lower suggests that the nondiseased adrenal's aldosterone production is actually fully suppressed [88]. Either the LI or the SI individually may be used to determine unilaterally of PA excess. The presence of an LI > 3 with SI < 1 is considered to be essentially diagnostic.

5.7.3.5 *Other laboratory measures that predict unilateral primary aldosteronism*

On occasion, even successful AVS may be hard to interpret if the LI or SI approaches the diagnostic definitions mentioned earlier (i.e., a LI of 2.7). In such cases, there are several validated biochemical and radiologic markers which also predict unilateral PA, including [89]:

- CT imaging showing adrenal mass >10 mm in size
- hypokalemia
- eGFR > 100 mL/min
- very high plasma aldosterone (>450 pmol/L)

In such cases, AVS may be interpreted as being consistent with probable unilateral adrenal disease when multiple other predictors are present.

KEY POINTS

After a biochemical diagnosis of PA is made the clinician should consider a "confirmatory" test in more borderline cases. Only if the patient is a surgical candidate should localizing studies be performed. The current gold standard for defining PA as unilateral (and therefore potentially surgically treatable) is adrenal vein sampling. This procedure is technically difficult, requires very careful laboratory handling, and labeling of tubes and proper reporting of results. Results should be interpreted by an experienced clinician given the complexity of numbers generated by the test.

5.7.4 HYPERTENSION WITH HIGH RENIN

When drawn in the absence of mineralocorticoid receptor antagonists, a plasma renin activity of >1.0 ng/mL/h or direct renin >10 mIU/L should essentially exclude any further consideration of primary (autonomous) aldosterone excess [79]. A nonsuppressed plasma renin found in the presence of multidrug resistant hypertension or hypertensive young women may indicate the presence of renovascular disease or very rarely, a renin-producing renal tumor. Renal vein renin sampling has been reported as a means to identify unilateral versus bilateral renal renin over-production but its predictive value for finding hemodynamically significant renal artery stenosis is poor and thus not recommended.

5.7.5 MINERALOCORTICOID HYPERTENSION WITH LOW ALDOSTERONE

The "syndromes of apparent mineralocorticoid excess" comprise individuals with hypertension and hypokalemia with appropriately suppressed renin but low or suppressed aldosterone. This rare constellation suggests the presence or action of a non-aldosterone mineralocorticoid. Potential explanations include:

- severe Cushing's syndrome
- Liddle's syndrome (activating mutation in epithelial Na + channel)
- 11-BHSD2 defect (defect in conversion of cortisol to cortisone)
- glycyrrhetinic acid (black licorice inhibition of 11-beta-hydroxysteroid dehydrogenase (BHSD))
- 11-deoxycorticosterone secreting adrenal tumor
- CYP11B1 and CYP17A1 mutations (with resulting high levels of mineralocorticoid precursors)
- activating mutations in the mineralocorticoid receptor, often worsened by pregnancy

Expert consultation with endocrinology should be sought to arrange the necessary specialized testing if one of the above syndromes is considered.

5.7.6 COMMON MISUSE OF THE LABORATORY IN THE INVESTIGATION OF PRIMARY ALDOSTERONISM

5.7.6.1 Failure to discontinue mineralocorticoid receptor antagonists 6 weeks prior to testing

Many drugs may affect the ARR, usually by increasing renin levels and thus decreasing the ARR (see Table 5.2). The impact of this change may be minimal in those with true PA where aldosterone over-secretion continues to suppress renin. However, aldosterone antagonists may result in a dramatic rise in plasma renin (and aldosterone) thus giving a normal ARR even in true PA. Spironolactone, the most common mineralocorticoid receptor antagonist, is metabolized to the biologically active mineralocorticoid blocker canrenone which has a half-life of 2 weeks. Discontinuation of spironolactone is thus necessary for at least 6 weeks prior to any measurement of either ARR or adrenal vein sampling.

5.7.6.2 Failure to consider the assays and reporting units in interpretation of results

With numerous assays and reporting units in use around the world, clinicians may be easily confused when trying to compare their local results with the numbers reported in studies from different centers. It is critical that the clinical chemist be familiar with the relative similarities and differences of the various renin measurements in order to advise the clinician as to the best interpretation or comparison to report diagnostic values.

5.7.6.3 Failure to recognize the fluidity of what is determined to be "abnormal" on the aldosterone–renin ratio

Considering the overlap of PA with essential hypertension the diagnostic cut-off for ARR interpretation may vary according to the locally accepted diagnosis of PA. If a very high cut-off defining abnormality is used in reporting the ARR, clinicians may be confused by situations where the clinical presentation is highly convincing for PA yet the biochemistry is apparently normal. Both clinicians and clinical chemists should recognize that the ARR abnormal cut-off may not be 100% sensitive for all clinical cases and should be interpreted according a disease spectrum comprising a variety of clinical signs and symptoms as well as other indicators.

5.7.6.4 Ordering adrenal vein sampling without understanding of appropriate result interpretation steps

AVS generates a large volume of both cortisol and aldosterone levels from both adrenals and IVC. At face value, one might think to simply compare the aldosterone levels between the two adrenals but this fails to account for the multiple steps in assuring correct catheter placement and sample dilution.

5.7.7 EXAMPLE WITH DISCUSSION

A 48-year-old thin woman is diagnosed with hypertension and started on a thiazide type diuretic. There is no family history of hypertension and she takes no other medications. Renal function is normal. Her doctor notices that the medication does not seem to control her blood pressure all that well and follow up bloodwork continues to show normal creatinine but low potassium (3.2 mmol/L). The thiazide medication is stopped in favor of an ACE-inhibitor drug which still does not really control the blood pressure well but at least the repeat potassium level is normal (3.6 mmol/L).

Question 1: *Should secondary hypertension be considered here?*

Answer: Yes, for several reasons. (1) She is thin and has no family history of hypertension—both of which suggest that this may not be primary hypertension. (2) Her blood pressure was not well controlled by either of the two first-line medications that were tried. (3) The thiazide induced hypokalemia is a "red flag," denoting the likely presence of aldosterone excess. PA is a strong possibility.

Question 2 What test should be ordered?

Answer: The ARR should be measured, after ensuring she is eukalemic and on a high salt diet for at least 2 days. Some would argue that the ACE-inhibitor should be discontinued or substituted with another noninterfering drug like an alpha blocker. However, if her ARR is found to be high despite use of an ACE-inhibitor, it actually adds to the specificity of the diagnosis (i.e., if anything, ACE-inhibitors could slightly lower an ARR result, thus a high ARR on ACE-inhibitor should be viewed as a likely true positive finding). In this case the ARR is 400% above the lab reference range and the diagnosis is confirmed. Additional testing will now include CT adrenals and possibly adrenal vein sampling to confirm any abnormal CT findings as consistent with unilateral adrenal adenoma.

5.8 SPECIFIC TESTS OF ADRENAL FUNCTION: SEX STEROIDS

5.8.1 GENERAL CONSIDERATIONS

5.8.1.1 Most disorders of excessive adrenal sex steroids are part of a larger adrenal pathology

Isolated hypersecretion of adrenal sex steroids with otherwise normal adrenal function is very rare. Such situations may include the pure androgen or estrogen secreting adrenal tumors. Cosecretion of adrenal androgens along with cortisol, which causes Cushing's syndrome, from a primary adrenal tumor is typical of malignant adrenocortical carcinoma. Hypersecretion of adrenal androgens in the setting of various steroidogenic enzyme deficiencies (CAH) is usually accompanied by deficiencies of glucocorticoids with or without deficiency of mineralocorticoids. This typically presents as ambiguous genitalia in 46XX females or precocious puberty in 46XY males. Milder forms of adrenal androgen excess may also present later in life among females with very mild defects of steroidogenesis (typically 21-hydroxylase deficiency), known as "non-classic congenital adrenal hyperplasia." Similar mild adrenal androgen excess may accompany ACTH-dependent (pituitary) CD, driven by ACTH simulation of steroidogenesis in the zona reticulosum.

5.8.1.2 Most pediatric disorders of deficient adrenal sex steroids are part of a generalized steroidogenic defect that also affects ovaries and testes

Some forms of CAH (such as CYP17A1 deficiency) may impair androgen (and thus estrogen) synthesis, causing sexual infantilism.

5.8.1.3 Adrenal sex steroid deficiency acquired in adulthood is of uncertain clinical significance

Adrenal androgen synthesis in males is dwarfed by testicular androgen production and therefore adrenal androgen excess or deficiency is largely clinically undetectable if they have normal testicular function. Adrenal androgens account for 50% of androgen synthesis in females (ovaries producing the other 50%). Thus deficiency of adrenal androgens is often not detected unless there is combined adrenal and ovarian failure. Anything that suppresses ACTH (such as exogenous steroids) may also decrease adrenal androgen synthesis. Clinical trials of adrenal androgen replacement in adrenal failure have not proved to be of meaningful clinical benefit.

5.8.2 LABORATORY CONSIDERATIONS

While any of the androgen synthesis pathway products may be measured, the most useful general measurement is DHEA-S, which is the sulfated form of DHEA. DHEA-S, while physiologically inert, has a long serum half-life and is thus best suited as a general marker of adrenal androgen synthesis. The common indications for measuring DHEA-S would include: investigation of female hirsutism/virilization/precocious adrenarche and investigation of incidental adrenal mass functionality. Use of DHEA-S to screen for adrenal insufficiency has been suggested but not widely validated. Owing to the lack of older age-specific reference ranges, the definition of low DHEA-S is still in question.

Despite standard reference ranges for both adult men and women, there is likely variation or decrease with age that may not be reflected in a single reference range. The clinical implications of finding a high DHEA-S are that adrenal imaging is needed to rule out a potentially malignant adrenal tumor. The clinical implications of a low DHEA-S are less understood. In some scenarios it may coexist with or even predate the appearance of adrenal insufficiency. However, low DHEA-S alone should not be taken as proof of adrenal insufficiency. At present there is no compelling indication to treat a patient with androgens on the basis of a low serum DHEA-S level.

KEY POINTS

Excessive (and to a lesser extent, deficient) adrenal androgens may be detected through measurement of DHEA-S. Adrenal androgens play an important part of the pathophysiology and clinical picture in congenital steroidogenic enzyme deficiencies as well as functional (often malignant) adrenal tumors. The clinical interpretation of apparently low DHEA-S levels in adults is uncertain in most cases and routine measurement of DHEA-S is discouraged.

5.8.3 MISUSE OF THE LABORATORY WITH ADRENAL ANDROGEN MEASUREMENTS

5.8.3.1 Failure to measure DHEA-S in men with incidentally found adrenal tumors

An adrenal tumor that produces androgens is often easily detected in women by clinical assessment. However, it may not produce obvious signs in men with normal testicular function other than possible testicular shrinkage. As androgen-producing adrenal tumors are often malignant, it is important to measure DHEA-S in all men with incidental adrenal tumors, even in the absence of any clinical signs.

5.8.3.2 Inappropriate measurement in adults with nonspecific symptoms

With increasing interest in postmenopausal hormone tests and treatments and the availability of "health promotion" clinics that do multiple hormone tests, there is increasing use of DHEA-S measurements. In many postmenopausal women the DHEA-S levels may be on the lower side. The clinical significance of this finding is unknown although often offered as an explanation for a variety of likely unrelated symptoms. Replacement of DHEA has not been proven to be either efficacious or safe, especially in the absence of a real diagnosis such as adrenal insufficiency.

KEY POINTS

Adrenal disorders are very rare diseases, but detection is crucial since the consequences of missed diagnosis can be fatal.

- Diagnosis of adrenal disorders requires several lines of investigation incorporating both clinical and laboratory data over time.
- Clinicians and laboratories should be aware that universal biochemical definitions of many adrenal disorders are lacking due to the absence of adequate gold-standard comparisons and consistency between laboratory methods that are used to drive clinical guidelines.
- Ongoing collaboration between the clinical team and the laboratory is needed for proper validation of assay parameters and for appropriate interpretation in light of the clinical picture.

REFERENCES

[1] Mantero F, Terzolo M, Arnaldi G, Osella G, Masini AM, Ali A, et al. A survey on adrenal incidentaloma in italy. Study group on adrenal tumors of the italian society of endocrinology. J Clin Endocrinol Metab 2000;85:637—44.

[2] Gimenez-Roqueplo AP, Favier J, Rustin P, Rieubland C, Crespin M, Nau V, et al. Mutations in the sdhb gene are associated with extra-adrenal and/or malignant phaeochromocytomas. Cancer Res 2003;63:5615—21.

[3] Bravo EL, Tagle R. Pheochromocytoma: state-of-the-art and future prospects. Endocr Rev 2003;24:539—53.

[4] Boscaro M, Barzon L, Fallo F, Sonino N. Cushing's syndrome. Lancet 2001;357:783—91.

[5] Mulatero P, Monticone S, Bertello C, Viola A, Tizzani D, Iannaccone A, et al. Long-term cardio- and cerebrovascular events in patients with primary aldosteronism. J Clin Endocrinol Metab 2013;98:4826—33.

[6] Venos ES, So B, Dias VC, Harvey A, Pasieka JL, Kline GA. A clinical prediction score for diagnosing unilateral primary aldosteronism may not be generalizable. BMC Endocr Disord 2014;14:94.

[7] Ayala AR, Ilias I, Nieman LK. The spectrum effect in the evaluation of cushing syndrome. Curr Opin Endocrinol Diab Obes 2003;10:272—6.

[8] Nanba K, Tsuiki M, Sawai K, Mukai K, Nishimoto K, Usui T, et al. Histopathological diagnosis of primary aldosteronism using cyp11b2 immunohistochemistry. J Clin Endocrinol Metab 2013;98:1567—74.

[9] Burt MG, Mangelsdorf BL, Rogers A, Ho JT, Lewis JG, Inder WJ, et al. Free and total plasma cortisol measured by immunoassay and mass spectrometry following acth(1)(-)(2)(4) stimulation in the assessment of pituitary patients. J Clin Endocrinol Metab 2013;98:1883—90.

[10] Handelsman DJ, Wartofsky L. Requirement for mass spectrometry sex steroid assays in the journal of clinical endocrinology and metabolism. J Clin Endocrinol Metab 2013;98:3971—3.

[11] Lenders JW, Duh QY, Eisenhofer G, Gimenez-Roqueplo AP, Grebe SK, Murad MH, et al. Pheochromocytoma and paraganglioma: an endocrine society clinical practice guideline. J Clin Endocrinol Metab 2014;99:1915—42.

[12] Sawka AM, Jaeschke R, Singh RJ, Young Jr. WF. A comparison of biochemical tests for pheochromocytoma: measurement of fractionated plasma metanephrines compared with the combination of 24-hour urinary metanephrines and catecholamines. J Clin Endocrinol Metab 2003;88:553—8.

[13] van Duinen N, Steenvoorden D, Kema IP, Jansen JC, Vriends AH, Bayley JP, et al. Increased urinary excretion of 3-methoxytyramine in patients with head and neck paragangliomas. J Clin Endocrinol Metab 2010;95:209—14.

[14] Hoy LJ, Emery M, Wedzicha JA, Davison AG, Chew SL, Monson JP, et al. Obstructive sleep apnea presenting as pseudopheochromocytoma: a case report. J Clin Endocrinol Metab 2004;89:2033—8.

[15] Little RA, Frayn KN, Randall PE, Stoner HB, Morton C, Yates DW, et al. Plasma catecholamines in the acute phase of the response to myocardial infarction. Arch Emerg Med 1986;3:20—7.

[16] Lenders JW, Willemsen JJ, Eisenhofer G, Ross HA, Pacak K, Timmers HJ, et al. Is supine rest necessary before blood sampling for plasma metanephrines? Clin Chem 2007;53:352−4.

[17] Eisenhofer G, Lattke P, Herberg M, Siegert G, Qin N, Darr R, et al. Reference intervals for plasma free metanephrines with an age adjustment for normetanephrine for optimized laboratory testing of phaeochromocytoma. Ann Clin Biochem 2013;50:62−9.

[18] Hickman PE, Leong M, Chang J, Wilson SR, McWhinney B. Plasma free metanephrines are superior to urine and plasma catecholamines and urine catecholamine metabolites for the investigation of phaeochromocytoma. Pathology 2009;41:173−7.

[19] Eisenhofer G, Lenders JW, Timmers H, Mannelli M, Grebe SK, Hofbauer LC, et al. Measurements of plasma methoxytyramine, normetanephrine, and metanephrine as discriminators of different hereditary forms of pheochromocytoma. Clin Chem 2011;57:411−20.

[20] McHenry CM, Hunter SJ, McCormick MT, Russell CF, Smye MG, Atkinson AB. Evaluation of the clonidine suppression test in the diagnosis of phaeochromocytoma. J Hum Hypertens 2011;25:451−6.

[21] Eisenhofer G, Huysmans F, Pacak K, Walther MM, Sweep FC, Lenders JW. Plasma metanephrines in renal failure. Kidney Int 2005;67:668−77.

[22] Algeciras-Schimnich A, Preissner CM, Young Jr. WF, Singh RJ, Grebe SK. Plasma chromogranin a or urine fractionated metanephrines follow-up testing improves the diagnostic accuracy of plasma fractionated metanephrines for pheochromocytoma. J Clin Endocrinol Metab 2008;93:91−5.

[23] Stewart PM. Is subclinical cushing's syndrome an entity or a statistical fallout from diagnostic testing? Consensus surrounding the diagnosis is required before optimal treatment can be defined. J Clin Endocrinol Metab 2010;95:2618−20.

[24] Klee GG. Interferences in hormone immunoassays. Clin Lab Med 2004;24:1−18.

[25] Krasowski MD, Drees D, Morris CS, Maakestad J, Blau JL, Ekins S. Cross-reactivity of steroid hormone immunoassays: clinical significance and two-dimensional molecular similarity prediction. BMC Clin Pathol 2014;14:33.

[26] Findling JW, Raff H. Screening and diagnosis of cushing's syndrome. Endocrinol Metab Clin 2005;34:385−402.

[27] Mericq MV, Cutler Jr. GB. High fluid intake increases urine free cortisol excretion in normal subjects. J Clin Endocrinol Metab 1998;83:682−4.

[28] Tamada D, Otsuki M, Kashine S, Hirata A, Onodera T, Kitamura T, et al. Obstructive sleep apnea syndrome causes a pseudo-cushing's state in japanese obese patients with type 2 diabetes mellitus. Endocr J 2013;60:1289−94.

[29] Petersenn S, Newell-Price J, Findling JW, Gu F, Maldonado M, Sen K, et al. High variability in baseline urinary free cortisol values in patients with cushing's disease. Clin Endocrinol (Oxf) 2014;80:261−9.

[30] Raff H, Auchus RJ, Findling JW, Nieman LK. Urine free cortisol in the diagnosis of Cushing's syndrome: is it worth doing and, if so, how? J Clin Endocrinol Metab 2015;100:395−7.

[31] Elias PC, Martinez EZ, Barone BF, Mermejo LM, Castro M, Moreira AC. Late-night salivary cortisol has a better performance than urinary free cortisol in the diagnosis of Cushing's syndrome. J Clin Endocrinol Metab 2014;99:2045−51.

[32] Findling JW, Raff H, Aron DC. The low-dose dexamethasone suppression test: a reevaluation in patients with cushing's syndrome. J Clin Endocrinol Metab 2004;89:1222−6.

[33] Nieman LK, Biller BM, Findling JW, Newell-Price J, Savage MO, Stewart PM, et al. The diagnosis of Cushing's syndrome: an endocrine society clinical practice guideline. J Clin Endocrinol Metab 2008;93:1526−40.

[34] Abou Samra AB, Dechaud H, Estour B, Chalendar D, Fevre-Montange M, Pugeat M, et al. Beta-lipotropin and cortisol responses to an intravenous infusion dexamethasone suppression test in Cushing's syndrome and obesity. J Clin Endocrinol Metab 1985;61:116−19.

[35] Jung C, Alford FP, Topliss DJ, Burgess JR, Long F, Gome JJ, et al. The 4-mg intravenous dexamethasone suppression test in the diagnosis of Cushing's syndrome. Clin Endocrinol (Oxf) 2010;73:78−84.

[36] Sharma ST, Yanovski JA, Abraham SB, Nieman LK. Utility of measurement of dexamethasone levels in the diagnostic testing for Cushing's syndrome. In Diagnosing and Treating Cortisol Excess and Deficiency 2014 Jun (pp. OR02-1). *Endocr Soc.*

[37] Yanovski JA, Cutler Jr. GB, Chrousos GP, Nieman LK. The dexamethasone-suppressed corticotropin-releasing hormone stimulation test differentiates mild Cushing's disease from normal physiology. J Clin Endocrinol Metab 1998;83:348−52.

[38] Erickson D, Natt N, Nippoldt T, Young Jr. WF, Carpenter PC, Petterson T, et al. Dexamethasone-suppressed corticotropin-releasing hormone stimulation test for diagnosis of mild hypercortisolism. J Clin Endocrinol Metab 2007;92:2972−6.

[39] Yanovski JA, Cutler Jr. GB, Chrousos GP, Nieman LK. Corticotropin-releasing hormone stimulation following low-dose dexamethasone administration. A new test to distinguish cushing's syndrome from Pseudo-Cushing's states. JAMA 1993;269:2232−8.

[40] Rollin GA, Costenaro F, Gerchman F, Rodrigues TC, Czepielewski MA. Evaluation of the ddavp test in the diagnosis of cushing's disease. Clin Endocrinol (Oxf) 2015;82:793−800.

[41] Moro M, Putignano P, Losa M, Invitti C, Maraschini C, Cavagnini F. The desmopressin test in the differential diagnosis between cushing's disease and pseudo-cushing states. J Clin Endocrinol Metab 2000;85:3569−74.

[42] Tirabassi G, Papa R, Faloia E, Boscaro M, Arnaldi G. Corticotrophin-releasing hormone and desmopressin tests in the differential diagnosis between cushing's disease and pseudo-cushing state: a comparative study. Clin Endocrinol (Oxf) 2011;75:666−72.

[43] Page-Wilson G, Freda PU, Jacobs TP, Khandji AG, Bruce JN, Foo ST, et al. Clinical utility of plasma pomc and agrp measurements in the differential diagnosis of acth-dependent Cushing's syndrome. J Clin Endocrinol Metab 2014;99:E1838−45.

[44] Grant P, Dworakowska D, Carroll P. Maximizing the accuracy of inferior petrosal sinus sampling: validation of the use of prolactin as a marker of pituitary venous effluent in the diagnosis of cushing's disease. Clin Endocrinol (Oxf) 2012;76:555−9.

[45] Carroll TB, Fisco AJ, Auchus RJ, Kennedy L, Findling JW. Paradoxical results after inadvertent use of cosyntropin [adrenocorticotropin hormone (1-24)] rather than acthrel (ovine corticotropin releasing hormone) during inferior petrosal sinus sampling. Endocr Pract 2014;20:646−9.

[46] Lacroix A, Bourdeau I, Lampron A, Mazzuco TL, Tremblay J, Hamet P. Aberrant g-protein coupled receptor expression in relation to adrenocortical overfunction. Clin Endocrinol (Oxf) 2010;73:1–15.

[47] Jung C, Ho JT, Torpy DJ, Rogers A, Doogue M, Lewis JG, et al. A longitudinal study of plasma and urinary cortisol in pregnancy and postpartum. J Clin Endocrinol Metab 2011;96:1533–40.

[48] Odagiri E, Demura R, Demura H, Suda T, Ishiwatari N, Abe Y, et al. The changes in plasma cortisol and urinary free cortisol by an overnight dexamethasone suppression test in patients with cushing's disease. Endocrinol Jpn 1988;35:795–802.

[49] Cardoso EM, Arregger AL, Budd D, Zucchini AE, Contreras LN. Dynamics of salivary cortisol in chronic kidney disease patients at stages 1 through 4. Clin Endocrinol (Oxf) 2016;85:313–19.

[50] Blondin MC, Beauregard H, Serri O. Iatrogenic cushing syndrome in patients receiving inhaled budesonide and itraconazole or ritonavir: two cases and literature review. Endocr Pract 2013;19:e138–41.

[51] Moalli PA, Pillay S, Weiner D, Leikin R, Rosen ST. A mechanism of resistance to glucocorticoids in multiple myeloma: transient expression of a truncated glucocorticoid receptor mrna. Blood 1992;79:213–22.

[52] Hinz LE, Kline GA, Dias VC. Addison's disease in evolution: an illustrative case and literature review. Endocr Pract 2014;20:e176–9.

[53] Bornstein SR, Allolio B, Arlt W, Barthel A, Don-Wauchope A, Hammer GD, et al. Diagnosis and treatment of primary adrenal insufficiency: an endocrine society clinical practice guideline. J Clin Endocrinol Metab 2016;101:364–89.

[54] Tordjman K, Jaffe A, Trostanetsky Y, Greenman Y, Limor R, Stern N. Low-dose (1 microgram) adrenocorticotrophin (acth) stimulation as a screening test for impaired hypothalamo-pituitary-adrenal axis function: Sensitivity, specificity and accuracy in comparison with the high-dose (250 microgram) test. Clin Endocrinol (Oxf) 2000;52:633–40.

[55] Arvat E, Di Vito L, Lanfranco F, Maccario M, Baffoni C, Rossetto R, et al. Stimulatory effect of adrenocorticotropin on cortisol, aldosterone, and dehydroepiandrosterone secretion in normal humans: dose-response study. J Clin Endocrinol Metab 2000;85:3141–6.

[56] Abraham SB, Abel BS, Sinaii N, Saverino E, Wade M, Nieman LK. Primary vs secondary adrenal insufficiency: acth-stimulated aldosterone diagnostic cut-off values by tandem mass spectrometry. Clin Endocrinol (Oxf) 2015;83:308–14.

[57] Limor R, Tordjman K, Marcus Y, Greenman Y, Osher E, Sofer Y, et al. Serum free cortisol as an ancillary tool in the interpretation of the low-dose 1-mug acth test. Clin Endocrinol (Oxf) 2011;75:294–300.

[58] Karpman MS, Neculau M, Dias VC, Kline GA. Defining adrenal status with salivary cortisol by gold-standard insulin hypoglycemia. Clin Biochem 2013;46:1442–6.

[59] Simsek Y, Karaca Z, Tanriverdi F, Unluhizarci K, Selcuklu A, Kelestimur F. A comparison of low-dose acth, glucagon stimulation and insulin tolerance test in patients with pituitary disorders. Clin Endocrinol (Oxf) 2015;82:45–52.

[60] Cho HY, Kim JH, Kim SW, Shin CS, Park KS, Jang HC, et al. Different cut-off values of the insulin tolerance test, the high-dose short synacthen test (250 mug) and the low-dose short synacthen test (1 mug) in assessing central adrenal insufficiency. Clin Endocrinol (Oxf) 2014;81:77–84.

[61] Hamrahian AH, Yuen KC, Gordon MB, Pulaski-Liebert KJ, Bena J, Biller BM. Revised gh and cortisol cut-points for the glucagon stimulation test in the evaluation of gh and hypothalamic-pituitary-adrenal axes in adults: results from a prospective randomized multicenter study. Pituitary 2016;19:332−41.

[62] Merke DP, Bornstein SR. Congenital adrenal hyperplasia. Lancet 2005; 365:2125−36.

[63] Lacey JM, Minutti CZ, Magera MJ, Tauscher AL, Casetta B, McCann M, et al. Improved specificity of newborn screening for congenital adrenal hyperplasia by second-tier steroid profiling using tandem mass spectrometry. Clin Chem 2004;50:621−5.

[64] Hamrahian AH, Oseni TS, Arafah BM. Measurements of serum free cortisol in critically ill patients. N Engl J Med 2004;350:1629−38.

[65] Torpy DJ, Bachmann AW, Grice JE, Fitzgerald SP, Phillips PJ, Whitworth JA, et al. Familial corticosteroid-binding globulin deficiency due to a novel null mutation: association with fatigue and relative hypotension. J Clin Endocrinol Metab 2001;86:3692−700.

[66] Nader N, Raverot G, Emptoz-Bonneton A, Dechaud H, Bonnay M, Baudin E, et al. Mitotane has an estrogenic effect on sex hormone-binding globulin and corticosteroid-binding globulin in humans. J Clin Endocrinol Metab 2006; 91:2165−70.

[67] Betterle C, Dal Pra C, Mantero F, Zanchetta R. Autoimmune adrenal insufficiency and autoimmune polyendocrine syndromes: autoantibodies, autoantigens, and their applicability in diagnosis and disease prediction. Endocr Rev 2002;23:327−64.

[68] Chen S, Sawicka J, Betterle C, Powell M, Prentice L, Volpato M, et al. Autoantibodies to steroidogenic enzymes in autoimmune polyglandular syndrome, Addison's disease, and premature ovarian failure. J Clin Endocrinol Metab 1996;81:1871−6.

[69] Riley WJ, Maclaren NK, Neufeld M. Adrenal autoantibodies and addison disease in insulin-dependent diabetes mellitus. J Pediatr 1980;97:191−5.

[70] Husebye ES, Allolio B, Arlt W, Badenhoop K, Bensing S, Betterle C, et al. Consensus statement on the diagnosis, treatment and follow-up of patients with primary adrenal insufficiency. J Intern Med 2014;275:104−15.

[71] Tanaka H, Perez MS, Powell M, Sanders JF, Sawicka J, Chen S, et al. Steroid 21-hydroxylase autoantibodies: measurements with a new immunoprecipitation assay. J Clin Endocrinol Metab 1997;82:1440−6.

[72] Holst JP, Soldin SJ, Tractenberg RE, Guo T, Kundra P, Verbalis JG, et al. Use of steroid profiles in determining the cause of adrenal insufficiency. Steroids 2007;72:71−84.

[73] Schimke KE, Greminger P, Brandle M. Secondary adrenal insufficiency due to opiate therapy—another differential diagnosis worth consideration. Exp Clin Endocrinol Diabetes 2009;117:649−51.

[74] Alvarez-Madrazo S, Padmanabhan S, Mayosi BM, Watkins H, Avery P, Wallace AM, et al. Familial and phenotypic associations of the aldosterone renin ratio. J Clin Endocrinol Metab 2009;94:4324−33.

[75] Shimamoto K, Ando K, Fujita T, Hasebe N, Higaki J, Horiuchi M, et al. The japanese society of hypertension guidelines for the management of hypertension (jsh 2014). Hypertens Res 2014;37:253−390.

[76] Montori VM, Schwartz GL, Chapman AB, Boerwinkle E, Turner ST. Validity of the aldosterone–renin ratio used to screen for primary aldosteronism. Mayo Clin Proc 2001;76:877–82.

[77] Funder JW, Carey RM, Mantero F, Murad MH, Reincke M, Shibata H, et al. The management of primary aldosteronism: case detection, diagnosis, and treatment: an endocrine society clinical practice guideline. J Clin Endocrinol Metab 2016;101:1889–916.

[78] Ahmed AH, Gordon RD, Taylor PJ, Ward G, Pimenta E, Stowasser M. Are women more at risk of false-positive primary aldosteronism screening and unnecessary suppression testing than men?. J Clin Endocrinol Metab 2011;96:E340–6.

[79] Rye P, Chin A, Pasieka J, So B, Harvey A, Kline G. Unadjusted plasma renin activity as a "first-look" test to decide upon further investigations for primary aldosteronism. J Clin Hypertens (Greenwich) 2015;17:541–6.

[80] Baudrand R, Guarda FJ, Torrey J, Williams G, Vaidya A. Dietary sodium restriction increases the risk of misinterpreting mild cases of primary aldosteronism. J Clin Endocrinol Metab 2016;101:3989–96.

[81] Ahmed AH, Gordon RD, Taylor PJ, Ward G, Pimenta E, Stowasser M. Effect of contraceptives on aldosterone/renin ratio may vary according to the components of contraceptive, renin assay method, and possibly route of administration. J Clin Endocrinol Metab 2011;96:1797–804.

[82] Jones JC, Carter GD, MacGregor GA. Interference by polar metabolites in a direct radioimmunoassay for plasma aldosterone. Ann Clin Biochem 1981;18:54–9.

[83] Rehan M, Raizman JE, Cavalier E, Don-Wauchope AC, Holmes DT. Laboratory challenges in primary aldosteronism screening and diagnosis. Clin Biochem 2015;48:377–87.

[84] Yin G, Zhang S, Yan L, Wu M, Xu M, Li F, et al. Effect of age on aldosterone/renin ratio (arr) and comparison of screening accuracy of arr plus elevated serum aldosterone concentration for primary aldosteronism screening in different age groups. Endocrine 2012;42:182–9.

[85] Williams B, MacDonald TM, Morant S, Webb DJ, Sever P, McInnes G, et al. Spironolactone versus placebo, bisoprolol, and doxazosin to determine the optimal treatment for drug-resistant hypertension (pathway-2): a randomised, double-blind, crossover trial. Lancet 2015;386:2059–68.

[86] Kline GA, Pasieka JL, Harvey A, So B, Dias VC. High-probability features of primary aldosteronism may obviate the need for confirmatory testing without increasing false-positive diagnoses. J Clin Hypertens (Greenwich) 2014;16:488–96.

[87] Kline GA, So B, Dias VC, Harvey A, Pasieka JL. Catheterization during adrenal vein sampling for primary aldosteronism: failure to use (1-24) acth may increase apparent failure rate. J Clin Hypertens (Greenwich) 2013;15:480–4.

[88] Kline GA, Chin A, So B, Harvey A, Pasieka JL. Defining contralateral adrenal suppression in primary aldosteronism: implications for diagnosis and outcome. Clin Endocrinol (Oxf) 2015;83:20–7.

[89] Kupers EM, Amar L, Raynaud A, Plouin PF, Steichen O. A clinical prediction score to diagnose unilateral primary aldosteronism. J Clin Endocrinol Metab 2012;97:3530–7.

Endocrine markers of diabetes and cardiovascular disease risk

CHAPTER 6

Erik Venos, MD[1] and Lawrence de Koning, PhD[2]

[1]Lecturer, Division of Endocrinology, Department of Medicine, Division of Endocrinology, University of Calgary, Calgary, Alberta, Canada

[2]Associate Professor, Department of Pathology and Laboratory Medicine, University of Calgary, and Clinical Biochemist, Calgary Laboratory Services, Alberta, Canada

CHAPTER OUTLINE

Endocrine Biomarkers. DOI: http://dx.doi.org/10.1016/B978-0-12-803412-5.00006-9

6.1 OVERVIEW/INTRODUCTION

Diabetes mellitus is defined by a persistent elevation in blood glucose, resulting from the reduction of insulin production by the beta cells of the pancreas (type 1), or the lack of physiologic response to insulin (type 2 and gestational diabetes mellitus (GDM)). Type 1 diabetes typically occurs in childhood and is characterized by an autoimmune response, culminating in the destruction of pancreatic beta cells. Persons with type 1 diabetes mellitus require insulin replacement therapy. Type 2 diabetes mellitus occurs mostly in adulthood, due to the presence of excess adipose tissue, resulting in resistance to insulin action. GDM is defined as elevated serum glucose values first occurring during pregnancy.

6.1.1 CLINICAL EPIDEMIOLOGY

- Diabetes mellitus is present 1 in 11 people worldwide, increasing to 1 in 10 by 2040 becoming the 7th leading cause of death worldwide [1].
- The prevalence of type 1 diabetes, comprising 10% of diabetes cases, is increasing globally at approximately 3% per year in children with 86,000 children developing this condition annually [2].
- Several environmental hypotheses have also been advanced to explain this increased prevalence including dietary changes, vitamin D status, or novel viruses, but none have been definitive [2].
- Type 2 diabetes is the most prevalent form of diabetes, accounting for 90% of cases. In the United States, 12%−14% of adults are affected but is steadily increasing—both in adults and children [3]. This is mainly due to an increasing prevalence of overweight and obesity.
- GDM is also increasingly common and is found in as much as 9% of pregnant women in the United States [3].
- Mature onset diabetes of the young (MODY) manifests in childhood or late adolescence and is generally includes monogenic disorders of pancreatic beta cell function. MODY makes up 1%−4% of pediatric diabetes [3].

KEY POINTS

Diabetes mellitus is one of the most common medical conditions. Its increasing prevalence is mostly due to increased type 2 diabetes from high rates of overweight and obesity. Type 1 diabetes is due to autoimmune conditions while GDM is due to the action of placental hormones.

6.1.2 PATHOGENESIS

6.1.2.1 Type 1 diabetes

- Type 1A diabetes is an autoimmune condition resulting in destruction of insulin-producing pancreatic beta cells by T-cell and B-cells over months or years [2].
- Type 1A diabetes, where autoantibodies are detected in the serum, comprises 70%—90% of type 1 diabetes. The remaining 10%—30% is considered type 1B diabetes, which does not include autoimmune involvement. Some of these patients may have unrecognized monogenic diabetes, which is caused by mutations in specific genes (e.g., HNF-4a) [2].
- Antigens targeted in type 1A diabetes include proinsulin, glutamic acid decarboxylase (GAD65), the tyrosine phosphatase-like protein IA-2, the islet-specific glucose-6-phosphatase catalytic subunit-related protein, and the cation efflux transporter (ZnT8) [2].
- There is a strong genetic component to type 1 diabetes. For example, human leukocyte antigen genotype (e.g., DQ8/DQ2) explains 40%—50% of inheritable type 1 diabetes risk [2].

6.1.2.2 Type 2 diabetes

- Type 2 diabetes is lifestyle related, where excess energy intake and decreased energy expenditure lead to adiposity, which causes insulin resistance and hyperglycemia. Core features also include oxidative stress, inflammation, and also beta cell dysfunction [4].
- Type 2 diabetes is linked to more than 60 genetic polymorphisms each associated with a small increase in risk [4].
- Pregnancy includes metabolic changes to provide adequate nutrition (i.e., glucose) to the developing fetus. While the exact mechanism is not known, placental hormones are likely responsible for antagonizing the action of insulin in mid-late gestation, leading to an increase in insulin resistance by 45%—70%—which can result in GDM [5].
- Risk factors for GDM include obesity, smoking, ethnic origin, polycystic ovarian syndrome, impaired fasting glucose (IFG) or glucose tolerance, history of fetal death, increased maternal age, and existing beta cell dysfunction [5].

> **KEY POINTS**
>
> Type 1 diabetes is characterized by an absolute insulin deficit (not enough made) whereas type 2 diabetes is characterized by a relative insulin deficit (not enough is usable). GDM is due to physiologic changes related to fetal nutrition.

6.1.3 CARDIOVASCULAR COMPLICATIONS OF DIABETES

- Diabetes induces the development of atherosclerotic plaques through sustained elevations of blood glucose, which causes (1) a characteristic atherogenic dyslipidemia, (2) endothelial damage and atherogenic modification of lipoproteins via glycation and glycoxidation reactions, (3) increased inflammation, and (4) smooth muscle cell proliferation [4].
- Microvascular complications include ischemic damage to the retina (retinopathy), the kidney (nephropathy), and the nervous system (neuropathy). Macrovascular complications include ischemic damage to the heart (myocardial infarction) and brain (stroke). Both can contribute to limb ischemia through peripheral artery disease, leading to amputation [6].
- For type 1 diabetes the relative risk for heart attack and stroke is 3−7 times higher than nondiabetics, with high risk in women versus men [6].
- Hemoglobin A1c (HbA1c) has a close relationship with diabetes-related complications. A 1% increase in glycosylated hemoglobin is associated with an increased risk of 11% for stroke, to 20% for peripheral arterial disease. Mortality risk increases with a 15% increase for every 1% increase in HbA1c [6].

> **KEY POINTS**
>
> Diabetes accelerates the development of atherosclerotic plaques which can lead to thrombosis, tissue ischemia, and necrosis.

6.1.4 DIAGNOSTIC OVERVIEW OF TYPE 1 AND TYPE 2 DIABETES

6.1.4.1 Biochemical definitions of diabetes mellitus

- Glucose detected in urine (glucosuria) by taste was considered the hallmark feature of diabetes. However, urine glucose does not become detectable until plasma glucose exceeds a value of ~10 mmol/L (180 mg/dL), which exceeds the capacity of the proximal convoluted tubule to reabsorb glucose.
- Glucosuria can be detected using semiquantitative dipstick urinalysis or a quantitative urine glucose test, though glucosuria is not used for diagnosis today as the test lacks sensitivity.

- As decreased insulin secretion (type 1) and suboptimal response to insulin (type 2) characterize diabetes, fasting, postload (e.g., 75 g glucose), and random plasma glucose concentrations are elevated in diabetes.
- Elevated HbA1c, a hemoglobin molecule modified by the nonenzymatic addition of a glucose molecule, may also be used to diagnose diabetes.
- Measurement of HbA1c yields a time-integrated measure of glycemic control and is measured in units of % glycated hemoglobin (GhB) versus total hemoglobin (National Glycohemoglobin Standardization Program (NGSP) units) or as mmol GhB/mol of total hemoglobin (International Federation of Clinical Chemistry (IFCC) units). HbA1c standardization was necessary to assure that all HbA1c assays identified the same part of the HbA1c molecule and reported results in similar units.
- Glucose cut points that define diabetes are derived from epidemiologic studies examining the relationship between plasma glucose, morbidity (e.g., micro- and macrovascular complications) and mortality but also controlled trials that examine the effects of lowering glucose past certain cut points.
- International discussions regarding the appropriateness of cut points occur regularly to address limitations of the data in which cut points are derived and to determine whether cut points are applicable in other regions with different ethnic compositions.
- Prediabetes is considered a risk factor for diabetes and is defined as IFG (5.6–6.9 mmol/L (100–125 mg/dL) or 6.1–6.9 mmol/L (110–125 mg/dL)), impaired glucose tolerance (IGT; 7.8–11.0 mmol/L (140–199 mg/dL)), or an HbA1c of 6.0%–6.4%.
- Diabetes is diagnosed at higher concentrations of plasma glucose (fasting $\geq$ 7.0 mmol/L (126 mg/dL); 2 h postload or random $\geq$ 11.1 mmol/L (200 mg/dL)) or an HbA1c of 6.5% or greater [3].
- To confirm a diagnosis, glucose or HbA1c elevations must be observed on two occasions, or on one occasion in the presence of clinical signs of hyperglycemia (e.g., polyuria (excess urination), polydipsia (thirst), polyphagia (hunger), and diabetic ketoacidosis (DKA)).
- For a random glucose elevation, confirmation with an alternate test is required.
- Type 1 diabetes is distinguished from type 2 diabetes by a family history of an autoimmune disorder, younger age of onset, rapidity of symptom onset, thinner body habitus, and evidence of autoantibodies.

6.1.4.2 Biochemical population screening for diabetes

- Population screening for diabetes may be done by (1) testing of all individuals above a specific age (e.g., 45 years), or (2) testing only individuals who possess a diabetes risk factor (e.g., family history of diabetes, obesity). Professional organizations such as the American Diabetes Association recognize the validity of both strategies and incorporate them into their testing guidelines.

- Population screening for type 1 diabetes is likely not cost-effective because of its low prevalence and because symptoms appear quickly after onset. Widespread autoantibody testing may also not be useful because of assay heterogeneity, and a lack of consensus on what to do after a positive autoantibody result.
- As type 2 diabetes can be mistaken for other conditions and appear silent for long periods of time, population wide screening for type 2 diabetes may identify many undiagnosed cases. However, screening is only useful if individuals identified by screening as having IFG or IGT do not progress to type 2 diabetes after treatment, or those identified as having type 2 diabetes do not develop morbidity or mortality when treated. At present, results from randomized trials do not fully support widespread population screening because patient outcomes have not significantly changed despite the added cost of screening [3].

KEY POINTS

Plasma glucose and HbA1c tests are used for diagnosis of type 1 and type 2 diabetes, which requires repeat testing in the absence of symptoms of hyperglycemia. Diagnostic cut points are defined as thresholds beyond which risk of morbidity and mortality significantly increase. Population screening for diabetes is currently being done; however, the cost-effectiveness of widespread screening is unclear.

6.2 SPECIAL CLINICAL SITUATIONS

6.2.1 OTHER TYPES OF DIABETES

- Individuals with hepatitis C may develop type 2 diabetes, which appears to be due to viral interference in insulin signaling.
- Conditions that affect exocrine pancreatic function (cystic fibrosis, pancreatitis, and hemochromatosis) can induce diabetes.
- Cushing's syndrome (hypercortisolism), glucagonoma (elevated glucagon), and pheochromocytoma (elevated catecholamines) increase the risk of diabetes.
- Medications can also induce diabetes: corticosteroids and antipsychotics affect feeding behavior and weight gain, whereas HIV protease inhibitors and immunosuppressants for solid organ transplantation can induce insulin resistance and type 2 diabetes.
- Monogenic diabetes (MODY) represents <5% of diabetes and is usually diagnosed before 25 years of age. The most common mutation associated with MODY results in decreased hepatocyte nuclear factor (HNF)-1a, which regulates insulin production. Other deficits include the enzyme glucokinase,

which converts glucose to glucose-6-phosphate, the metabolism of which stimulates insulin secretion by beta cells. Less common forms of MODY result from mutations in other transcription factors, including HNF-4a, HNF-1b, insulin promoter factor-1, and NeuroD1.

- Diabetes occurring in the first 6 months of life is often not related to autoimmune activity and is separately classified as neonatal diabetes [3].

> **KEY POINTS**
>
> Diabetes has many causes, including medications, endocrine conditions, and individual genetic variants. The clinician should look for clues of these conditions and test appropriately.

6.2.2 PREEXISTING DIABETES IN PREGNANCY

- Those with diabetes prior to pregnancy benefit from better glycemic control to reduce the risk of harm to the offspring. For women with type 1 diabetes the target HbA1c is <7%.
- Management of diabetes in pregnancy aims to achieve a glucose profile as normal as possible, realizing that the principal limitation of this approach is hypoglycemia.
- Patients with type 1 diabetes have higher rates of preeclampsia compared with those with type 2 diabetes. The preeclampsia rate can be as high as 50% for those with type 1 diabetes complicated by nephropathy.
- Preeclampsia is the leading cause of perinatal morbidity and mortality worldwide and is a future risk factor for maternal cardiovascular (CV) disease [7].

> **KEY POINTS**
>
> Diabetes before pregnancy is a serious condition that increases risk to both mother and infant. Type 1 diabetes is far more serious than type 2 in pregnancy.

6.2.3 GESTATIONAL DIABETES

- GDM is defined as glucose intolerance first recognized in pregnancy and is associated with increased risk of intrauterine death in the final 4−8 weeks of gestation.
- The hyperglycemia and adverse pregnancy outcomes (HAPOs) established a linear relationship between fasting, 1 h and 2 h post-75 g glucose, and adverse birth outcomes (birthweight >90% (macrosomia/large for gestational age),

primary cesarean section, neonatal hypoglycemia, and cord C-peptide levels >90%) [8].

- GDM is also associated with increased risk of maternal diabetes after pregnancy and may increase the risk of prediabetes and diabetes in offspring.
- Controlled trials indicate that treating GDM results in improved neonatal outcomes, including a reduced rate of shoulder dystocia and cesarean sections.
- Screening and diagnostic testing for GDM is best done between 24 and 28 weeks of gestational age, the time of peak insulin resistance.
- Initially, diagnostic criteria were designed to identify women deemed to be at risk for GDM from factors such as obesity, previous GDM, and advanced maternal age. This screening process missed some women with GDM. Guidelines now recognize that screening may be performed earlier if women are deemed to be at high risk, which includes belonging to specific ethnic groups.
- GDM can be diagnosed using a one-step or a two-step tolerance test procedure. In the one-step procedure the patient is required to fast and is given a 75-g glucose load. If glucose cut-offs are exceeded in either fasting, 1 h or 2 h plasma tests, a diagnosis of GDM can be made.
- In the two-step procedure a 1-h 50-g glucose tolerance test is first administered without fasting for screening purposes. GDM can be diagnosed by a sufficiently high 1-h plasma glucose (e.g., ≥ 11.1 mmol/L (200 mg/dL)). Patients with intermediate results (e.g., 7.8−11.0 mmol/L (140−199 mg/dL)) can then be given a 75 g or 100 g glucose tolerance test, which requires fasting. As with the one-step procedure, if glucose cut-offs are exceeded in fasting, 1 h or 2 h samples, a diagnosis of GDM can be made.
- In either procedure, diagnosis by a sufficiently high fasting value eliminates the need for the patient to undergo the tolerance test—which is unpleasant and exposes pregnant patients to a risk of vomiting and hyperglycemia. Stopping the tolerance test also eliminates unnecessary phlebotomy and time spent in an outpatient laboratory.
- Some guidelines (International Association of Diabetes and Pregnancy Study Groups (IADPSG), American Diabetes Association (ADA)) favor the one-step procedure for screening and diagnosis, whereas others recommend diagnostic testing via oral glucose tolerance only in the presence of risk factors (e.g., advanced age, obesity, previous glucose intolerance) (National Institutes of Health and Care Excellence (NICE), UK) or the two-step procedure (Canadian Diabetes Association (CDA)).
- Those in favor of population wide screening and diagnosis of all GDM cases likely advocate the one-step procedure as it performs the diagnostic test (oral glucose tolerance test) on all patients at risk. However, this procedure may be more costly than the two-step procedure and provide the same diagnostic power. This is likely due to the higher number of patients receiving the tolerance test versus the simpler and less expensive screening test.

- There is also controversy on what diagnostic cut points should be used. IADSPG guidelines favor cut points that convey an odds ratio of 1.75 for HAPOs whereas CDA guidelines favor cut points that convey an odds ratio of 2.0. Lower cut points associated with a lower odds ratio will result in more patients treated—and additional cost. However, because there is a linear relationship between glycemia and adverse birth outcomes, disagreement regarding the ideal procedure and cut points used is likely to continue [7].

> **KEY POINTS**
>
> GDM is considered to be diabetes newly diagnosed during pregnancy. GDM is associated with significant risk to both mother and offspring. Procedures and cut points for GDM screening and diagnosis are controversial and are decided based on sensitivity and specificity for adverse health outcomes, and cost to the healthcare system.

6.2.4 HYPOGLYCEMIC DISORDERS

6.2.4.1 Clinical presentation

- The cardinal symptoms of hypoglycemia are classified as both neuropenic and adrenergic.
- Symptoms include nausea, fatigue, lightheadedness, hunger, trembling, and ataxia as well as anxiety and depression. Severe hypoglycemia can cause unconsciousness and death.
- An important diagnostic component of hypoglycemia is Whipple's Triad, which includes low blood glucose, symptoms of hypoglycemia, and resolution of symptoms with correction of the low blood glucose.
- Hypoglycemic symptoms with normal plasma glucose suggest that symptoms are not due to a true hypoglycemic disorder.
- Both insulin use and hypoglycemic agents such as sulfonylureas increase the chance of symptomatic hypoglycemia in patients.
- Those with type 1 diabetes suffer more frequent and severe hypoglycemia episodes than those with type 2 diabetes. This difference is more pronounced in those receiving intensive insulin therapy. Persons with type 1 diabetes also exhibit impaired glucagon and epinephrine secretion in response to hypoglycemia, and those who experience frequent hypoglycemic events may develop hypoglycemic unawareness, which makes them more prone to severe hypoglycemia.

6.2.4.2 Nondiabetic hypoglycemia

- In a healthy individual with low glucose and high serum insulin levels, insulinoma or nesidioblastosis (high insulin secretion and hypoglycemia due from extremely active beta cells) may be suspected. In particular, nesidioblastosis may be related to changes in pancreatic beta cells secondary to gastric bypass surgery.

- Rarely, two forms of autoimmune hypoglycemia can occur. The first involves formation of insulin autoantibodies (IAAs), which change insulin kinetics and favor formation of highly reactive insulin complexes. The second involves formation of autoantibodies to the insulin receptor, which activate glucose uptake.

- Clinicians also need to be mindful of the possible surreptitious use of an insulin secretagogue such as a sulfonylurea as this may biochemically mimic an insulinoma.

- In a less well appearing individual, drugs such as pentamidine and quinidine can cause hypoglycemia. However, among oral medications, hypoglycemic agents are the most prevalent cause of hypoglycemia.

- Other important causes including hypoglycemia include heavy alcohol use (inhibition of gluconeogenesis and glycogen depletion), adrenal insufficiency (glucocorticoid deficiency), hepatic failure (decreased glucose production), renal failure (decreased insulin clearance and degradation, decreased renal gluconeogenesis, poor nutrition), sepsis (inflammatory state, insulin treatment), or a nonislet cell tumor such as a sarcoma-producing IGF2.

- Neonatal hypoglycemia is common because of low glycogen stores with risk factors including prematurity and maternal risk factors such as GDM [9].

KEY POINTS

Symptoms of hypoglycemia occur when plasma glucose is driven below the lower limit of normal by increased cellular uptake or decreased production. This can occur through exogenous insulin use, medications or disorders that increase insulin production or change insulin kinetics—such as in activation of the insulin receptor or modification of insulin by autoantibodies.

6.3 LABORATORY CONSIDERATIONS FOR THE ASSESSMENT OF DIABETES

6.3.1 PLASMA GLUCOSE

- Measuring plasma glucose is an essential part of the diagnostic workup for diabetes. While plasma and serum are often described interchangeably, serum has a lower (5%) glucose concentration than plasma. This is because serum is produced by allowing whole blood to clot for 30 min at room temperature. During this time, cells present in the specimen metabolize glucose.

- Conversely, plasma is produced from anticoagulated whole blood, which is centrifuged immediately. As preparing plasma does not require 30 min of clotting which can affect glucose results, plasma is the specimen of choice for glucose testing and will be referred to throughout this section.

- Blood samples for plasma testing can be collected after an overnight fast, a glucose load (e.g., 75 g) or completely at random.
- Plasma glucose measured in fasting samples is not affected by short-term variation in diet and can be used to diagnose diabetes. However, glucose measured after a tolerance test (glucose load) is generally regarded as a more relevant assessment because it evaluates a physiologic response to glucose.
- Plasma glucose tested in randomly collected samples is not the assessment of choice, and elevations compatible with diabetes should be confirmed using an alternate method [10].

6.3.1.1 Preanalytical issues

- Care must be taken to ensure that blood samples for glucose analysis are not allowed to sit unseparated for long. Glucose decreases approximately 5%−7% (0.3−0.6 mmol/L (5−10 mg/dL)) per hour in unseparated blood at room temperature due to cellular glycolysis.
- Unexpectedly low-values can therefore be encountered in a specimen that is not promptly separated, which is magnified by high red or high white cell count. High hematocrit can be encountered in neonates (polycythemia of the newborn), where hematocrit rises to 60% or higher due to factors such as intrauterine oxygen deprivation, intrauterine growth restriction, and maternal diabetes. A high white cell count can be encountered in patients with leukemia or infections.
- Drawing blood in gray-top tubes will limit the decrease in plasma glucose as the fluoride oxalate anticoagulant inhibits glycolysis.
- High glucose can sometimes be observed when blood is drawn from an intravenous line that has been used for a dextrose infusion and insufficiently rinsed. In these situations, blood for plasma glucose analysis should be drawn after rinsing the IV line with saline and drawing blood samples that are discarded [10].

6.3.1.2 Analytical issues

- Plasma glucose can be measured on many different core laboratory analyzers. Methods utilize hexokinase, glucose oxidase, or glucose dehydrogenase enzyme systems.
- The hexokinase method is the most common and detects glucose via generation of NADPH, which absorbs light at 340 nm.
- Interferences can occur when using hemolyzed, icteric, or lipemic samples or the presence of certain medications.
- Glucose oxidase methods are specific to glucose molecules and generate hydrogen peroxide that results in a colored indicator which is proportional to glucose concentrations [10].

6.3.1.3 Postanalytical issues

- It is most important that the reference interval (set of result values expected in healthy individuals) used is appropriate for the specimen type (e.g., plasma vs. whole blood vs. serum) and the patient conditions at the time of collection (fasting, tolerance testing, or random sampling).
- Tolerance test and random samples have higher upper limits of normal than do fasting samples.
- Age is also a consideration. Fasting glucose is lowest in the neonatal period and gradually increases across the lifespan as insulin resistance becomes more common. Reference intervals are therefore lower in neonates versus adults.
- Interestingly, because diagnosis of prediabetes and diabetes relies on medical decision points, the upper limit of normal for plasma glucose is usually ignored. However, the lower limit of normal is frequently used when evaluating possible hypoglycemia. However, hypoglycemia is not defined by a laboratory result alone as it must accompany symptoms of hypoglycemia [10].

KEY POINTS

Plasma glucose decreases with cellular metabolism, necessitating rapid separation of plasma from cells. This is why serum glucose is slightly lower than plasma glucose. While numerous testing methodologies exist for plasma glucose, the hexokinase method is the most common. Reference intervals for plasma glucose are not used like they are for other analytes. On the higher end of concentration they are replaced with medical decision points for diagnosis of prediabetes and diabetes. On the lower end, they are used to help diagnose hypoglycemia.

6.3.2 POINT OF CARE GLUCOSE TESTING

- Point of care testing (POCT) refers to testing in close proximity to the patient where a medical decision can be made immediately based on the result.
- POCT glucose testing employs small handheld units (glucose meters or glucometers) which use testing strips to analyze whole blood. This requirement eliminates the need for centrifugation, thereby decreasing time from specimen collection to result verification.
- Testing is simple (1- or 2-step testing process), fast (1−2 min per test), does not require a highly trained operator, uses integrated calibration and quality control, does not require refrigeration to maintain meters and reagents (e.g., test strips), and produces results similar to main laboratory analyzers.
- POCT glucose testing forms the basis of self-monitoring blood glucose in type 1 diabetics and should be performed multiple times daily.
- In patients with type 2 diabetes who do not take insulin the use of POCT testing should not be used frequently on the account of expense and the lack of clinical meaningful difference in outcomes compared to using measures such as the HbA1c [10].

KEY POINTS

Point of care glucose testing uses small portable analyzers (meters) to test whole blood. They are simple to operate and are mostly used for monitoring of patients with type 1 diabetes.

6.3.2.1 Preanalytical issues

- Whole blood glucose is approximately 10% lower than plasma glucose because of its lower water content. However, it can be as much as 25% higher than plasma glucose following a meal because capillary blood closely resembles arterial blood.
- Hematocrit has a complex effect on glucose meters. As red cells contain glucose (but at a lower concentration than plasma) an increase in hematocrit results in lower whole blood glucose. High hematocrit can also block sensors in glucose meters, resulting in low results. Some glucose meters automatically account for the effect of hematocrit.
- Patients on lipid emulsions can have falsely high or low results due to a scattering of light by lipid droplets on reflectance photometry-based assays. Triglycerides from lipid emulsions can also compress the water phase of a sample, resulting in low whole blood glucose.
- Low oxygen concentration (present at high altitudes) and low temperature can cause falsely high glucose results when using glucose oxidase electrode-based meters.
- Testing strips have a finite lifetime, and storing them at high temperature or humidity can damage them.
- If patient extremities are cold or have poor perfusion, glucose values can be falsely low as blood spends more time in contact with tissue.
- High uric acid, as found in patients with gout or tumor lysis syndrome, can be oxidized in glucose oxidase electrode-based methods leading to false high results.
- The presence of other sugars and sugar alcohols in the blood (galactose, xylose, and maltose) can cause false positives in glucose dehydrogenase-based meter assays.
- Several medications can cause interferences on glucose oxidase−based meters, including acetaminophen (false negative), L-DOPA (variable), and ascorbic acid (variable). Icodextrin, which is sometimes added to peritoneal dialysate fluids, can cause large positive interferences on glucose dehydrogenase-based meters.
- Highly viscous samples (e.g., hyperproteinemia, high ketones) can cause false low results [11].

KEY POINTS

Point of care glucose testing is affected by many interferences. Incorrect results can be obtained with extremes in hematocrit, oxygen, temperature, as well as high triglycerides, proteins, and uric acid—depending on the method. Several medications and other sugars can also cause interferences, but these are method dependent.

6.3.2.2 Analytical issues

- Glucose meters are required by the International Standards Organization (ISO) Standard 15197 to have a minimum accuracy such that 95% of results are ± 20% for results >4.2 mmol/L (76 mg/dL), and <0.83 mmol/L (15 mg/dL) for results below 4.2 mmol/L.
- The American Diabetes Association recommends an acceptable coefficient of variation of no >5%.
- Accuracy is affected by environmental factors and patient-related factors mentioned earlier.
- Approximately 50% of variation in results is due to operator error.
- The principal test methods for glucose meters include (1) glucose oxidase, (2) glucose dehydrogenase, and (3) hexokinase.
- Glucose oxidase–based methods can be based on either reflectance photometry or electrochemistry. In reflectance photometry-based assays a colored product is generated by a peroxidase reaction in the presence of glucose. In the electrochemical method the glucose is oxidized to gluconic acid with production of hydrogen peroxide, which causes either a peroxidase reaction to oxidase a color-bearing compound, or release of electrons that is detected amperometrically. In the glucose dehydrogenase reaction, glucose is dehydrated to form gluconolactone. The hexokinase method is identical to those in core laboratory analyzers, where glucose is phosphorylated to 6-phosphogluconate. Many of the limitations of these assays are discussed earlier.
- In general, glucose meters are unreliable at very high and very low glucose concentrations [11].

KEY POINTS

Glucose meters produce results that are usually within 20% of core laboratory glucose assays. Their imprecision should be close to 5%. Glucose meters operate by several analytical principles which are affected by different interferences. It is important to discuss the strengths and weaknesses of the meters in use with the laboratory.

6.3.2.3 Postanalytical issues

- Glucose meters do not have the same analytical performance as core laboratory chemistry analyzers; thus they should not be used for diagnosis.
- However, glucose meter results are appropriate for self-monitoring of glucose in diabetic patients.
- Glucose meters can also be used as screening devices to identify individuals with grossly high or low glucose values that may require confirmatory testing by better methods [11].

KEY POINTS

Glucose meters should not be used for diagnosis. However, they can be used for monitoring and as screening tools.

6.3.3 URINE GLUCOSE

- Measuring urine glucose is not required to diagnose diabetes, but it can be helpful to identify potential cases for further testing.

6.3.3.1 Preanalytical issues

- Urine specimens must be analyzed immediately or preserved and refrigerated to reduce the impact of bacterial metabolism. Urine glucose can decrease by 40% in specimens stored at room temperature for 24 h. Individuals with urinary tract infections will have more bacteria present, which increases the rate of glucose metabolism.
- Common preservation techniques include freezing but also addition of preservatives such as, glacial acetic acid, sodium benzoate, chlorhexidine, sodium nitrate, and benzethonium chloride. Preservatives can cause significant interferences with other tests so they should be chosen depending on what other tests may be ordered on the same specimen. For example, acids can cause precipitation of uric acid—making it impossible to measure this analyte. Salt-based preservatives can cause false high urine electrolyte concentrations [10].

KEY POINTS

Urine glucose is consumed by bacteria and therefore specimens must be analyzed promptly or preserved.

6.3.3.2 Analytical issues

- Quantitative measurements of urine glucose can be made using any of the plasma/serum tests provided they have been evaluated for use in urine. However, the quantitative glucose oxidase method should not be used for urine because it is susceptible to interferences from numerous substances in urine such as uric acid and antioxidants such as vitamin C.
- Qualitative or semiquantitative measurement of urine glucose can be made using Benedict's test for reducing sugars (e.g., glucose, fructose, lactose), or urine dipsticks containing glucose oxidase test pads.
- Dipstick tests are resistant to many interfering substances in urine but are still susceptible to oxidants (e.g., bleach, oxygen in air following exposure for

24 h), which can cause false positives, or reducing agents (e.g., vitamin C, ketones, salicylates), which can cause false negatives [10].

KEY POINTS

Urine glucose analysis is susceptible to numerous interferences which are more important if quantitative measurement is required. Oxidants or antioxidants can cause positive or negative results if the glucose oxidase method is used.

6.3.3.3 Postanalytical issues

- Reference intervals for urine glucose should have very low upper limits of normal as glucose is not normally present in urine.
- Urine glucose in 24 h collections should be calculated appropriately by multiplying glucose concentration in the sample by the total volume of urine collected.
- Qualitative and semiquantitative tests should report out results in "bands" appropriate to the calibration and precision of the test as well as need for appropriate clinical interpretation (e.g., positive/negative, $+$, $++$, $+++$, or concentration ranges) [10].

KEY POINTS

Urine glucose concentration is much lower than plasma glucose. Reporting must account for the type of testing (e.g., 24 h urine collection, or semiquantitative dipstick testing).

6.3.4 HEMOGLOBIN A1C

- HbA1c, often referred to as GhB, has become a popular diagnostic test for diabetes as it does not require fasting or ingestion of a glucose load and is relatively stable over time.
- Glycation of the hemoglobin molecule occurs via nonenzymatic addition of a glucose molecule to the N-terminal valine of the beta chain, which requires an Amadori rearrangement.
- Modification of hemoglobin from other chemicals can also occur from renal failure (carbamylation), lead poisoning, heavy alcohol intake, or salicylate poisoning (acetylated hemoglobin).
- HbA1c assess the relative proportion (% or mmol per mol) of hemoglobin that has a glycated beta chain, which increases with duration and extent of glycemia and the lifespan of red blood cells (mean 120 days).
- In individuals with normal red blood cell lifespans, HbA1c integrates changes in plasma glucose over the preceding 8−12 weeks. Changes in glucose over

the preceding month explain 50% of variation in HbA1c, whereas changes in the preceding 60−120 days explain only 25% of variation [12].

- Owing to the relatively stable nature of HbA1c, clinical practice guidelines almost universally recommend a retesting interval of 90 days even in patients who are experiencing changes in treatment or who have poorly controlled glucose levels [13].
- Retesting at 6 months is acceptable if treatment goals are met and glycemic control is steady.
- Some have argued for more frequent measurement of HbA1c among pregnant or women at child-bearing age (18−45 years) patients with diabetes because of early and sustained hypoglycemia, and because red blood cells have a shorter lifespan.
- In patients with type 1 diabetes and newly diagnosed type 2 diabetes, clinicians aim to target a HbA1c of <7% (53 mmol/mol) as these population have evidence that reducing glucose values lower than this level is associated with lower rates of macrovascular and microvascular complications [14].
- On the other hand, achieving this target in a patient with long-standing diabetes or abundant complications tends to be less effective in reducing events and may also lead to higher rates of side effects from therapy, such as hypoglycemia; avoiding symptomatic hyperglycemia, typically achieved with an HbA1c of 8% (64 mmol/mol), is a better option for this population [14].

KEY POINTS

Long-term glycemic control is reflected in the proportion of hemoglobin this is glycated to form HbA1c—which is associated with an increased risk of vascular disease. HbA1c is formed by nonenzymatic addition of glucose to the hemoglobin molecule. Owing to the rate of glycation, lifespan of red blood cells, and studies of clinical improvement according to testing frequency, numerous clinical practice guidelines recommend that retesting be done at 90 days. HbA1c is used as a target for therapy—with clinicians opting for lower values in patients newly discovered to have diabetes in order to minimize complications, and higher values in patients with long-standing diabetes or who are at higher risk of complications from treatment.

6.3.4.1 Preanalytical issues

- There are no special patient requirements for specimen collection for HbA1c. However, as HbA1c is measured from hemoglobin in red blood cells, unseparated anticoagulated whole blood (lavender top tube containing dipotassium EDTA) is required.
- Patients with reduced red blood cell lifespans such as those with hemolytic disease will have lower HbA1c than normal patients, whereas patients with iron deficient anemia often have high HbA1c due to red blood cells with longer lifespans.

- Recent transfusion can change HbA1c values, usually to much lower values than what would be expected in a diabetic patient.
- Thalassemias, hemoglobin variants (e.g., Hb J, Hb J-Baltimore, Hb F), can alter measurements of HbA1c; however, their impacts are assay specific and vary according to disease severity.
- Renal failure can lead to the creation of carbamylated hemoglobin, formed by reaction with urea, which can cause false positive results with some methods [10].

KEY POINTS

HbA1c is the test of choice for diagnosing and monitoring diabetes because it does not require fasting or consumption of a glucose drink. However, clinicians must be aware that disorders affecting red blood cell lifespan can increase or decrease HbA1c, changing its interpretation. Patients possessing hemoglobin variants, thalassemias, or renal failure may also present with altered HbA1c.

6.3.4.2 Analytical issues
- HbA1c can be measured using several methods that yield slightly different results.
- Some manufacturers have immunoassay analyzers with very high throughput that are ideal for large laboratories with very high HbA1c volumes.
- Common methods include ion exchange mini-columns, boronate affinity chromatography, high-performance liquid chromatography (HPLC), and immunoassays.
- Ion exchange mini-columns are usually temperature controlled and use a negatively charged cation-exchange resin to bind positively charged hemoglobin from patient samples. A buffer that makes GhBs less positive than other forms of hemoglobin is used to selectively elute HbA1c, which is often separated further using HPLC. Detection is by spectrophotometry.
- Ion exchange methods are susceptible to positive or negative interferences from hemoglobin variants (e.g., Hb F, Hb C, Hb S) and positive interferences from chemically modified forms of hemoglobin that elute at the same time as HbA1c [20].
- A labile precursor molecule to HbA1c may also interfere with ion exchange methods unless samples are chemically pretreated.
- Affinity chromatography separates HbA1c using boronic acid attached to a solid matrix in a column. When patient specimens are applied to the column, boronic acid binds to glucose molecules attached to hemoglobin. Once bound, nonglycated forms of hemoglobin can be washed away, after which GhB is eluted with sorbitol and detected by spectrophotometry.

- While affinity chromatography detects not just HbA1c but all glycated forms of hemoglobin, it is unaffected by labile precursor to HbA1c or other hemoglobin variants.
- Results from affinity chromatography are also reported in HbA1c equivalent values determined using more specific methods.
- HPLC is another popular method, which can be set up to be specific to only HbA1c. Separation is accomplished using cation-exchange columns and buffer gradients. HPLC analysis is also very precise, rapid, and requires small sample volumes.
- Immunoassays utilize antibodies specific to glucose and the first few amino acids of the hemoglobin beta chain. These assays are comparable to HPLC in that they are not generally affected by the labile precursor, many hemoglobin variants, or chemically modified hemoglobin molecules.
- Owing to the great variety in HbA1c assays the NGSP was initiated in the late 1990s to bring all results to be equivalent to those from the diabetes control and complications trial (DCCT), which associated HPLC-measured HbA1c to retinopathy and nephropathy.
- The NGSP helps manufacturers calibrate their GhB assays so that results are comparable to those produced in the DCCT. This program has greatly improved agreement between assays.
- The IFCC employed a different accuracy-based strategy, implementing a purified reference material and a pretreatment step combined with reference method of either HPLC with mass spectrometry or capillary electrophoresis.
- IFCC traceable methods report results in mmol/L HbA1c versus mol Hb.
- Even though the standardization has been responsible for increasing the agreement between HbA1c assays, minor differences are still expected.
- POCT is also available for HbA1c, which can accelerate clinical evaluation of the efficacy of interventions. This can help in the more rapid lowering of future HbA1c values in patients. These benefits must be balanced against the greater per test cost of POCT HbA1c versus those from core laboratory analyzers [15].
- POCT HbA1c testing is performed by similar, albeit miniaturized, methods as core laboratory analyzers; however, affinity chromatography and immunoassay are the most popular methods.
- As POCT assays use similar methods as core laboratory analyzers, they are susceptible to similar interferences.
- While POCT assays are certified under the NGSP, their performance is not the same as that of the core laboratory analyzers. Greater imprecision, expressed as a higher coefficient of variance on repeated testing of the same samples, is commonly observed with POCT HbA1c. Small proportional and constant biases are also seen relative to core laboratory analyzers.

> **KEY POINTS**
>
> Standardization programs such as the NGSP have brought HbA1c assays into closer agreement. Assay methods include affinity chromatography, ion exchange chromatography, immunoassays, HPLC, and capillary electrophoresis. POCT is also available.

6.3.4.3 Postanalytical issues

- Reference intervals defined as the central 95% of results from a healthy population are, in practice, rarely used for HbA1c. This is because a value of 6.5% or greater is considered diagnostic for type 2 diabetes, and that absolute change in HbA1c is important to monitor glycemic control in diabetes (a 1% increase in HbA1c = a 1.9 mmol/L (34 mg/dL) increase in long-term glycemic control).
- Use of diagnostic thresholds and change values in HbA1c interpretation is similar to how results from plasma glucose tests are used.
- As minor performance differences are expected between assays, a patient may receive slightly different results if tested at two separate laboratories. Discussions with clinical chemists regarding the assay performance may help to clarify why these differences have occurred [10].

> **KEY POINTS**
>
> HbA1c is interpreted by comparing individual results to diagnostic thresholds and evaluating the clinical significance of change across multiple results. Despite standardization of HbA1c assays, comparison of assays by clinical laboratories is needed to uncover minor differences in performance which can cause interpretational challenges.

6.3.5 INSULIN, C-PEPTIDE, PROINSULIN, GLUCAGON, AND SERUM KETONES

- Insulin and c-peptide are specialized tests that should not be used outside of very select clinical scenarios.
- Insulin measurement is best reserved for patients with fasting hypoglycemia, but it can also be used for evaluating insulin resistant women with polycystic ovarian syndrome or for classifying type of diabetes and directing appropriate treatment.
- A typical hypoglycemia protocol involves a supervised fast for up to 72 h. First the purpose of the test is to determine whether reported symptoms are truly explained by hypoglycemia (Whipple's Triad) and second, should hypoglycemia be proved, to determine the mechanism of disease. The fast is done to induce the symptoms of concern or until blood glucose drops to

2.5 mmol/L (45 mg/dL). During the test, blood is repeatedly drawn and tested for insulin, c-peptide proinsulin and glucose to see how well the body is handling low glucose. In patients with insulinoma, insulin and c-peptide are elevated (i.e., outside the reference interval) despite low glucose.

- C-peptide can be used to confirm whether hypoglycemia is caused by exogenous insulin. C-peptide is not found in exogenous-administered insulin and would therefore not be elevated after insulin injection, whereas insulin would be elevated.
- Glucagon levels are helpful for identifying glucagonoma, an alpha-cell tumor (although this presents as diabetes mellitus, not hypoglycemia).
- Proinsulin is most useful for identifying beta cell tumors as in many cases only proinsulin and not insulin and c-peptide are elevated. It can also be used to identify a familial form of hyperproinsulinemia, or to identify the degree of cross-reactivity of insulin assays.
- Plasma/serum ketones are normally measured as beta-hydroxybutyrate, which is present at six times greater than acetoacetate in diabetic ketoacidosis. Results for beta-hydroxybutyrate correlate better with acid–base status in diabetics than other ketones (acetoacetate, acetone) [10].
- During an observed hypoglycemic fast, if hyperinsulinism is the underlying cause, there will be no observed rise in beta-hydroxybutyrate and this may be a useful clue to the functional presence of insulin.

KEY POINTS

Insulin, c-peptide, proinsulin, and serum ketones are highly specialized tests that should not be ordered frequently.

6.3.5.1 Preanalytical issues

- An 8-h fast may be necessary to evaluate baseline levels of insulin, c-peptide, proinsulin, and glucagon as they all respond to glucose, unless a stimulation test requiring postprandial levels is required. Other technical requirements may be assay specific (e.g., minimum volume, anticoagulant, etc.).

6.3.5.2 Analytical issues

- Insulin immunoassays are notorious for poor performance. They have multiple problems including imprecision, cross-reactivity to proinsulin, poor agreement with other assays, and susceptibility to positive or negative interference by IAAs.
- As a result, each testing laboratory must extensively evaluate the assays they introduce, monitor their performance, and compare results to other assays using common calibrators.

- Insulin assays also have varying cross-reactivity with insulin analogs, which can cause false elevations when testing insulin. Consultation with the clinical chemist is important to determine the specificity of each assay.
- Proinsulin assays also suffer from some drawbacks, especially cross-reactivity with both c-peptide and insulin. They may also be affected by IAAs.
- While c-peptide assays are not affected by autoantibodies, they have problems with cross-reactivity and can produce very different results.
- Glucagon tests are frequently set up as radioimmunoassay. Owing to the poor performance and relatively large differences between assays, it is especially important for central laboratories to thoroughly evaluate them and document their ongoing performance. It is critical for clinicians to consult with clinical chemists when ordering these tests.
- Serum ketone assays are usually designed to be specific only to beta-hydroxybutyrate; however, some are sensitive to interferences from other ketones such as acetoacetate.
- The most common measurement principally utilizes the enzymes beta-hydroxybutyrate dehydrogenase and diaphorase to generate a colored compound.
- Ketones can be measured using the nitroprusside reaction in urine dipsticks; however, this reaction does not detect beta-hydroxybutyrate [10].

KEY POINTS

Assays for insulin, c-peptide, proinsulin, and glucagon have multiple problems that require close consultation with the laboratory to understand their performance. Plasma ketone assays are usually specific only for beta-hydroxybutyrate.

6.3.5.3 Postanalytical issues

- Owing to varying and poor performance for these assays, it is recommended that each laboratory establishes reference intervals based on their own patient populations.
- The recommended approach is to verify manufacturer's recommended reference intervals by using 20 or 40 specimens from relatively healthy individuals (not suffering from diabetes) who come from the laboratory's normal patient population. The same procedures should be followed each time a new method is implemented in the laboratory.
- If the new method is a "home-made" method developed in the laboratory, reference intervals from the literature can be verified by the new method.
- Alternatively, 120 randomly selected relatively healthy volunteers (from laboratory's usual patient populations) can be used to establish the first reference intervals for the new method. This reference interval should be monitored each year to ensure that it is a true representation of the "normal/ disease free" population.

- As some of these tests are run during stimulation tests (e.g., glucose tolerance test), a separate reference interval may be necessary [10].
- In many clinical scenarios, it is critical that several related tests (e.g., glucose, c-peptide, insulin) be interpreted as a whole, rather than as individual parts according to otherwise unrelated reference ranges. A "normal" insulin level in the setting of "normal range C-peptide" but concomitant hypoglycemia is not normal at all; it is inappropriate and indicative of pathologic endogenous hyperinsulinism.

6.3.6 AUTOIMMUNE MARKERS OF DIABETES

- Autoantibodies can be detected long before patients develop hyperglycemia; however, the presence of autoantibodies combined with clinical symptoms is important for confirming type 1 diabetes.
- Common autoantibodies include islet cell cytoplasmic antibodies (ICAs), IAAs, GAD65 antibodies, and antibodies directed against insulinoma-associated antigens (IA-2A, IA-2Aβ).
- Approximately 70% of patients with type 1 diabetes have 3 or 4 different autoantibodies.
- ICAs react with cytoplasmic proteins in pancreatic cells and are present in 75%−85% of patients with type 1 diabetes.
- IAAs have a high sensitivity for detecting type I diabetes in children and may predict the appearance of other autoantibodies in the future. IAAs may also develop in patients treated with insulin, but rarely exert an effect because they remain at low titers.
- GAD65 that is found in beta cells is one of the proteins contributing to ICA positivity.
- GAD65 autoantibodies are present in 70%−80% of patients with newly diagnosed type 1 diabetes, may help identify patients who will progress from type 2 to type 1 diabetes, and have the highest sensitivity for predicting the presence of other autoantibodies.
- Antibodies to IA-I2A and IA-2βA react with tyrosine phosphatases and are present in 32%−75% of newly diagnosed type 1 diabetes.
- The Diabetes Antibody Standardization Program (DASP) revealed that GAD65 had the highest sensitivity for type 1 diabetes [16].

KEY POINTS

Demonstrating the presence of autoantibodies is important to confirm the diagnosis of type 1 diabetes, of which GAD65 autoantibodies have the highest sensitivity.

6.3.6.1 Preanalytical issues

- There are very few patient and collection requirements for these assays other than technical requirements specified by the manufacturer (e.g., anticoagulant, minimum volume).
- Fasting is not generally required for these tests; however, several commercial assays recommend serum from red-top collection tubes to eliminate interferences from gel barriers.

> **KEY POINTS**
>
> Assays have different requirements for specimen collection, and so staff should consult with the laboratory to determine what is required.

6.3.6.2 Analytical issues

- ICAs are tested using a frozen section of human pancreas that is exposed to patient serum. Immunofluorescence microscopy is then used to visually detect the presence of antibodies binding to the tissue.
- This test requires considerable technical skill to run, and so assays and their results vary substantially from laboratory to laboratory.
- IAAs are generally tested using competitive radio ligand assays.
- Several immunoassay formats exist for measuring GAD65 autoantibodies, including ELISA and Western blot. Developing monoclonal antibody and recombinant GAD65 standards, resulting in several commercial assays, has reduced assay variability.
- ELISA- and radioimmunoassay-based methods are available for detecting IA-2A and IA-2βA [16].

> **KEY POINTS**
>
> ICAs have historically been tested using a manual process involving visual interpretation of stained tissue, whereas other autoantibodies are tested using immunometric assays. Monoclonal antibodies have helped to reduce assay variability.

6.3.6.3 Postanalytical issues

- The greatest issue in the analysis of autoantibodies is the lack of assay standardization, which means that specimens tested for the same autoantibody on different platforms may yield disparate results.
- While the Centers for Disease Control (CDC) is working towards the standardization of these assays via the DASP, this program is not yet widely

available and so clinicians must remain in close communication with testing laboratories to understand the performance and interpretation of these tests [16].

KEY POINTS

Owing to a lack of assay standardization, test results from different laboratories running different assays should be evaluated with the assistance of a clinical chemist.

6.3.7 URINE ALBUMIN EXCRETION (BIOMARKER FOR DIABETIC KIDNEY DISEASE)

- As albumin is too large to enter glomerular filtrate, elevated urinary albumin indicates microvascular damage to the glomeruli.
- Elevated urine albumin excretion is a major risk predictor for end-stage renal disease, occurring in 33% of persons with type 1 diabetes and represents a significant predictor for CV disease.
- Urine albumin excretion in combination with estimated glomerular filtration is used for staging of chronic kidney disease (CKD).
- Urine albumin concentration gradually rises in those with diabetes, becoming detectable by routine methods (e.g., dipstick urinalysis) in overt diabetic nephropathy (clinical albuminuria).
- Urine albumin below the detection limit of routine methods (e.g., dipstick urinalysis) but above what is considered normal (i.e., 30−300 mg/day) has in the past been termed *microalbuminuria.* This term refers to the low concentration of urine albumin, not a smaller version of the albumin molecule.
- In those with elevated (30−300 mg/day) urine albumin the risk of coronary heart disease is 50% higher than those with normal urine albumin and is two times greater in those that have clinical albuminuria (>300 mg/day) [17].
- The American Diabetes Association recommends annual quantitative urine albumin assessment in patients who have had type 1 diabetes for 5 years, or at diagnosis and yearly thereafter in those with type 2 diabetes [18].

KEY POINTS

Elevated urine albumin excretion is a marker of glomerular and microvascular damage and indicates increased risk for end-stage renal disease and CV disease. Quantitative urine albumin methods are needed for albumin assessment in type 1 diabetics because dipstick urinalysis is not sensitive enough to detect small but clinically meaningful elevations.

6.3.7.1 Preanalytical issues

- Random, first-morning void, overnight, hourly, or 24-h urine specimens may be collected depending on the test ordered.
- 24 h collections are difficult for patients to follow and challenging for the central laboratory to verify that they were collected adequately.
- First-morning void specimens are most appropriate for measuring albumin: creatinine ratio, which should be assessed on three separate days due to high biological variation. Protein preservatives such as sodium azide may be added to specimens for albumin testing; however, these can interfere with some assays.
- Urine albumin excretion may be increased by exercise, upright posture, and diuresis following an acute fluid load, pregnancy, urinary tract infections, acute illness, or surgery [10].

KEY POINTS

Several different urine collections are appropriate for albumin quantification; however, first-void specimens are best for measuring albumin:creatinine ratios. Albumin excretion may be temporarily increased by several conditions unrelated to diabetes, and clinicians should be aware of the presence of these conditions.

6.3.7.2 Analytical issues

- Urine albumin can be measured by semiquantitative dipstick assays; however, these are insufficiently sensitive to rule out the presence of renal disease and cannot correct for dilute urine.
- Quantitative immunoassays in several formats (e.g., RIA(radioimmunoassay), ELISA, immunoturbidimetry) are available and all have similar performance.
- Immunoturbidimetry is the most popular method as it is easily adapted for high volume testing in automated analyzers.
- Serum-based assays that use dye binding methods do not have sufficient sensitivity for urine protein analysis.
- Normalizing quantitative albumin concentration to urine creatinine in the albumin:creatinine ratio is a convenient method increasingly preferred because it accounts for urine dilution and flow rate [10].

KEY POINTS

The most popular method for measuring urine albumin is dipstick urinalysis; however, this does not have sufficient sensitivity to detect clinically important elevations in urine albumin among type 1 diabetes. Quantitative assays for high volume analysis are available and most use immunoturbidimetric methods. Albumin:creatinine ratio is the preferred measurement of urine albumin because it corrects for dilution.

6.3.7.3 Postanalytic issues

- Urine albumin excretion can be expressed as mass units per min (mcg/min), per 24 h (mg/24 h), or as a albumin:creatinine ratio (mg/g creatinine). Each has separate reference intervals, classifying individuals as normal, having increased albumin excretion or clinical albuminuria [18].
- The coefficient of variation in urinary albumin excretion is up to 40%, so small changes in this laboratory value should not immediately change clinical management.
- Kidney Disease Improving Global Outcomes (KDIGOs) guidelines define prognostically important stages of CKD as increasing from <30 mg/g (<3 mg/mmol), $30-300$ mg/g ($3-30$ mg/mmol), and >300 mg/g (>30 mg/mmol) creatinine.

> **KEY POINTS**
>
> Owing to different metrics for reporting urine albumin, attention should be paid to use of the appropriate reference interval.

6.3.8 FRUCTOSAMINE

- Fructosamine refers to glycated serum proteins, of which albumin is the most abundant.
- Fructosamine is formed by a similar process as HbA1c via nonenzymatic glycation of protein amino groups.
- The main advantage of this test is that it reflects more short-term ($2-3$ weeks) changes in glucose control than HbA1c because albumin has a much shorter half-life of ~ 20 days.
- Fructosamine may be useful in patients with hemoglobin variants that interfere with HbA1c assays, or diseases that affect red blood cell lifespan (e.g., thalassemia).
- At present, fructosamine is not used for diagnosis of diabetes; however, it is recognized as being useful in patients with the above conditions. As well, fructosamine may be helpful in patients undergoing multiple changes in treatment or who are pregnant [12].

> **KEY POINTS**
>
> Fructosamine is analogous to HbA1c except that it involves glycation of mostly albumin and integrates glycemic changes over a much shorter time span. Fructosamine may be useful in patients who have conditions which interfere with HbA1c measurement.

6.3.8.1 Preanalytical issues

- Any disease that chronically decreases the concentration of albumin can result in low fructosamine.
- Diseases that result in lower albumin include liver failure, nephrotic syndrome, analbuminemia, or acute inflammatory processes.
- Hyperthyroidism can also accelerate the catabolism of albumin and result in lower than expected fructosamine.
- Specimen collection and other requirements are assay specific. However, long-term storage of specimens may not be advisable because glycation reactions continue as glucose remains in contact with serum proteins [19].

> **KEY POINTS**
>
> Conditions that cause albumin loss or degradation invalidate the use of fructosamine assays.

6.3.8.2 Analytical issues

- Several different methods are available, which share similarities to HbA1c assays. These include affinity chromatography, HPLC, and those requiring chemical pretreatment of specimens to detect colored reaction products. These methods may yield different results and so discussion with the central laboratory regarding their performance is essential.

> **KEY POINTS**
>
> There are many kinds of fructosamine assays.

6.3.8.3 Postanalytical issues

- Reference intervals for fructosamine have been determined among healthy populations. However, the appropriateness of these intervals should be verified on the method in use by the central laboratory [20].
- Fructosamine levels have not been used in clinical studies testing their ability to diagnose diabetes or as a suitable marker for clinical decision making regarding glycemic control. Therefore clinical interpretation of a fructosamine level is often performed by making a comparison to the "expected" HbA1c since it is the clinical standard for management.
- Fructosamine is related to HbA1c by the following formula; however, the relationship is not perfectly linear ($R^2 = 0.78$):

$$\text{Fructosamine } (\mu mol/L) = (\%HbA1c\text{-}1.61)/0.017$$

KEY POINTS

Owing to variety in the types of fructosamine assays available, clinicians should discuss specifics and performance with the central laboratory. Fructosamine reference intervals should be validated by the laboratory and may be used to estimate the corresponding HbA1c.

6.4 LABORATORY CONSIDERATIONS FOR THE ASSESSMENT OF CARDIOVASCULAR DISEASE RISK

6.4.1 PRIMARY AND SECONDARY CLINICAL HYPERLIPIDEMIAS

6.4.1.1 Basic physiology of lipid metabolism

- Many primary (genetic) lipid disorders exist and involve isolated or combined elevations in cholesterol and triglycerides from specific lipoprotein particles, including low-density lipoproteins (LDLs), very low-density lipoproteins (VLDLs), intermediate-density lipoproteins (IDLs), or chylomicrons.

- The extrinsic lipid pathway is responsible for transport and processing of chylomicrons, which are produced during digestion of dietary fat. Briefly, triglyceride-rich chylomicrons, carrying surface apolipoprotein (Apo) A-I and apo B-48 are created by intestinal enterocytes and transported into the circulation. By interacting with HDL particles, surface apo A-I is exchanged for apo E, and apo C-II is transferred to the chylomicron. At the same time, membrane phospholipids and membrane free cholesterol are transferred to the HDL particle. Chylomicrons activate endothelial cell lipoprotein lipase with newly acquired apo C-II, which hydrolyzes triglycerides into fatty acids. These are taken up by adipocytes and myocytes after binding with albumin. The chylomicron remnant then interacts with hepatic receptors for apo B-48 and apo E, causing it to be internalized and broken down into lipid building blocks [21].

- The endogenous lipid pathway is responsible for regulating production and elimination of lipoprotein particles production by the liver. Briefly the liver packages large quantities of synthesized triglycerides and small amounts of cholesterol into VLDL particles, which carry surface apo B-100, apo C-II, and apo E. As with chylomicrons in the intrinsic pathway, VLDL particles interact with endothelial cell lipoprotein lipase via Apo C-II and release fatty acids into adipocytes and myocytes. This leaves behind VLDL remnants, which bind to hepatic receptors and are metabolized, but also IDL particles—some of which also bind to hepatic receptors and are metabolized. IDL particles also transfer apo C-II, apo E, membrane phospholipids, and membrane free cholesterol to HDL particles in exchange for cholesterol esters. This leaves a cholesterol-rich LDL particle carrying small amounts of triglycerides as well as surface apo B-100. The LDL particle interacts with hepatic receptors for apo B-100, causing it to be internalized and broken down into lipid building blocks. In addition to offloading cholesterol esters to LDL, HDL interacts directly with hepatic HDL and remnant receptors to transfer cholesterol esters to hepatocytes for further processing. Transfer of cholesterol esters from HDL back to the liver is called "reverse cholesterol transport" [21].

6.4.1.2 Clinical classification of common hyper- and hypolipidemias

- Fredrickson classifications of hyperlipidemias include:
  - Type I (high chylomicrons)
  - Type II (IIA: high LDL, IIB: high LDL and VLDL)
  - Type III (high IDL and chylomicron remnants)
  - Type IV (high VLDL)
  - Type V (high chylomicrons and VLDL) [22].
- Tangier disease is a heritable dyslipidemia specifically resulting in extremely low HDL.
- Plasma from patients with dyslipidemias involving elevated chylomicrons (I and V) have a cloudy triglyceride-rich upper layer (supernatant), whereas those with elevated VLDL can have a turbid lower layer (infranatant).
- The most common disorder is familial hypercholesterolemia (Fredrickson type II) associated with a mild elevation in LDL cholesterol (LDL-C; cholesterol from LDL particles) in the heterozygous form and marked elevation in LDL-C in the homozygous form. These conditions are most often due to mutations in the LDL receptor while other defects in the apo B-100 molecule or in PCSK 9 can be responsible. Polygenic forms of the condition also exist.
- The most common primary cause of elevated VLDL and LDL levels, manifesting as an elevated LDL-C and triglycerides, is familial combined hyperlipidemia, a polygenic condition (Fredrickson type IV).
- Deficits in lipoprotein lipase and apo C can lead to increases in chylomicrons and VLDL (Fredrickson types I and V).

6.4.1.3 Lipid disorders arising from other medical conditions (secondary dyslipidemia)

- Secondary etiologies of LDL-C elevation include hypothyroidism and nephrotic syndrome. Type 2 diabetes, obesity, cholestatic liver disease, glucocorticoids, and antipsychotic drugs can also give rise to this pattern.
- Elevations in triglycerides may be caused by type 2 diabetes, alcohol abuse, metabolic syndrome, high-dose beta blockers, and thiazide diuretics.
- The dyslipidemia resulting from type 2 diabetes is characterized by low HDL cholesterol (HDL-C) and high triglycerides, which is mostly due to increased HDL catabolism, increased VLDL production and decreased VLDL catabolism. Other changes also occur, including a shift towards small, dense, highly atherogenic LDL. The actual mechanism responsible for diabetic

dyslipidemias is not well understood, but is likely due to the relative and absolute insulin deficits associated with diabetes [23].

6.4.1.4 Clinical approach to patients with elevated low-density lipoprotein cholesterol

- People with a markedly elevated total cholesterol level (e.g., 7 mmol/L; 270 mg/dL) merit comprehensive CV risk reduction at an early age as despite their low 10-year risk estimated by CV risk calculators as their lifetime CV risk is markedly elevated.
- Modifiable risk factors such as smoking, hypertension, and obesity should be avoided where possible.
- The principle means of LDL-C lowering is by statin therapy, which enhance uptake of LDL-C by inhibiting the enzyme HMG-CoA(3-hydroxy-3-methyl-glutaryl-coenzyme A) in the liver, which increase LDL receptor expression on the liver [24].
- Other options for LDL-C lowering include inhibiting intestinal absorption of LDL-C by medications such as ezetimibe and binding resin or promoting the retention of the LDL receptor on the surface of the liver by using proprotein convertase subtilisin-kexin type 9 (PCSK9) inhibitors [25].

6.4.1.5 Clinical approach to patients with elevated triglycerides

- An important distinction in patients with elevated triglycerides is based on the degree of elevation.
- For patients with only mild elevations in triglyceride levels ($\leq$10 mmol/L; 886 mg/dL) a statin is preferred because of its ability to prevent CV disease.
- For marked elevations in triglycerides ($>$10 mmol/L; 886 mg/dL) the risk of pancreatitis increases, necessitating the use of fibrates, which promote triglyceride reduction by agonizing peroxisome proliferator-activated receptor alpha receptors, in addition to statins.
- Other therapies would include the use of pharmacologic doses of fish oil, a very low fat diet, and insulin therapy [26].

6.4.1.6 Approach to the patient with low high-density lipoprotein

- These patients are considered at higher risk of CV disease.
- The primary pharmacological therapy again includes statin therapy with the goal of lowering LDL-C as low as possible.
- Pharmacological therapies designed to increase total levels of HDL-C through pharmacologic doses of the vitamin niacin or through the use of cholesterol ester transport protein inhibitors have not proven successful, suggesting that HDL function and not HDL concentration may be the more important means of its HDL modification of CV risk [27].

KEY POINTS

Clinical dyslipidemias are caused by the accumulation of lipoprotein particles which can be due to either increased production or decreased clearance. The exogenous pathway is responsible for distributing dietary fat as triglyceride-rich chylomicrons throughout the body, whereas the endogenous pathway distributes synthesized cholesterol and triglycerides in VLDL particles which are gradually relieved of their triglyceride content. This leaves cholesterol-enriched particles, the most important of which is LDL. HDL particles participate in reverse cholesterol transport, where cholesterol esters from other lipoprotein particles and tissues are transferred to the liver. Statin therapy is the mainstay of lipid modifying therapy as it has the most clinical evidence. Newer therapies such as PCSK9 inhibitors hold promise in clinical trials. Therapies that primarily modify triglyceride or HDL levels have proven less successful in clinical trials.

6.4.2 CARDIOVASCULAR RISK ESTIMATION AND IMPACT OF DIABETES

- There are at least 100 clinical risk calculators for estimating CV risk in patients. However, many have not been fully validated across populations, nor do they consistently agree with each other. This leaves clinicians with some challenges in deciding which calculator is the best one for them to use.
- Of the calculators available the most well-known is the Framingham risk calculator. Based on associations of patient characteristics with 10-year risk of CV disease (e.g., coronary death, myocardial infarction, angina) in a long-term cohort study of healthy participants from Framingham, MA, Framingham risk is determined using a scoring system that awards higher points for high-risk characteristics such as advanced patient age, smoking, presence of diabetes, high blood pressure, decreased HDL-C, and increased total cholesterol. Other factors such as body mass index (BMI) have been added over time.
- The traditional way of translating findings from risk scores to practice has been to classify patients into low (<10%), moderate (10%−20%), and high (>20%) risk of CV disease and treating appropriately. That is, patients at highest risk receive aggressive pharmacotherapy and lifestyle modification whereas those at low risk receive no treatment or at the minimum diet and lifestyle modification.
- In recent years there has been substantial interest in treating apparently low- and intermediate-risk patients with low doses of statins and other CV therapeutics (e.g., beta blockers, diuretics) to further reduce lifetime risk of CV disease.
- There are approximately 45 CV risk calculators that account for type 2 diabetes. Given the very large CV risk associated with diabetes, one could argue from a prevention perspective that lifestyle and pharmacologic interventions should be directed to most patients as part of a process of shared informed decision making [28].
- The practical use of these calculators is discussed in Section 6.6.

> **KEY POINTS**
>
> There are numerous CV risk calculators that award higher points for the presence and degree of risk-related traits (e.g., diabetes) in patients. The total score calculated is proportional to the individual risk of CV disease or CV mortality. The Framingham risk score is one of the most well-known calculators and their use is endorsed by several clinical practice guidelines.

6.4.3 LIPID PANEL

- The lipid panel consists of LDL-C, HDL-C, total cholesterol, and triglycerides.
- Lipid measurement is essential for estimating CV risk not only because HDL-C and total cholesterol are components of the Framingham risk score, but also because LDL-C is the primary target for statin therapy and lifestyle interventions.
- LDL particles lodge under arterial endothelium to initiate self-perpetuating inflammatory lesions characterized by cholesterol-engorged macrophages called foam cells, which constitute the bulk of advanced atherosclerotic plaques.
- HDL particles are thought to slow or reverse atherogenesis because they remove cholesterol and transport it back to the liver [29].

> **KEY POINTS**
>
> LDL particles promote atherosclerosis whereas HDL particles protect against it. However, rather than measuring the number of particles, it is customary to measure the amount of cholesterol attributed to each particle type.

6.4.3.1 Preanalytical issues

- The most important historical requirement has been fasting. Recent consumption of dietary fat can elevate blood triglycerides and disrupt the relatively stable relationship between triglycerides and VLDL-C, which is used in the estimation of LDL-C. However, several studies now show that duration of fasting has a much smaller usual impact on triglycerides than originally thought.
- Recent food intake has little impact on total cholesterol and HDL-C except through hemodilution. As a result, fasting is increasingly viewed as unnecessary, except perhaps if nonfasting triglycerides are significantly elevated (e.g., 5 mmol/L, 443 mg/dL).

- While lipid levels can change due to factors such as acute illness, lipid lowering medications or pregnancy—these are not random fluctuations and can be taken into consideration during testing.
- Owing to biological variation in lipid levels, several measures have been recommended to assess "usual" or steady state values in patients, ideally 1 week apart.
- Anticoagulants in collection tubes can also affect lipids levels as well, for example citrate (royal blue top tube) and fluoride oxalate (grey top tube) anticoagulants cause water to leave cells and enter plasma, resulting in a dilution of approximately 10%.
- Biological variation in total cholesterol is 5%−10% whereas for triglycerides is 5%−25%—mainly because triglyceride levels are strongly affected by recent diet. However, triglycerides tend to peak in the morning regardless of recent diet.
- In women, cholesterol and triglycerides are highest at the midpoint of the menstrual cycle and are elevated in pregnancy and increase in menopause to be similar to that of men. Women generally have lower lipid results than men (except HDL) because of the effect of estrogen on lipid metabolism.
- LDL-C can decrease and HDL-C can increase after weight loss, long-term switch to diet rich in plant-based foods, or after aerobic exercise. Alcohol intake tends to increase HDL concentrations by increasing the synthesis and transport of its structural apolipoprotein (apo A) without changing the catabolic rate [30].

KEY POINTS

Cholesterol measures are relatively stable; however, triglyceride levels vary significantly with recent diet. Men generally have higher lipid concentrations than women; however, in both sexes lipids can be modified by lifestyle modifications and drug therapy.

6.4.3.2 Analytical issues

- Many methods exist to measure blood lipids. However, the most common utilize coupled enzyme systems to measure cholesterol and triglyceride concentration by generating colored products via oxidation reactions.
- Most are affected by the presence of interfering colored products or substances that can participate in redox reactions.
- Hemoglobin, for example, contains heme iron which has peroxidase activity can be oxidized from Ferrous to Ferric iron by an oxidant to methemoglobin. Vitamin C is an antioxidant that quenches reactions induced by oxidants.
- Indirect measurement of LDL-C involves chemical measurement of total cholesterol, HDL-C and triglycerides and then estimation of LDL-C using the

Friedewald equation (LDL-C (mmol/L) = total cholesterol (mmol/L)—HDL-C (mmol/L)—VLDL-C (triglycerides; mmol/L/2.17) OR LDL-C (mg/dL) = total cholesterol (mg/dL)—HDL-C (mg/dL)—VLDL-C (triglycerides; mg/dL/5)). The relatively stable relationship between triglycerides and VDL-C is disrupted when triglyceride concentrations are above 4.5 mmol/L (400 mg/dL).

- A popular chemical method to measure HDL-C involves the precipitation of positively charged non-HDL lipoproteins with negatively charged polyanions (e.g., dextran sulfate) and divalent cations (e.g., Mg^{2+}). Precipitated lipoproteins can then be removed by centrifugation or magnetic beads, leaving only HDL particles which can be tested for HDL-C using a cholesterol assay.
- HDL-C can also be measured directly using homogeneous assays that protect non-HDL lipoproteins from detection by the cholesterol assay with antibodies, polymers, or detergents. The main advantage of these assays is that, unlike precipitation methods, non-HDL particles do not need to be physically removed.
- Polyanion precipitation is also used in beta quantification, an indirect ultracentrifugation-based method which separates plasma and serum into lipoprotein fractions based on density. In this method, low-density chylomicrons and VLDL particles float to the top of the specimen, whereas the lower layers become enriched by IDL, HDL, LDL, and also Lp(a). Total cholesterol measured in the lower layers followed with treatment with polyanions allows for HDL-C and LDL-C to be separately measured (LDL = total cholesterol$_{lower\ layer}$ − HDL-C), whereas VLDL-C is measured by testing total cholesterol in the entire specimen, and subtracting total cholesterol in the lower layer (VLDL = total cholesterol$_{specimen}$ − total cholesterol$_{lower\ layer}$).
- Beta quantification has frequently been used to determine LDL-C in patients with high triglycerides. As with HDL-C, LDL-C can also be measured using direct methods. These include antibody capture of non-LDL particles followed by testing for LDL-C using a cholesterol assay, or using components that block or solubilize non-LDL lipoproteins in order to mask them from the cholesterol assay. These homogeneous assays generally meet performance targets set by the national cholesterol education program (NCEP); however, the specificity and performance of each assay should still be assessed prior to clinical use.
- Standardization of lipid assays was achieved in the 1990s by having reagent manufacturers participate in the cholesterol reference method laboratory network (CRMLN), which set assay calibrator values to agree with results from reference methods for lipids.
- Non-HDL-C (i.e., total cholesterol—HDL-C) can be easily determined by any of these methods, representing cholesterol from atherogenic and potentially atherogenic apo-B-100 containing lipoproteins (LDL, IDL, VLDL, Lp(a)).

For these reasons, non-HDL-C is more strongly associated with the development of CV disease than LDL-C.

- As calculating non-HDL-C is computationally simple, laboratories should implement automatic calculation of this measure in their laboratory information system.
- Nuclear magnetic resonance spectroscopy can also be used to measure the number and size of lipoprotein particles. While this is a flexible research tool, it has yet to evolve to accommodate high volume analysis in a clinical laboratory.
- In patients with intrahepatic biliary obstruction a lack of intrahepatic bile flow causes lipoproteins and bile to back up and enter plasma, reacting with serum proteins to form modified LDL particle called lipoprotein X. These patients present with a falsely elevated LDL-C concentration. Lipoprotein electrophoresis is needed to distinguish between the lipoprotein X band and unmodified LDL particles. Lipoprotein X does not express apo B and is not considered to be atherogenic [21].

> **KEY POINTS**
>
> In most laboratories, LDL-C is calculated indirectly using the Friedwald formula. This formula does not work well when triglyceride concentrations are high, which may necessitate the use of direct assays or other methods. Cholesterol and triglyceride assays have been standardized for many years and have very low imprecision. Non-HDL-C is increasingly recognized as an easy-to-calculate lipid measure that is strongly associated with CV risk.

6.4.3.3 Postanalytical issues

- Interpretation of lipid results is governed by clinical practice guidelines for the management of hyperlipidemias such as those from American College of Cardiology and Canadian Cardiovascular Society, therapeutic decision points rather than reference intervals are used.
- However, premenopausal women have lower total cholesterol and LDL-C but higher HDL-C than men due to the role of estrogen. Regardless of sex, total cholesterol, LDL-C, and triglycerides increase with age, whereas HDL decreases with age [21].
- Lipid reference intervals also exist for pediatric patients; however, there is great controversy on whether these should be used for general population screening or CV risk assessment of children with a family history of hyperlipidemias. Some guidelines (e.g., Canadian Clinical Practice Guidelines on the Management and Prevention of Obesity Clinical Guidelines) recommend testing of children only who are overweight or obese.

KEY POINTS

Like with HbA1c, lipid levels are interpreted according to medical decision points rather reference intervals. In general, women have healthier lipid profiles than men; however, this changes at menopause.

6.4.4 APOLIPOPROTEINS

- In addition to measuring lipoprotein cholesterol, surface proteins called apolipoproteins can be measured to estimate lipoprotein particle number, which can provide further information about CV risk.
- Apo B refers to apolipoprotein B-100, which is synthesized in the liver and found on LDL, IDL, VDL, and Lp(a) particles. Only one apo B molecule is found on these lipoproteins, which means that their concentration is equivalent to a particle count.
- The function of apo B is to interact with LDL receptors to facilitate cholesterol transport into cells. Apo B has been endorsed as a superior marker than LDL-C as it is not affected by recent food intake and provides a direct count of the number of potentially atherogenic particles [21].
- However, a similar argument could be made regarding non-HDL-C. Only when the cholesterol content of apo B containing particles varies could apo B and non-HDL have variant associations with CV risk. This is best illustrated in the highly atherogenic small dense LDL phenotype typical of diabetes. In this phenotype, apo B particles are elevated relative to LDL-C.
- Several prospective cohort studies have found that apo B has stronger independent associations with CV risk than non-HDL-C. However, other studies have not, including a meta-analysis of randomized trials of lipid lowering therapy examining CV risk according to changes in apo B and non-HDL-C.
- Both apo B and non-HDL-C are potential targets for lipid lowering therapy.
- Apo A-I is the major apolipoprotein found on HDL particles. Unlike apo B, more than one apo A-I is present on HDL which means that a measurement of Apo-AI concentration is not necessarily equivalent to a particle count.
- Apo A-I is a cofactor for lecithin cholesterol acyltransferase, which esterifies cholesterol as it is transported out of cells into HDL particles.
- Apo A-I concentration may be more strongly associated with CV risk than HDL-C. In a meta-analysis of randomized trials of lipid lowering therapy, increased apo A-I but not HDL was associated with decreased risk of CV disease. This may be due to the unique biochemical role of apo A-I [31].

KEY POINTS

Apolipoproteins can be measured instead of the cholesterol contained within lipoprotein particles. There is a controversy as to whether these are superior measures of risk compared with traditional lipid measures. Measuring apolipoproteins in combination with cholesterol subfractions may improve detection of the atherogenic small dense LDL particles.

6.4.4.1 Preanalytical issues

- There are few variables that must be rigidly controlled for in apolipoprotein testing. Fasting is not required, as both apo B and apo A-I are not present on chylomicrons, and apo B and A-I containing lipoproteins are not significantly affected by food intake.
- The only other conditions that must be met are assay specific, such as minimum specimen volumes, anticoagulants, and patient medications. For example, some laboratories maintain the traditional requirement of serum from a red-top or yellow-top serum separator tube, whereas others require plasma from a lavender top (dipotassium EDTA) tube.

> **KEY POINTS**
>
> Fasting is not required for apolipoprotein testing.

6.4.4.2 Analytical issues

- Apo B and apo A-I are generally measured using immunoassay techniques.
- Reference materials provided by the IFCC, approved by the World Health Organization (WHO), and distributed through the CDC have significantly improved the agreement between different manufacturers of apo B and A-I assays.

> **KEY POINTS**
>
> Apolipoproteins are measured using different methods than components of the lipid panel. These are frequently immunoassays.

6.4.4.3 Postanalytical issues

- While reference intervals and therapy decision points are available for apo B and apo A-I, interpretation of results is not routine because of disagreement on how to use these tests. Clinical practice guidelines in Canada now suggest apo B as an alternative target for therapy [32].
- Given that calculated levels of LDL-C cannot be determined with moderate-to-severe hypertriglyceridemia, measurement of apo B in this setting may provide the clinician with an alternate way of knowing the atherogenic profile that accompanies the hypertriglyceridemia.

> **KEY POINTS**
>
> At best, apolipoproteins currently remain an alternative treatment target for most patients.

6.4.5 LIPOPROTEIN (A)

- Lipoprotein(a) (Lp(a)) is a LDL-like particle that is highly variable in size. It contains apo B-100 that is covalently bound to the plasminogen-like glycoprotein apolipoprotein(a).
- Discovered in 1963, its biological effects are uncertain.
- Lp(a) is both atherogenic and prothrombotic, and testing for it may be warranted in certain high-risk patients.
- A meta-analysis of prospective cohort studies found that Lp(a) was significantly associated with an increased risk of CV disease, which was independent of other lipid measures such as LDL-C.
- Incorporating Lp(a) into routine risk profiles has proved challenging as there are few therapies that can reduce Lp(a) other than niacin.
- Lp(a) is generally resistant to changes in diet or lifestyle and other therapies tend to lower LDL-C as well as Lp(a), making it difficult to determine the independent impact of lowering Lp(a). However, Lp(a) may be successfully lowered using monoclonal antibodies to PCSK9 [21].

KEY POINTS

Lp(a) is an LDL-related particle bound to a plasminogen-like glycoprotein. It represents an independent risk factor for CV disease.

6.4.5.1 Preanalytical issues

- Lp(a) shares similar requirements with apo B and apo A-1. Fasting does not appear to be required for Lp(a), but is occasionally recommended if a standard lipid panel is ordered.
- Other preanalytical requirements are assay specific.

KEY POINTS

Preanalytical requirements for Lp(a) are similar to those of apo B.

6.4.5.2 Analytical issues

- Lp(a) immunoassays assays are highly variable partly due to the structural heterogeneity and variation in size of Lp(a).
- The apo(a) molecule is comprised of numerous kringle-type polypeptide domains, some of which are repeated numerous times. Multiple possible antigenic sites exist for binding with antibodies, with stable positions being preferred.

- It is important to check with the testing laboratory, which sites are targeted and whether monoclonal or polyclonal antibodies are used in the assay in order to make valid comparisons between results from different laboratories.
- While the IFCC has developed a primary reference material to help manufacturers create assays that agree with each other, this has not been completely successful [33].

KEY POINTS

Lp(a) immunoassay assays can be highly variable depending on what antigenic site antibodies are directed against.

6.4.5.3 Postanalytical issues

- Lp(a) can be reported either as mass of Lp(a) particles or Lp(a) protein per unit volume. If antibodies are directed against an invariant, single antigen of the molecule, immunoassays can report a particle count in SI units (e.g., nmol/L).
- There is some ethnic variation in Lp(a) levels, with African Americans having much higher Lp(a) levels than Caucasians, which historically has not been correlated to excess risk in this group [21].

KEY POINTS

Lp(a) can be reported in either mass or SI units.

6.4.6 HIGH-SENSITIVITY C-REACTIVE PROTEIN

- C-reactive protein (CRP) is a pentameric protein produced by the liver as part of an acute phase response.
- First identified in 1930s, CRP was found to bind the bacterial cell wall C-polysaccharide of *Streptococcus pneumoniae* although it was later discovered to bind and agglutinate numerous other organisms as well.
- While CRP is a sensitive marker to inflammatory processes, especially infection, desire to detect subclinical inflammatory processes led to the development of high-sensitivity C-reactive protein (hsCRP) assays. These assays have a limit of detection one order of magnitude lower than normal assays (e.g., <0.3 mg/L) [34].
- The concentration of hsCRP is associated with development of CV disease and type 2 diabetes. In the Justification for the Use of Statins in Primary Prevention: an Intervention Trial Evaluating Rosuvastatin (JUPITER),

individuals who achieved both a reduction of hsCRP and LDL-C experienced greater benefit than those who only achieved a reduction in LDL-C [21].

- This and other findings have prompted some clinical practice guidelines to include hsCRP as a secondary therapeutic target.
- However, controversy remains whether hsCRP is a causal factor in the development of these diseases or is simply a marker of existing inflammation attributed to diabetes and atherosclerosis.

KEY POINTS

hsCRP is marker of subclinical inflammation but is likely not a causal factor in CV disease.

6.4.6.1 Preanalytical issues

- hsCRP can be elevated by numerous patient conditions, many of which are unrelated to inflammation associated with diabetes and CV disease.
- Any tissue damage, infection, or autoimmune disease can elevate hsCRP. Other factors associated with elevated hsCRP include advanced age, female sex, obesity, smoking, and hormone replacement therapy in women.
- Antiinflammatory medications, statins, and antioxidant drugs can reduce hsCRP. Provided patients are in good health, hsCRP levels do not vary significantly. They are not affected by diet, posture, or circadian rhythms. However for cardiac risk assessment, two hsCRP measurements taken 2 weeks apart are recommended.

KEY POINTS

While hsCRP is more specific to inflammation than other markers of inflammation (e.g., erythrocyte sedimentation rate), it can still be elevated by numerous factors unrelated to the underlying inflammatory process in atherosclerosis.

6.4.6.2 Analytical issues

- There are now dozens of hsCRP assays in production. Most use an enhanced immunoturbidimetry principle to detect hsCRP; however, assays differ substantially in their performance, which has resulted in the launch of a CDC-sponsored standardization program using a certified reference material to assign assay calibrator values.
- As with all assays, it is important to check with the central laboratory to understand how well it performs (e.g., sensitivity, specificity, reporting limits, interferences, etc.).

> **KEY POINTS**
>
> Most hsCRP assays use immunoturbidimetry principles. A standardization program has been launched to help assays agree with each other.

6.4.6.3 Postanalytical issues

- Expected hsCRP values have been determined in the United States population, from which values associated with low (<1 mg/L), average (1–3 mg/L), and high risk (>3 mg/L) have been determined. These correspond with tertile ranges of hsCRP.
- Those in the high-risk group have roughly a 2× higher risk of developing CV disease than those in the low risk group.

> **KEY POINTS**
>
> The upper limit of normal for hsCRP corresponds to the top tertile of hsCRP values.

6.4.7 HOMOCYSTEINE

- Homocysteine is a sulfur-containing amino acid and a risk factor for CV disease.
- This association was first made in the 1960s through careful observation of young patients with homocystinuria, a metabolic disease characterized by deficiencies in cystathionine beta synthase and plasma total homocysteine concentrations >100 μmol/L. These patients frequently die of atherothrombotic disease [35].
- Smaller elevations of homocysteine occur in deficiencies of vitamin cofactors (e.g., folic acid, vitamins B6 and B12) used for disposal of homocysteine, and may increase the risk of CV disease in otherwise healthy individuals [28].
- Testing for homocysteine may be important in patients who develop CV disease before the age of 40, or those who are high risk of CVD.
- Fortification of flours with folic acid in North America was successful in lowering average homocysteine levels.

> **KEY POINTS**
>
> Homocysteine is a risk factor for CV disease and testing for it may be revealing in patients with premature CV disease. Flour fortification with folic acid has lowered population wide-levels of homocysteine, lessening the importance of this risk factor.

6.4.7.1 Preanalytical issues

- Patients are frequently required to fast prior to testing because homocysteine can be affected by amino acid intake.
- In blood, homocysteine is present as protein-bound, reduced or oxidized species which individually are unstable. This issue was solved in commercial assays by converting all species to the reduced form (total homocysteine).
- Glycolysis will artificially increase homocysteine, which can be reduced by rapid blood collection and separation of cells from plasma/serum using refrigerated centrifuges.
- Addition of glycolysis inhibitors (e.g., fluoride oxalate anticoagulant) will also decrease *in vitro* generation of homocysteine.
- Total homocysteine is relatively stable within individuals, and a single measurement is a good representation of average homocysteine over at least 1 month [28].

KEY POINTS

Homocysteine is not a stable analyte and so specimens collected for it must be centrifuged promptly.

6.4.7.2 Analytical issues

- HPLC-, immunoassays-, and gas chromatography-based methods are available.
- Most yield similar results; however, assays have not technically been standardized. Newborn screening programs generally use tandem mass spectrometry to identify neonatal cases of homocystinuria by testing blood spots for methionine, which is elevated in patients with homocystinuria [36].
- HPLC was the gold standard for this analyte and was used in several clinical trials. LC−MS/MS is now the gold standard and other methods should be evaluated against it.

KEY POINTS

As several assays exist for homocysteine, it is important to check with the laboratory to understand the performance of the assay.

6.4.7.3 Postanalytical issues

- Both adult and pediatric reference intervals are available for fasting total homocysteine.
- Males generally have higher homocysteine levels than females, and children have lower levels than adults.

> **KEY POINTS**
>
> As with several CV risk factors, males have higher levels of homocysteine than women.

6.5 COMMON ERRORS IN LABORATORY TESTING WITH DIABETES AND CARDIOVASCULAR RISK

- Not checking the patient's medication list, which could contain drugs that interfere with test results (e.g., sulfonylurea and insulin levels).
- Asking subject to fast (for glucose, for example) prior to laboratory testing where the fasting result will not change the management or will cause those taking insulin to experience unnecessary hypoglycemia.
- Drawing blood from an insufficiently flushed intravenous line that was recently used for dextrose infusion (may cause false high glucose).
- Allowing whole blood specimens to sit at room temperature for too long before testing for plasma glucose (may cause false low glucose).
- Making a diabetes diagnosis using a glucose meter.
- Using a glucose meter that does not correct for hematocrit on a neonate.
- Comparing patient results to a reference interval when the standard practice is to compare patient results to diagnostic thresholds (e.g., HbA1c).
- Not confirming a random plasma glucose test in the diabetic range with a second test when symptoms are absent.
- Not following up a random glucose result in the diagnostic range for diabetes with a different test (e.g., fasting glucose).
- Giving a glucose drink unnecessarily after a diabetes diagnosis has been made by a fasting test.
- Adding an electrolyte order on a urine specimen containing salt-based preservatives.
- Not verifying the level of completeness of a 24-h urine collection using volume collected and creatinine normalized values.
- Using a urine dipstick to test for the presence ketones in diabetic ketoacidosis (will not detect beta-hydroxybutyrate).
- Calculating LDL using the Friedewald formula when triglycerides are >4.5 mmol/L (>400 mg/dL).
- Routinely testing for esoteric markers of CV risk such as apolipoproteins, Lp(a), and homocysteine instead of the lipid panel in average-risk patients.
- Ordering HbA1c too close to the last test (e.g., <90 days), or among those who have disorders affecting red blood cell lifespan.
- Ordering plasma insulin without accompanying c-peptide, or not ordering either on presentation of hypoglycemia.
- Ordering autoantibody tests in patients without confirmed type 1 diabetes.

6.6 CASE STUDIES

6.6.1 DIABETES DIAGNOSIS

A 23-year-old previously healthy man presents to the emergency room with an episode of the flu. On further questioning, he reports 3 months history of a 10-pound weight loss where he was not obese before; excessive urination, voiding five times per night; along with the consumption of 5 L of water per day. He has a history of autoimmune hypothyroidism treated with levo-thyroxine. There is no family history of type 1 diabetes. On examination, his vital signs are within normal limits with the exception of a temperature of 38.3°C. He has signs of mild volume depletion and no suggestion of insulin resistance such as acanthosis nigracans. POCT glucose testing indicates a random glucose of 16.5 mmol/L.

Question 1: What diagnosis should be explored? What testing should be performed in the emergency room and what treatment should be offered?

Answer: The previous history of another autoimmune condition; the cardinal symptoms of diabetes such as weight loss, polyuria, and polydipsia; and elevated glucose well above normal should lead to the clinician to conclude that the most likely diagnosis is type 1 diabetes. A physiologic increase in glucose would not be expected to occur at this level without a large deficit in insulin secretion. Type 2 diabetes would be suggested against as the patient is both young and not obese with any signs of insulin resistance. The absence of a strong family history argues against a diagnosis of monogenic diabetes.

Laboratory testing indicates an anion gap of 20, a plasma glucose value of 20 mmol/L, and both glucosuria and ketonuria on urinalysis. A provisional diagnosis of type 1 diabetes is made and 10 units of regular insulin along with 1 L of normal saline are administered in the emergency room. The patient's serum glucose value subsequently becomes 12 mmol/L, his anion gap returns to normal, and he is discharged from the hospital with subcutaneous insulin.

Question 2: As part of this patient's initial follow-up visit and routine care, what testing should be arranged?

Answer: Soon after, he should receive comprehensive diabetes education and be initiated on a multiple daily injection regimen of insulin. He will be given a glucometer to allow him to monitor his blood glucose four times per day. He should go the laboratory at least 2 h after eating to ensure whether his glucometer is accurate. In this case, his glucometer reads 8.5 mmol/L from a capillary sample while his venous plasma glucose from the central laboratory is 9 mmol/L. This represents a difference of 5.5%, which meets the accuracy standard and the patient can continue to use the meter with retesting recommended every 1−2 years or if there is concern regarding accuracy of the meter.

GAD65 antibody testing can be ordered if there is diagnostic uncertainty regarding type 2 diabetes or monogenic diabetes as a positive test result has high specificity for type 1 diabetes. In this case the patient requested that this test be performed. The result is significantly elevated at three times the upper limit of normal range, confirming a diagnosis of type 1 diabetes. Testing for hypothyroidism (present in up to 10% of patients) is recommended annually while other autoimmune conditions are recommended to be evaluated should symptoms or laboratory abnormalities (such as hyperkalemia in adrenal insufficiency) arise.

As part of routine care for his type 1 diabetes, a lipid panel was ordered. Absolute risk of a CV episode was calculated as "low" given his young age and short duration of type 1 diabetes. In this case, lipid panel results were within healthy limits. Serum creatinine revealed no obvious kidney abnormalities. The development of proteinuria in type 1 diabetes typically occurs 5 years after diagnosis, and so testing is not required at diagnosis. After 5 years an albumin to creatinine ratio is recommended annually.

HbA1c will be ordered every 3–6 months, allowing him and his diabetes care team to determine his long-term diabetes control. In this case, it was initially 11.4% (101.1 mmol/mol), which declined to 7.2% (55.2 mmol/mol) after 3 months of treatment. He and his care team will aim for a value of 7% (53 mmol/mol) or less, as this range is associated with lower microvascular complications over 10 years or more.

This target value would be modified upward in clinical situations where advanced complications of diabetes or cognitive impairment are present or where frequent hypoglycemic episodes have resulted in hypoglycemia unawareness. In the latter case the goal of a higher HbA1c would be to allow for the patient to regain hypoglycemia awareness by avoiding hypoglycemia. In this case an HbA1c up to 8.0% may be permitted, to allow room for lower rates of hypoglycemia.

6.6.2 CARDIOVASCULAR RISK ASSESSMENT AND TREATMENT

A 64-year-old woman of South Asian heritage presents to her primary care provider for annual examination. She has an 8-year history of type 2 diabetes, which was preceded by a diagnosis of impaired fasting glucose. Her only current medication is metformin, which is administered at a dose of 1000 mg twice daily. Her 56-year-old sister has recently had a myocardial infarction and she is interested in discussing her CV risk.

Question 1: What other features from her clinical history should be sought to determine her risk for CV disease?

Answer: CV risk estimation can be divided into modifiable and nonmodifiable risk factors. Nonmodifiable risk factors include age, gender, ethnic background, and family history. Modifiable risk factors include hypertension, diabetes, cholesterol, obesity, and smoking.

With this in mind, it is noted she is a nonsmoker but has significant family history (defined as a male or a female first degree family member having a CV event before the ages of 55 and 65, respectively). Her BMI is 31 kg/m^2 (obese) with a blood pressure of 150/95 mmHg. Laboratory testing in anticipation for the current appointment was the following: total cholesterol = 5.4 mmol/L (209 mg/dL), LDL-C = 3.3 mmol/L (128 mg/dL), HDL-C 1 mmol/L (39 mg/dL), and triglycerides of 3 mmol/L (266 mg/dL). This was done in a nonfasting state. Her serum electrolytes and creatinine were within normal limits.

Her albumin:creatinine ratio was 2 mg/mmol and HbA1c was 6.8% (51 mmol/mol). An ECG did not suggest a prior myocardial infarction. Ordering novel markers of CV risk was decided against as they would not change immediate management of this patient.

Question 2: How is clinical laboratory information used to determine this patient's CV risk?

Answer: Several CV risk calculators can give a reasonable estimate of CV risk, allowing shared informed decision making about risk reduction behaviors and medication usage. The American College of Cardiology (ACC)/American Heart Association (AHA) risk calculator places her 10-year risk of a cardiac event at approximately 24%. A low dose statin is initiated, which may reduce her risk of a CV event by 25%. In addition a blood pressure lowering medication is prescribed, which may reduce her risk of a CV event by up to 50%. She is informed that her metformin may also reduce her CV risk. All medications are tolerated well.

Question 3: What further laboratory testing should be ordered to manage her CV risk?

ACC guidelines suggest that that serial cholesterol levels may not be needed to monitor the success of therapy. There may be a role in measuring serial cholesterol levels based on patient preference or if concerns exist regarding adherence. In this case, her statin medication was expected to lower her total cholesterol by approximately 50% and follow-up testing indicates values of total cholesterol was 2.8 mmol/L (108 mg/dL), LDL-C = 1.8 mmol/L (70 mg/dL), HDL-C = 1.1 mmol/L (43 mg/dL), and triglycerides = 1.8 mmol/L (159 mg/dL). She remained free of CV disease after 3 years of treatment.

REFERENCES

[1] International Diabetes Federation. About Diabetes; 2015. Available from: http://www. idf.org/about-diabetes [accessed 01.02.16].
[2] Atkinson MA, Eisenbarth GS, Michels AW. Type 1 diabetes. Lancet 2014;383:69–82.

[3] American Diabetes A. 2. Classification and diagnosis of diabetes. Diabetes Care 2016;39(Suppl. 1):S13−22.

[4] Mazzone T, Chait A, Plutzky J. Cardiovascular disease risk in type 2 diabetes mellitus: insights from mechanistic studies. Lancet 2008;371:1800−9.

[5] Chen P, Wang S, Ji J, Ge A, Chen C, Zhu Y, et al. Risk factors and management of gestational diabetes. Cell Biochem Biophys 2015;71:689−94.

[6] de Ferranti SD, de Boer IH, Fonseca V, Fox CS, Golden SH, Lavie CJ, et al. Type 1 diabetes mellitus and cardiovascular disease: a scientific statement from the American Heart Association and American Diabetes Association. Diabetes Care 2014;37:2843−63.

[7] American Diabetes A. 12. Management of Diabetes in Pregnancy. Diabetes Care 2016;39(Suppl. 1):S94−8.

[8] Metzger BE, Lowe LP, Dyer AR, Trimble ER, Chaovarindr U, Coustan DR. Hyperglycemia and adverse pregnancy outcomes. N Engl J Med 2008;358 1991−2002.

[9] Cryer PE, Axelrod L, Grossman AB, Heller SR, Montori VM, Seaquist ER, et al. Evaluation and management of adult hypoglycemic disorders: an Endocrine Society Clinical Practice Guideline. J Clin Endocrinol Metab 2009;94:709−28.

[10] Sacks DB. Carbohydrates. In: Burtis CA, Ashwood ER, Bruns DE, editors. Tietz textbook of clinical chemistry and molecular diagnostics. 4th ed. St. Louis: Elsevier Saunders; 2006.

[11] Ginsberg BH. Factors affecting blood glucose monitoring: sources of errors in measurement. J Diabetes Sci Technol 2009;3:903−13.

[12] Goldstein DE, Little RR, Lorenz RA, Malone JI, Nathan DM, Peterson CM, et al. Tests of glycemia in diabetes. Diabetes Care 2004;27(Suppl. 1):S91−3.

[13] Canadian Agency for Drugs and Technologies in Health. HbA1c testing frequency: a review of the clinical evidence and guidelines. Rapid response report: Summary with critical appraisal; Ottawa, Sept 26 2014:1−21.

[14] Rodriguez-Gutierrez R, Montori VM. Glycemic control for patients with type 2 diabetes mellitus: our evolving faith in the face of evidence. Circ Cardiovasc Qual Outcomes 2016;9:504−12.

[15] Whitley HP, Yong EV, Rasinen C. Selecting an A1C point-of-care instrument. Diabetes Spectr 2015;28:201−8.

[16] Bingley PJ. Clinical applications of diabetes antibody testing. J Clin Endocrinol Metab 2010;95:25−33.

[17] Bakris GL, Molitch M. Microalbuminuria as a risk predictor in diabetes: the continuing saga. Diabetes Care 2014;37:867−75.

[18] American Diabetes A. 9. Microvascular complications and foot care. Diabetes Care. 2016;39(Suppl. 1):S72−80.

[19] Balland M, Schiele F, Henny J. Effect of a 6-month storage on human serum fructosamine concentration. Clin Chim Acta: Int J Clin Chem 1994;230:105−7.

[20] Cohen RM, Holmes YR, Chenier TC, Joiner CH. Discordance between HbA1c and fructosamine: evidence for a glycosylation gap and its relation to diabetic nephropathy. Diabetes Care 2003;26:163−7.

[21] Rifai N, Warnick GR. Lipids, lipoproteints, apolipoproteins and other cardiovascualr risk factors. In: Burtis CA, Ashwood ER, Bruns DE, editors. Tietz textbook of clinical chemistry and molecular diagnostics. St. Louis: Elsevier Saunders; 2006.

[22] Winter WE, Harris NS. Lipoprotein disorders. Contemporary practice in clinical chemistry [Internet]. Washington, DC: AACC Press; 2006. p. 251−8, Chapter 21.

[23] Verges B. Pathophysiology of diabetic dyslipidaemia: where are we? Diabetologia 2015;58:886−99.

[24] Sniderman A, McQueen M, Contois J, Williams K, Furberg CD. Why is non-high-density lipoprotein cholesterol a better marker of the risk of vascular disease than low-density lipoprotein cholesterol?. J Clin Lipidol 2010;4:152−5.

[25] Everett BM, Smith RJ, Hiatt WR. Reducing LDL with PCSK9 inhibitors—the clinical benefit of lipid drugs. New Engl J Med 2015;373:1588−91.

[26] Berglund L, Brunzell JD, Goldberg AC, Goldberg IJ, Sacks F, Murad MH, et al. Evaluation and treatment of hypertriglyceridemia: an Endocrine Society clinical practice guideline. J Clin Endocrinol Metab 2012;97:2969−89.

[27] Siddiqi HK, Kiss D, Rader D. HDL-cholesterol and cardiovascular disease: rethinking our approach. Curr Opin Cardiol 2015;30:536−42.

[28] Garg UC, Zheng ZJ, Folsom AR, Moyer YS, Tsai MY, McGovern P, et al. Short-term and long-term variability of plasma homocysteine measurement. Clin Chem 1997;43:141−5.

[29] Allan GM, Garrison S, McCormack J. Comparison of cardiovascular disease risk calculators. Curr Opin Lipidol 2014;25:254−65.

[30] Hansson GK. Inflammation, atherosclerosis, and coronary artery disease. New Engl J Med 2005;352:1685−95.

[31] Sniderman A, Couture P, de Graaf J. Diagnosis and treatment of apolipoprotein B dyslipoproteinemias. Nat Rev Endocrinol 2010;6:335−46.

[32] Walldius G, Jungner I. Apolipoprotein A-I versus HDL cholesterol in the prediction of risk for myocardial infarction and stroke. Curr Opin Cardiol 2007;22:359−67.

[33] Kronenberg F. Lipoprotein(a): there's life in the old dog yet. Circulation 2014;129:619−21.

[34] Guyton JR, Dahlen GH, Patsch W, Kautz JA, Gotto Jr AM. Relationship of plasma lipoprotein Lp(a) levels to race and to apolipoprotein B. Arteriosclerosis 1985;5:265−72.

[35] Ridker PM, Danielson E, Fonseca FA, Genest J, Gotto Jr AM, Kastelein JJ, et al. Reduction in C-reactive protein and LDL cholesterol and cardiovascular event rates after initiation of rosuvastatin: a prospective study of the JUPITER trial. Lancet 2009;373:1175−82.

[36] McCully KS. Vascular pathology of homocysteinemia: implications for the pathogenesis of arteriosclerosis. Am J Pathol 1969;56:111−28.

Pituitary disorders

7

Bernard Corenblum, MD[1] and Ethan A. Flynn, MD[2]

[1]*Professor, Department of Medicine, Division of Endocrinology*
University of Calgary, Calgary, Alberta, Canada
[2]*Associate Professor, Department of Pathology and Laboratory Medicine,*
University of Calgary, and Section Chief of General Pathology,
Calgary Laboratory Services, Calgary, Alberta, Canada

CHAPTER OUTLINE

Endocrine Biomarkers. DOI: http://dx.doi.org/10.1016/B978-0-12-803412-5.00007-0

7.1 OVERVIEW OF PITUITARY PROBLEMS

The pituitary gland consists of an anterior lobe and posterior lobe. Each has its unique hormone production/release, and unique anatomical and physiological control mechanisms. Every pathogenetic cause may affect this gland, in isolation or as part of systemic disease. Pituitary disease may result in hyposecretion of one or more hormones, hypersecretion of one (and less likely more than one) hormone, mass effects resulting in neurological concerns, or any combination of these.

7.1.1 DESCRIPTION

The pituitary gland, also called hypophysis, weighs about 1 g. Hypophysis means "under the brain," where it lies outside the blood brain barrier. It is separated into two lobes (anterior and posterior) with different embryological origins and resulting different anatomical relations. It sits in the bony sella turcica, above the sphenoid sinus, the venous cavernous sinus is on each side, and it is below the optic chiasm and hypothalamus. Each cavernous sinus contains cranial nerves III, IV, VI, V1, and V2, as well as the internal carotid artery. Knowledge of the surrounding anatomy is important for understanding the various clinical presentations that arise from pituitary tumors.

7.1.2 CLINICAL EPIDEMIOLOGY

As with all endocrine glands the two basic groups of pituitary disorders are decreased function (hypopituitarism) and tumor formation, with tumor formation the most common cause of hypopituitarism. A tumor may secrete one or more pituitary hormones.

- Surveillance studies report the annual incidence of new hypopituitarism to be 4.2 cases per 100,000 population, with a prevalence rate of 45 cases per 100,000 [1]. It is thought by most clinicians that these reported rates very much underestimate the incidence, and with a higher index of suspicion and laboratory testing, the clinical diagnosis rate is higher. The incidence may be increasing with greater recognition of the effects of traumatic brain injury and cranial radiation in causing abnormal pituitary function.
- Pituitary adenomas are almost always benign, but can be invasive, and rarely malignant. They are arbitrarily divided into microadenomas (<1 cm) and macroadenomas (>1 cm).
- Large cancer registries list pituitary adenomas as 230 per 100,000 population [2]. Most clinically identified (symptomatic) pituitary adenomas are managed by endocrinologists, do not go to neurosurgery, and may go unreported to tumor registries. This is especially true with prolactinomas, the most common secretory tumor.
- Intensive screening for specific pituitary disease in populations at risk finds a higher incidence rate [2]. For example, acromegaly is more common in a population of patients with type 2 diabetes, compared to the general population.
- A small percentage of pituitary adenomas are associated with familial-related syndromes, such as multiple endocrine neoplasia type 1 (MEN1), Carney's complex, and familial-isolated pituitary adenoma. Described genetic alterations in pituitary adenomas include somatic activating mutations in the GNAS gene found in 35% of sporadic growth hormone–secreting adenomas [2]. MEN1 and CDKN1B germ line mutations occur in MEN1 and MEN4, respectively [2].
- AIP (aryl hydrocarbon receptor-interacting protein) functions as a tumor suppressor gene. AIP gene mutations are found in some families with familial-isolated pituitary adenomas and in younger patients with GH- or prolactin-secreting macroadenomas that are large and invasive, require more surgery, and are often poorly responsive to somatostatin agonists [2]. AIP mutations are more likely in familial pituitary tumors, any tumor found before age 18, and any macroadenoma found before age 30 [3].
- Multiple autopsy series have found the incidence of pituitary adenomas to be about 20% [2]. Most are small microadenomas. The widespread use of MRI (and CT to a lesser extent) does find unsuspected changes suggestive of pituitary adenomas in 10%–20% (incidentalomas), usually microadenomas [2].

7.1.3 BASIC PITUITARY PHYSIOLOGY

The anterior lobe (adenohypophysis) has five cell types secreting six separate peptide hormones. All but prolactin have their secretion controlled by hypothalamic stimulatory and/or inhibitory hormones and by feedback inhibition from target gland hormone secretion. Hypothalamic control of secretion by the adenohypophysis is via the vascular route from the portal vein.

The anterior pituitary hormones form five separate control and secretory systems that are best looked at as separate entities. The clinical and laboratory approach is to take each one by itself, the whole being a combination of the parts.

7.1.3.1 Prolactin

- Prolactin is under tonic dopamine inhibition, which gives unique characteristics and clinical applications (see Section 7.2). There is no target gland hormonal feedback, but prolactin does have a short-loop feedback upon itself.
- Prolactin is needed for lactation, but otherwise its other functions are minor and vestigial. It is not generally needed, so its secretion is under inhibition by hypothalamic dopamine. There is a short-loop feedback to its own hypothalamic control. There may be yet to be determined secretory factor(s) to facilitate secretion when needed, such as the response to suckling. The pituitary gland doubles in size during a normal pregnancy due to hyperplasia of the prolactin-secreting cells. This is from increased estrogen secretion and it is to prepare the breast for lactation.

7.1.3.2 Adrenocorticotrophic hormone

- Adrenocorticotropic hormone (ACTH) stimulates all three layers of the adrenal cortex, but mainly the cortisol-secreting layer, which enters the feedback control onto the hypothalamus and the pituitary corticotroph cell that secretes ACTH. The major hypothalamic stimulatory control of ACTH is by the hypothalamic hormone corticotropin-releasing hormone (CRH), and to a lesser degree by the posterior pituitary hormone arginine vasopressin (AVP). Higher centers in the central nervous system modulate this feedback control and result in a diurnal variation to secretion (highest upon awakening), and marked increase in ACTH-cortisol secretion under various forms of stress.

7.1.3.3 Thyroid-stimulating hormone

- Thyroid-stimulating hormone (TSH) activates all steps in thyroxine synthesis and secretion from the thyroid gland. Hypothalamic stimulatory control is with thyrotropin-releasing hormone (TRH) with feedback control by the thyroid hormones T4 (levothyroxine) and T3 (triiodothyronine) at the hypothalamus as well as modification of TRH action at the thyrotroph cell of the pituitary.

7.1.3.4 Gonadotrophic (sex-steroid stimulating) hormones

- Luteinizing hormone (LH) and follicle-stimulating hormone (FSH) are secreted by the same gonadotroph cell of the pituitary, the relative amounts being modified by gonadotropin-releasing hormone (GnRH) from the hypothalamus and sex steroid and inhibin feedback from the gonad. The ovary responds with the menstrual cycle, whereas the testicle responds in a more steady state with testosterone secretion (a diurnal rhythm does occur, highest upon awakening) and sperm production. Prolactin has an inhibiting action on the central control of GnRH secretion.

7.1.3.5 Growth hormone/somatotropin

- Growth hormone (GH) is under complex multifactorial control from hypothalamic, intrapituitary, and systemic influences which require the presence of thyroxine. The major controls are dual stimulation and inhibition from the hypothalamus with GH-releasing hormone (GHRH) and somatostatin, and negative inhibition from hepatically produced insulin-like growth factor-1 (IGF-1 (somatomedin-C)) in response to GH. Gastric production of a GH stimulatory peptide, ghrelin, has an adjunctive stimulatory role.
- GH is normally secreted in a pulsatile manner, being very low to undetectable most of the waking day. It is high when fasting in the morning, and during REM sleep. Its metabolic actions are mainly from GH itself, whereas its anabolic actions are mainly via IGF-1. IGF-1 is not pulsatile, so in the nourished state it does quantify net GH secretion. The malnourished catabolic state does decrease IGF-1. Obesity decreases both basal and stimulated GH but not IGF-1.

Anterior pituitary hormone secretion is affected by stage of life (growing child, adult, aging), gender, pregnancy, comorbid diseases, and various medications.

7.1.3.6 Posterior pituitary/arginine vasopressin

- The posterior pituitary lobe stores and secretes antidiuretic hormone (ADH), also known as AVP, and oxytocin, which are produced in the paraventricular and supraoptic nuclei of the hypothalamus. The axons from these nuclei traverse the pituitary stalk and directly connect to the posterior pituitary. Secretion of AVP that is stored in the posterior lobe is stimulated by hypothalamic response to increased serum osmolality and/or decreased arterial blood volume. AVP acts on the renal tubular vasopressin-2 receptor and via aquaporin-2 (AQP-2); it increases water reabsorption to lower the serum osmolality and increase intravascular volume (along with sodium). Thirst is also stimulated to facilitate water ingestion. Thus a physiological feedback loop is established between total body water content, osmolality, and AVP secretion.

7.1.3.7 Pituitary physiology as the key to interpretation of pituitary hormone measures

Although most pituitary hormone (ACTH, TSH, LH, FSH, GH) measurements are often reported by the clinical laboratory with reference to their "normal ranges," it is critically important to recognize that use of the reference range alone is wholly insufficient for clinical interpretation, even with the clinical picture in mind.

Pituitary hormone reference ranges are often developed through population measurements in normal subjects and each hormone may thus have its own unique reference range. However, pituitary hormones do not operate in isolation and thus can only be interpreted in light of the full pituitary hormone feedback loop.

For example, random ACTH levels are meaningless, even if they are found to be above or below the reference range for a random measure. ACTH secretion is under the exquisite control of both CRH and cortisol. Although CRH is rarely measured, cortisol represents the "end-organ" effect of any given ACTH being made by the pituitary. Therefore an ACTH level can only be understood as being truly "normal/physiologic" or pathologic when viewed in context of the concomitant serum cortisol level (and clinical history).

Therefore interpretation of pituitary hormones nearly always requires simultaneous measurement of the target hormone as well:

- ACTH levels require cortisol levels
- TSH levels require thyroxine levels
- LH/FSH levels require estrogen/testosterone (or clinical determination of sex steroid manifestations)
- GH levels require IGF-1 levels

In many cases it is not quite as simple as just measuring both pituitary and target organ hormone together; the reader is referred to the hormone/target gland specific sections of the book for details. Rather the principle is that pituitary hormone measurements are often difficult or impossible to interpret without both the clinical history and knowledge of the related target organ function.

7.1.3.8 The concept of "Inappropriately Normal" in pituitary hormone measurements

One of the most fundamental yet least understood concepts in pituitary medicine is that of the "inappropriate" normal. Many misdiagnoses may result when this concept is ignored.

As outlined in Section 7.1.3.7, pituitary hormone measurements may only be interpreted in light of knowledge of the target organ function either through biochemical or clinical means.

Consider three possible interpretations of a pituitary hormone such as ACTH which is measured and found to be within the normal reference range:

- The ACTH may be normal because the patient is normal and has no disease.

- If the patient is suspected of having primary adrenal failure (hypocortisolism), the expected normal pituitary response should be to produce high levels of ACTH as it tries to stimulate the adrenal to make cortisol; if the ACTH level in this setting is actually normal, then the clinician will be alerted to the fact that *this is inappropriate and thus indicative of abnormal pituitary function.* In such a case the actual underlying diagnosis may be ACTH underproduction from a pituitary tumor.
- If the patient is suspected of having primary adrenal cortisol over-production (Cushing's syndrome), the normal pituitary ACTH production should be fully suppressed by all the excess cortisol; if the ACTH level in this setting is actually normal, then the clinician will be alerted to the fact that this is also inappropriate and thus indicative of abnormal pituitary function. In this case the actual underlying diagnosis may be ACTH over-production by a pituitary tumor *even if the actual ACTH level remains within the "normal range."*

From the above example, there should be two key principles now seen:

1. Target organ dysfunction cannot be diagnosed through isolated measures of pituitary hormones (with some exceptions, discussed in later chapters).
2. A "normal" pituitary hormone may be either truly normal or inappropriately "normal" based upon the function of the target hormone-producing organ and the expected behavior of the pituitary as known from the classical target organ—pituitary—hypothalamus feedback loops.

7.1.3.9 The concept of "central/secondary" versus "target gland/primary" in endocrine dysfunction

Although these terms are commonly used by endocrinologists, there is often confusion among other practitioners as to exactly what is meant by primary/secondary forms of endocrine dysfunction.

Most commonly the qualifier term "primary" is used to indicate that the disease process exists within the final hormone-producing organ of interest.

- "primary" hypothyroidism pertains to disease of the thyroid gland that causes hypofunction
- "primary" adrenal insufficiency pertains to disease of the adrenal gland that causes hypofunction

In contrast the qualifier term "secondary" is used to indicate that the disease process exists within the endocrine organ that usually controls the secretion of the target organ.

- "secondary" hypothyroidism pertains to disease of the pituitary gland that results in deficient signaling to the otherwise normal thyroid gland
- "secondary" adrenal insufficiency pertains to disease of the pituitary that results in deficient signaling to the otherwise normal adrenal gland

Sometimes, clinicians may use the term "tertiary" to refer to disease of the hypothalamus, which controls the pituitary that controls the target organ. We recommend avoidance of this term which is often misunderstood. Hypothalamic disease may be simply combined with pituitary disease as "secondary."

The term "central" is often used interchangeably with "secondary"—both terms indicate hypothalamic/pituitary disease.

The major point of confusion is often related to the fact that in medicine, the term "primary" is often used to denote "congenital" whereas "secondary" can mean "acquired." Therefore speakers should take care to ensure their listeners understand which use of the word "primary" is intended. For this chapter and this book, we will use the terms to indicate whether it is pituitary or target gland that is concerned.

7.1.4 OVERVIEW OF PITUITARY DISEASE

- Pituitary tumors may present with any combination of hormonal hypersecretion, isolated or complete pituitary hormone hyposecretion, or mass symptoms from a macroadenoma with upward extension (loss of peripheral visual field from compression of the optic chiasm), lateral extension (double vision, altered facial sensation from impingement on cranial nerves), or downward extension (postnasal drip from cerebrospinal fluid (CSF) leaking into the sphenoid sinus), and headache.
- Laboratory investigation needs to take into account the secretion of each relevant pituitary hormone, and identification of any hormonal hypersecretion.
- Assessment of mass effects involves formal visual field assessment, while magnetic resonance imaging (MRI) is the test of choice to visualize the pituitary and surrounding anatomy.
- Increased mortality has repeatedly been described in the hypopituitary population [4]. Causes include increased cardiovascular disease, cerebrovascular disease (especially if pituitary radiotherapy was used), parasellar malignant disease (especially if radiotherapy was used), and hypocortisolism [4]. It is not known if this reflects the underlying disease process that resulted in hypopituitarism and its treatment, nonphysiologic hormone replacement therapy, or absent replacement therapy, such as with sex steroid or GH.
- Laboratory monitoring for optimal replacement therapy is limited to free thyroxine, testosterone, and IGF-1.

7.1.4.1 Pituitary hyposecretion

- Hyposecretion may be isolated, may involve one, several, or all anterior pituitary hormones, with or without associated posterior pituitary hyposecretion (resulting in the clinical disorder known as diabetes insipidus). Isolated posterior pituitary hyposecretion may occur. The clinical presentation is defined by the particular target gland hyposecretion, symptoms and signs of the underlying cause (such as a pituitary tumor), and will be modified by age, gender, and comorbid disease. For example, symptoms of GH deficiency are dramatic in a growing child, but subtle or clinically nonexistent in an adult. Gonadotropin deficiency would manifest in pubertal and reproductive aged women, but not in prepubertal or postmenopausal women.

- Clinical presentation may be sudden and dramatic, but hypopituitarism is usually chronic, insidious, and nonspecific in presentation.
- There are many pathophysiological causes of pituitary hypofunction, the most common being pituitary adenoma. Clinical clues for other etiologies may indicate traumatic brain injury, pituitary hemorrhage (apoplexy), infectious or inflammatory causes, symptoms of a locally compressive/invasive sellar mass, etc. Recently described is the effect of the oncology agent ipilimumab-inducing hypophysitis and pituitary hyposecretion [5].
- An apparently nonfunctioning pituitary mass has a differential diagnosis of all tissue types in and around the sella turcica, including metastases from distant malignancy. Most nonfunctioning pituitary adenomas (null cell adenomas) are composed of gonadotropin cells that make fragments of LH and FSH, but without sufficient biological activity to result in clinical manifestations.

7.1.4.2 Pituitary hypersecretion

- Hypersecretory states almost always result from functioning pituitary adenomas, the most common being prolactin hypersecretion, then GH hypersecretion (acromegaly and gigantism) including mixed prolactin and GH hypersecretion, then ACTH hypersecretion (Cushing's disease).
- Hypersecretion rarely may result from ectopic or local production of a hypothalamic-releasing hormone, such as GHRH. In fact, GHRH was first detected and isolated from an ectopic source [6].

KEY POINTS

The diagnosis of pituitary hyposecretion is based upon knowledge of the physiological control of pituitary function.

- ACTH, TSH, and LH/FSH hyposecretion are determined by testing the target gland function. Low level of the target gland hormone without elevation of the pituitary tropic hormone is diagnostic in the setting of pituitary disease.
- Pituitary hormone hypersecretion is tested independently for each one.
- In settings of both hyper- and hyposecretion of pituitary hormones, a "normal" level of the pituitary hormone is physiologically abnormal and therefore considered as "inappropriately" normal.
- Remembering the concept of "inappropriately normal" pituitary hormones will help to avoid missing important endocrine diagnoses despite the presence of apparently "normal" pituitary hormone levels, according to standard reference ranges.

7.2 PROLACTIN

7.2.1 CLINICAL PRESENTATION

- Prolactin is produced from the lactotrophs of the pituitary and inhibited by type 2 dopamine receptor activation by dopamine secreted from the hypothalamus into the portal vein. The prolactin secretion is pulsatile and can also increase with sleep, food, chest wall irritation/stimulation, and stress, all of which may cause minor elevations in measured serum levels.

- Prolactin should be measured in the morning, fasting. Elevations less than twice the upper range of normal should be repeated as the majority of times they are normal on repeat.
- The incidence of hyperprolactinemia is about 1% of women (especially age 25−34) and 0.1% of men. It is elevated in 5%−10% of women with amenorrhea (not pregnant) and 75% of women with amenorrhea and galactorrhea [7].
- Multiple body sites contain prolactin receptors, but the main action of prolactin is on the developed breast to produce milk.
- In excess, it acts centrally, directly, or indirectly (via kisspeptin-1) [7], to inhibit GnRH secretion resulting in some degree of hypogonadism, and acts centrally to reduce libido, independent of the fall of sex steroid.
- The hyperprolactinemic reproductive age woman would variably manifest galactorrhea, oligo-amenorrhea, infertility, and decreased libido.
- The only symptom in a male would be decreased libido (rarely galactorrhea), until long-term effects of hypogonadism arise such as infertility or fragility fractures.
- The underlying disease may give symptoms, such as headaches, loss of peripheral vision, or double vision resulting from a large pituitary tumor, and this may be the initial presentation in children, men, and postmenopausal women.

Hyperprolactinemia is a biochemical measurement that must be interpreted. Identification of the underlying cause and establishing the clinical problems and objectives must occur.

- A common error is to "treat" the biochemical finding without first meeting these clinical objectives.

7.2.2 CAUSES OF HYPERPROLACTINEMIA

- The serum prolactin normal range is skewed to the right, with more normal outliers than usually seen with typical bell-shaped assays [8]. Mild consistent elevation without clinical symptoms or any underlying cause of hyperprolactinemia is likely due to this, and further investigation and "treatment" should be avoided in this case.
- Physiological states of pregnancy and lactation must always be considered as the most common explanation for high prolactin levels.
- Transient mild elevations due to high protein meals, stress, intercourse, exercise, physical examination, or a spurious pulsatile peak; these must be considered if the prolactin level is less than twice the upper limit of normal. Thus the test should be repeated two more times, fasting, in the morning before concluding that a disease is necessarily present.
- Prolactin is under dopaminergic inhibition, possibly serotoninergic stimulation; therefore drugs may elevate the serum prolactin:
  - dopamine agonist mediated: neuroleptics, metoclopramide, domperidone, methyldopa, opiates, cocaine

- serotonin mediated: tricyclic antidepressants, some selective serotonin reuptake inhibitors (SSRIs)
- other: verapamil
- unknown effect of multiple over-the-counter medications/herbs/food supplements
- Prolactin-secreting pituitary tumors—microadenomas and macroadenomas, mixed secretory tumors.
- Any mass or infiltrate in the pituitary–stalk–hypothalamus may interfere with dopaminergic effect; a nonsecreting tumor may raise the serum prolactin mildly or several times above normal (referred to as disconnection hyperprolactinemia) but there is an inappropriately low level of the elevated serum prolactin than would be expected for the size of the mass if it were a prolactinoma. For example, a 2 cm mass with serum prolactin 50 times above normal is consistent with a macroprolactinoma; the same mass with serum prolactin two times above normal is likely a nonsecreting mass that is not a macroprolactinoma (not comprised of lactotroph cells).
- The suckling reflex is a physiological stimulation for prolactin secretion, so any chest wall irritating/painful state from T4 to T6 dermatomes may simulate this physiology.
- Chronic primary hypothyroidism, possibly due to the mass effect of longstanding, thyrotroph hyperplasia.
- Idiopathic—a diagnosis of exclusion, with the presence of symptomatic hyperprolactinemia. It is unknown if this is a true entity or a pituitary adenoma not found by MRI. There tends to be a more favorable natural history in that more patients return to normal, so this may suggest that some are a distinct entity [9].

7.2.3 RARE SITUATIONS

- Seizure activity—postictal state, or less obvious such as temporal lobe epilepsy. Serum prolactin can separate true seizure from pseudoseizure or malingering, where it is not elevated.
- Ectopic secretion from malignancy (ovarian dermoid cyst, bronchogenic carcinoma, renal cell carcinoma), very rare.
- Late stage chronic renal failure and chronic hepatic failure.
- Familial idiopathic hyperprolactinemia resulting from an inactivating mutation of the prolactin receptor [10].

7.2.4 DIAGNOSTIC APPROACH

- Serum prolactin should only be measured when it is clinically indicated: symptoms of oligo-amenorrhea, galactorrhea, delayed puberty, decreased libido (both sexes), or findings of a sellar mass on radiological study.
- Marked elevation over 150 µg/L in symptomatic patients almost always indicates a prolactinoma; the greater the elevation of serum prolactin, generally the greater the size of the prolactinoma.

- Modest elevations of serum prolactin <100 μg/L in the presence of a pituitary macroadenoma usually indicates a nonsecreting pituitary tumor or other mass, not a prolactinoma.
- Mild elevations <2 times upper limit of normal range need repeating two more times, and only considered a problem if consistently elevated and associated with clinical symptoms. Undergoing further investigations and treatment without repeating the test is a common clinical error.
- Obvious causes are first considered, such as drug ingestion, chest wall irritation, clinical signs of a sellar mass, and measurement of serum TSH (to rule out primary hypothyroidism) and creatinine. If no cause is found, then visualization of the pituitary (MRI) is needed.
- Interpretation is needed. For example, if a woman has breast pain and the serum prolactin is found to be mildly elevated, it is likely the elevated prolactin is resulting from the breast pain as hyperprolactinemia by itself does not usually cause breast pain.
- In asymptomatic patients, macroprolactinemia may be considered. See section 7.2.5.1, "Falsely high prolactin levels" below.

7.2.5 LABORATORY CONSIDERATIONS TO ASSESS HYPO- OR HYPERPROLACTINEMIA-ASSOCIATED DISEASE

7.2.5.1 Falsely high prolactin levels

- Prolactin may occur in vivo in different forms. The majority is monomeric; however, glycosylated ("big" prolactin) and inactive complexed ("big–big" prolactin) forms constitute a smaller but significant proportion of circulating prolactin variations. Different prolactin immunoassay methods may show different biases based on sensitivities for these prolactin forms.
- Macroprolactinemia consists of either prolactin-immunoglobulin complexes, polymeric complexes of prolactin, or both; forming molecules in excess of 100 kDa. These prolactin forms are biologically inert, relatively common, and are found exclusively in the intravascular space (since these molecules form outside of the pituitary and are too large to cross into the CSF). Macroprolactinemia is detected by most prolactin assays, leading to potential confusion in assay interpretation; and not infrequently leading to false diagnoses of hyperprolactinemia (partly because the kidneys clear the larger, IgG-bound prolactin more slowly), particularly in asymptomatic patients. Laboratories can add polyethylene glycol to the serum, causing precipitation of macroprolactin and enabling detection of the biologically active, nonpolymeric, noncomplexed prolactin (the clinically relevant fraction); or IgG-bound prolactin may be separated by gel chromatography (the former method is more widely used).
- Serum versus plasma prolactin measurement may be assay- or lab-dependent. Specimens are typically stable if stored at 4°C for 24 or more hours; however,

if analysis is delayed for longer than 24 h, freezing the samples is recommended. The lab's reference interval may be predicated on a particular time of collection, since prolactin concentrations are highest in the early morning hours. It is noteworthy that different states stimulate prolactin secretion, including protein-rich diet, exercise, and emotional stress; therefore collection in a resting patient after an overnight fast is most reliable.

7.2.5.2 Falsely low prolactin levels

- Particularly in patients with large pituitary adenomas (macroadenomas), caution should be used in avoiding a falsely low prolactin measurement due to the hook effect of very high prolactin levels. This is a phenomenon related to the principles of the immunoassay procedure. Normally the immunoassay involves an antibody−antigen−antibody complex ("sandwich") formation, such that the amount of complex formation is proportional to the concentration of analyte (in this case prolactin). However, in cases where the concentration of prolactin is extremely high (vastly greater than amount of reagent antibody), the usual cross-linking of antibody−antigen−antibody "sandwich" complexes (which would normally result in analyzer detection by means of a tagged dye, fluorescent tag, or isotope tag) does not occur as would normally be expected, and so analyzer detection of the bound complex does not occur, or occurs only minimally, yielding a falsely low or negative result. Therefore patients with newly diagnosed macroadenomas who may have a prolactinoma should have prolactin assay performed with serial dilutions in order to avoid an erroneously low or negative prolactin measurement due to high-dose hook effect.
- Communication between the endocrinologist and the laboratory is critical, in cases where there is a discrepancy between a large macroadenoma on imaging and relatively low prolactin levels, because laboratories do not routinely perform manual dilution studies (since it is not cost-effective laboratory practice to do so).

7.2.6 HYPOPROLACTINEMIA

This may occur from congenital transcription defects, or with structural pituitary disease of all pathogeneses. With pituitary disease, hypoprolactinemia occurs late in the course and suggests extensive pituitary disease with hyposecretion. The only clinical effect is inability to lactate in the postpartum state.

7.2.7 COMMON ERRORS IN USE OF PROLACTIN LAB TESTS

- *Measurement of prolactin without an appropriate clinical scenario.* Measuring serum prolactin when it is not clinically indicated sets off an inappropriate diagnostic and therapeutic pathway if it is elevated for any reason; "screening" for prolactinomas is not recommended unless there are clear and legitimate clinical clues to the diagnosis already present.

- *Failure to view prolactin levels as a signal to disease rather than a definite disease entity in all cases.* Deciding that an elevated prolactin is by itself a disease, and thus responsible for whatever symptoms the patient may lead away from the real cause. There are many causes of mildly elevated serum prolactin that have little or no pathologic significance. Patients who present with unusual or unrelated symptoms may be found to have a slight prolactin elevation by chance or by association with drug effects, etc. Attribution of atypical symptoms to mild (and unrelated) prolactin elevations is inappropriate and diverts medical attention from the actual underlying diagnosis.
- *Failure to consider the skewed nature of a normal reference range.* Not recognizing that the normal range of prolactin as defined by the lab may also include outliers that are "normal" but outside the 95% confidence interval, and therefore flagged as "high," for example, and so clinically interpreted as mild hyperprolactinemia. For example, a woman with unexplained infertility may have a mild elevation of serum prolactin wrongfully thought to be the cause. Thus even though hyperprolactinemia is a known cause of infertility, the finding of a slightly high prolactin level does not, by itself, secure the diagnosis.
- *Treating the hyperprolactinemia without first determining the diagnosis of the hyperprolactinemia.* The treatment is still of the underlying disease process. Initiation of medical therapy (dopamine agonists) to the hyperprolactinemic patient without determining the underlying cause may lead to serious missed diagnosis of pituitary tumor with associated other hormonal hyposecretory states.
- *Failure to consider "macroprolactinemia" in patients with high prolactin levels for whom the clinical scenario does not fit.* Macroprolactinemia is a laboratory phenomenon and not a disease process; appropriate early recognition of this problem may prevent extensive and unnecessary additional investigations or "therapy."
- *Failure to measure prolactin when it is indicated, such as after discovery of an incidental pituitary mass.* Many pituitary tumors are treated by surgical resection but prolactinomas are an extremely important exception. Prolactinomas are treated with medical therapy alone in virtually all cases, and failure to recognize a prolactinoma by failure to measure serum prolactin could result in unnecessary (and potentially unsuccessful) pituitary surgery.

7.2.8 EXAMPLE CASE WITH DISCUSSION

A 32-year-old woman complained of chronic nonspecific headaches and breast discharge. Her last pregnancy was 3 years ago, menses returned postpartum but were heavier, and breast milk production persisted to a minor degree. She also complained of nonspecific arthralgias, and rheumatological workup failed to find any cause. Because of her headaches, a CT scan of her head was done, and a homogeneous 1.5 cm mass was noted in her sella, with some suprasellar exten-sion. Clinically, she did not have acromegalic or Cushingoid features.

Question 1: Would you measure the serum prolactin?

Answer: Measurement of a serum prolactin is indicated in the presence of galactorrhea as well as the incidental finding of a pituitary mass. The breast discharge needs to be looked at under the microscope for fat droplets, to ensure it is true galactorrhea and not some other breast discharge. In this case, fat was present. The serum prolactin was 38 µg/L, twice normal. This would be high enough to promote galactorrhea, but not high enough to disrupt the menstrual cycle, which is her presentation.

Question 2: Is this a prolactinoma?

Answer: This is unlikely a prolactinoma as the serum prolactin would be expected to be several times higher, in keeping with the size of the presumed tumor. Generally the larger the prolactinoma, the higher the prolactin. A nonsecreting mass in the pituitary would likely raise the serum prolactin to this level by interfering with dopaminergic transmission to areas of the normal pituitary.

Question 3: Should there be a screen for macroprolactinemia?

Answer: The mild elevation in serum prolactin is likely biologically active due to the presence of galactorrhea, so a search for nonbiologically active measurable prolactin is not needed.

Question 4: What further investigation is needed?

Answer: TLook for other causes of the high prolactin: careful drug history, question chest wall irritation, rule out primary hypothyroidism, rule out renal failure, and look for clinical clues to the etiology of the mass. Her serum TSH was 0.240 mIU/L (normal 0.4–4.0 mIU/L). Subsequent measurement of serum free thyroxine was 7.0 pmol/L (4.0 ng/dL) (normal 8–25 pmol/L or 4.5–12.5 ng/dL).

hus she has chronic mild hypothyroidism, likely dating back to the last pregnancy; most of her other symptoms could be ascribed to this mild hypothyroidism.

Question 5: What is the pituitary mass?

Answer: Longstanding target organ failure results in hyperplasia of the tropic cells in the pituitary. The mass is likely thyrotroph hyperplasia and not a tumor. It is likely that autoimmune thyroid disease was brought out by recovery from the pregnancy, and persisted for 3 years until it was discovered.

Question 6: Should she have surgical removal of the pituitary mass?

Answer: No. She was given a thyroxine replacement and by 2 months her serum TSH was normal, by 4 months the serum prolactin was normal and at that time repeat MRI showed marked resolution of the mass.

KEY POINTS

Hyperprolactinemia is the most common laboratory abnormality of the hypothalamic-pituitary axis.

- Because prolactin is under tonically inhibited secretion, an increase of serum prolactin occurs with most organic diseases of the hypothalamus–pituitary. It is analogous to ESR in this regard, in that it may be a nonspecific indicator, rather than pointing to a particular cause, unless markedly elevated.
- Prolactin should only be measured when clinically indicated: relevant symptoms or discovery of a sellar mass.
- Mild elevation needs to be repeated two more times, fasting in the morning.
- There is a wide differential diagnosis for causes of symptomatic prolactin elevation.
- Hyperprolactinemia may occur without any clinical symptoms.

7.3 GROWTH HORMONE

7.3.1 OVERVIEW OF GROWTH HORMONE EXCESS

7.3.1.1 Clinical picture of growth hormone excess

- GH excess in adults after epiphyseal fusion results in a clinical syndrome called acromegaly, and before bony epiphyseal fusion, (i.e., childhood), results in the much rarer gigantism. The diagnosis is made by nonsuppressible serum GH and/or increased IGF-1 levels.
- The same tests are used for most clinical and therapeutic decisions over the chronic course of GH excess.
- Patients under age 40 have a higher risk of an identifiable germline mutation causing their pituitary GH-producing tumor.
- Incidence was previously thought to be about 6 per million, prevalence 50 per million. Increased awareness and screening has possibly increased this to 80–240 per million [11].

The onset of signs and symptoms is slowly progressive, usually over a decade before diagnosis. More aggressive tumor secretion and growth lead to a shorter course to clinical diagnosis. Excessive growth in a child is more obvious, with tall stature and increased growth velocity.

- The phenotype of the physical/facial changes of acromegaly is noted by the patient, friends/family/strangers, or physician. Physical changes include bony changes of coarse facial features with prominent mandible and supraorbital ridges, widening of the jaw with spreading of the lower teeth with an underbite, and widening of the hands and feet (acral enlargement). Soft tissue anabolic changes include skin thickening, skin tags, macroglossia, and carpal tunnel syndrome. Other features include complaints of increased sweating/

body odor, arthralgias, and snoring. There may be symptoms/signs of a sellar mass and/or pituitary insufficiency due to the mass effect on the normal pituitary.

- Acromegaly may be noticed when the patient presents with complications such as carpal tunnel syndrome, type 2 diabetes, hypertension, sleep apnea, osteoporosis and fracture, large joint arthritis, or abnormal dentition.
- Increased cardiovascular disease related to GH excess includes atherosclerotic disease, increased arterial stiffness, increased aortic root diameter, and cardiomyopathy. Because of the cardiovascular risk, other cardiovascular risk factors need to be evaluated and treated. The most common malignancy associated with acromegaly is thyroid cancer. Increased colon cancer risk is controversial.

It is recommended that all incidentally discovered sellar masses be screened for acromegaly as the phenotype may not yet be clinically obvious. GH excess may present with hyperprolactinemia and is in the differential diagnosis of hyperprolactinemia.

Acromegaly, if untreated, has an increased ($1.72 \times$) mortality rate compared to the normal population [12], due to cardiovascular disease, type 2 diabetes, and from complications of therapy. Normalization of the GH and IGF-1 levels reduces the mortality rate back to normal [13]. A partial biochemical improvement is associated with a better but still higher mortality rate than normal.

7.3.1.2 Causes of acromegaly/gigantism

GH-secreting pituitary adenoma is the underlying cause of acromegaly/gigantism in 95%−99%. Tumor size is variable; a high proportion not macroadenomas although this rate is falling, likely due to increased awareness and earlier diagnosis, and improvements in diagnostic techniques and imaging. Tumors may cosecrete prolactin in 30% or be plurihormonal in secretion, with other pituitary hormones such as TSH or alpha subunit being expressed although this usually is found only on tissue immunostaining. Pituitary carcinomas are rare.

7.3.1.3 Special situations

Some patients with pathological GH excess have no symptoms or any of its phenotypic changes.

Patients with common presenting symptoms, and individuals with coarse facial features who are otherwise normal, need to have GH excess ruled out by specific testing.

- *Special consideration: pregnancy*
  Pregnancy is associated with placental secretion of GH that is almost identical to pituitary GH, with resultant increase in IGF-1. It is difficult to first diagnose acromegaly in pregnancy (demonstration that GH secretion remains pulsatile suggests it is pituitary GH, not apulsatile placental GH that is raising the IGF-1), so the diagnosis is best delayed until after delivery. For the same

reasons, it is difficult to assess the status of a woman already being followed for suspected acromegaly, during a pregnancy. A normal pregnancy in an otherwise normal woman may be associated with acromegalic-like soft tissue changes.

- *Special consideration: falsely low IGF-1*

  Serum IGF-1 is less reliable in states when it is normally suppressed, such as malnutrition, chronic disease such as uncontrolled diabetes, liver and renal disease, and hypothyroidism. Recovery is generally needed, and previously normal ranges of serum IGF-1 may then become elevated.

- *Special consideration: estrogen therapy*

  Oral estrogen decreases the hepatic IGF-1 secretion in response to GH, so it may have to be stopped to detect an elevated IGF-1 level.

7.3.1.4 Rare causes of acromegaly

- GH-secreting pituitary adenomas may be part of a genetic syndrome, such as MEN1, Carney complex, McCune—Albright syndrome, and familial-isolated pituitary adenoma (the latter associated with larger and more aggressive and poorly responsive tumors).
- A rare cause is secretion of GHRH locally, such as a hypothalamic hamartoma, or ectopically from malignancies such as pancreatic islet cell carcinoma. Serum GHRH is measurable in the blood. These rare causes should be considered if no pituitary adenoma is visualized.
- Rarely, GH-deficient patients treated with exogenous GH have demonstrated acromegalic phenotypic changes [14].

7.3.2 DIAGNOSIS OF ACROMEGALY/GROWTH HORMONE EXCESS

7.3.2.1 Growth hormone levels and growth hormone suppression tests

- GH is episodically secreted and responsive to stress, so random GH levels are not useful.
- Demonstration of lack of physiological suppression of GH and/or elevated serum IGF-1 is utilized.
- Glucose normally suppresses GH, so failure to suppress to an oral glucose load (oral glucose tolerance test, OGTT) has long been the standard diagnostic test.
- Because the GH assay has become more sensitive over time, the lower limit of normal GH suppression has changed correspondingly. Presently, this nadir is $<1\,\mu g/L$, and most centers are using $<0.4\,\mu g/L$. Rarely, some patients do suppress with OGTT, but have elevated serum IGF-1.

7.3.2.2 Serum insulin-like growth factor-1

- It is important to measure IGF-1 with a reliable assay.
- It is important to measure IGF-1 and GH for diagnosis and follow-up of treatment in the same assay/laboratory, due to interassay variations.
- In the follow-up of patients thought to be cured, there may be a discrepancy between the serum IGF-1 and postglucose load serum GH. Although

controversial, a normal GH level to suppression but elevated serum IGF-1 is thought to be evidence of residual disease.

- Provided the assay is robust and reference range accurate, there are very few causes of false elevations in serum IGF-1. Reported causes include renal failure and hyperthyroidism.
- IGF-1 should be distinguished from IGF-2 (important in gestation as a growth-promoting hormone; may be produced by islet cell tumors, potentially causing hypoglycemia); this is important in that IGF-2 may cross-react in some IGF-1 laboratory assays (see "IGF Assays" and "Other IGF" later).

7.3.2.3 Other diagnostic considerations with growth hormone excess

In the presence of a macroadenoma, normal pituitary function needs to be assessed, which is done by assessing the target glands: AM serum cortisol, serum-free thyroxine, and AM serum testosterone (males) or estradiol (females). Because of the frequency of hyperprolactinemia, serum prolactin is measured.

Assess cardiovascular risk factors: smoking, blood pressure, body-mass index (BMI), lipids, diabetes screen.

KEY POINTS

GH excess produces a clinical phenotype, but this is not always obvious.

- Screening tests are serum IGF-1 or GH response to glucose load.
- Random GH should not be used as a screening test for GH excess.
- The same testing is used to monitor response to therapy and for long-term clinical follow-up. The same sensitive assay should be used for monitoring GH and IGF-1.
- Focus on the entire patient, not the GH and IGF-1. Any improvement in elevated GH/IGF-1 levels is of long-term benefit, even if they are not completely normalized.

7.3.3 OVERVIEW OF GROWTH HORMONE DEFICIENCY

7.3.3.1 Clinical picture of growth hormone deficiency

Growth hormone deficiency (GHD) may be child-onset or adult-onset.

7.3.3.1.1 Childhood GHD

- In children the annual incidence is 1 in 30,000 [15]. Presentation is by disorder of growth. The GH Research Society consensus for GH deficiency in children older than 2 years of age includes the following criteria:
  - severe short stature more than 3SD below the mean or
  - height more than 2SD below the mean and growth velocity over a year more than 1SD below mean or
  - a decrease of growth velocity more than 0.5 SD over 1 year.
- There may be more extensive hypofunction in other pituitary hormones, or signs of a sellar/CNS mass, or insult to the pituitary such as previous radiation or trauma.

- Hypothyroidism is first excluded, then GH deficiency is diagnosed by failure to respond to a stimulation test and measurement of serum IGF-1 and/or insulin-like growth factor binding protein 3 (IGFBP-3).
- To diagnose isolated GH deficiency, two abnormal stimulation tests are required. In the absence of known pituitary disease, it is difficult to separate mild GH deficiency in children, from children with short stature who are not GH deficient. Diagnosed GH deficiency leads to neuroimaging to evaluate for pituitary mass/lesion, and if normal then perhaps a search for a genetic causation.
- Not all cases of childhood-onset idiopathic GH deficiency continue to be GH deficient as adults. Presumably, deficient GHRH secretion in childhood does not persist into adult life.

7.3.3.1.2 Adult GHD

- Adults may have GH deficiency persisting from childhood, or acquired for the first time.
    - There are associated symptoms, signs, and laboratory tests which form a recognizable clinical syndrome in adults with GH deficiency.
    - GH (and IGF-1) action is normally on muscle (anabolic), bone (remodeling), adipose tissue (lipolytic), kidney (antinatriuretic), liver, central nervous system, and cardiovascular. Thus GH deficiency is associated with altered body composition (less muscle and bone, and more fat, especially visceral fat), decreased exercise capacity, poorly tolerated quality of life changes (mood, generalized well-being, social interaction), and changes in metabolic status involving lipids and carbohydrates [16].
    - There is increased fracture risk from low bone turnover and low bone mineral density. Some of the lean body mass loss, decreased exercise ability, and strength may in part be related to decreased body water and sodium that occurs in GH deficiency. The increased abdominal adiposity gives propensity to insulin resistance and all the features of the metabolic syndrome, and increased atherogenesis. Dyslipidemia, increased arterial intima-media thickness, and arterial stiffness are risk factors for premature atherogenesis. The decreased longevity associated with hypopituitarism may be in part related to untreated GH deficiency although to date GH replacement therapy has not yet been demonstrated to change this [17].
    - Decreased quality of life is best detected by specifically designed questionnaires. Quality of life does tend to improve with GH replacement [18].

GH deficiency is often associated with other pituitary hormone deficiencies, and the symptoms usually overlap. Only when the remaining pituitary hormones are found to be normal or are adequately replaced can potential GH deficiency symptoms/signs be evaluated.

7.3.3.2 Causes

Childhood causes include isolated idiopathic GH deficiency, associated with genetic syndromes, or related to causes of hypopituitarism in general.

Causes in adults are of those which result in hypopituitarism. Most commonly, GH deficiency results from pituitary adenoma, craniopharyngioma or other parasellar tumors and their treatment. Other causes include infiltrative/inflammatory disease, apoplexy, cranial radiation, and traumatic brain injury blast injury/subarachnoid hemorrhage, or sport injury (severe and/or repetitive head trauma). GH deficiency may be transient. The greater the number of pituitary hormones that are deficient, the greater is the likelihood of also having GH deficiency.

7.3.3.3 Special situations

Isolated GH deficiency in adults is controversial and is questionable. Nonspecific complaints in obese patients (BMI > 32) can be associated with poor GH response to stimulation testing, which is reversible with weight loss [19]. Obesity does not decrease IGF-1.

The GH stimulation tests lack specificity. Only patients with high pretest probability of GH deficiency should be tested.

7.3.3.4 Diagnosis of growth hormone deficiency

7.3.3.4.1 Pediatric testing

- Other than in the neonatal period, random GH measurements are not useful.
- GH is secreted in a pulsatile manner, mostly overnight. Most of the day the GH level is normally very low.
- Physiological tests such as exercise have been replaced by stimulation tests to separate normal from GH deficiency.
- The choice of stimulation tests may be based on patient factors, but also local, regional, or national preferences.
- Tests used are insulin-induced hypoglycemia (IIH), glucagon, levodopa, clonidine, arginine, and GHRH with or without arginine. GHRH for testing is difficult to obtain in many countries.
- All the above-mentioned tests have a false positive rate, problems with reproducibility, and unique side effects [20]. Different tests have different criteria of normal response. One problem is the somewhat arbitrary nature of the separation point of normal and abnormal [20].
- If adolescents are to have continued GH replacement as adults, repeated GH testing is needed, unless there are proven mutations or irreversible pituitary disease.

7.3.3.4.2 Additional considerations

In the presence of known pituitary disease, one stimulation test is used; to diagnose idiopathic GH deficiency, two abnormal tests are usually required. If there is an anticipated problem in initiating GH therapy (such as cost), it would not be of use to undergo the diagnostic process.

Sex steroids increase GH secretion, best seen at puberty. In prepubertal children the short-term use of estrogen or testosterone prior to GH stimulation testing has been shown to decrease the false positive rate. This approach of "priming" with sex steroid is variable in clinical practice (personal experience), and usually not done in adults.

GH acts on the liver to produce IGF-1, which circulates bound to binding proteins, mainly IGFBP-3, both of which are more stable in serum than pulsatile GH. Low levels of IGF-1 and IGFBP-3 may support the diagnosis of GH deficiency. They are not sensitive in younger children and are decreased by hypothyroidism, malnutrition, chronic disease, and diabetes.

7.3.3.4.3 Adult testing

- Adults with known or suspected pituitary disease can have the clinical syndrome of GH deficiency. Such adults who are thought to have acquired GH deficiency require GH stimulation testing. This should only be done in those with a probability of having the disease, due to lack of specificity of the tests (a false positive test would result in a person inappropriately treated with GH).
- The choice of test depends on patient factors and preference of the physician. IIH, glucagon, and GHRH (if available) are the most common, while others such as arginine and levodopa do not adequately separate GH deficiency from normal.
- In most centers, IIH is the gold standard, and glucagon is the alternative. Both of these also measure ACTH-cortisol responses, with IIH being the more reliable of the two in this regard.
- Measurement of serum IGF-1 is not useful as a screening test in adults as there is overlap in IgF-1 levels between GH deficiency and normal state, thus IGF-1 has poor diagnostic sensitivity.
- IGF-1 may be normal in those with GH deficiency or low in normal adults with other causes such as malnutrition.
- A very low serum IGF-1 may be diagnostic if in the presence of more generalized pituitary hypofunction. Most third party payers require more diagnostic evidence than a low serum IGF-1 before permitting human growth hormone (hGH) therapy reimbursement.

7.3.3.5 Special situations

- GH neurosecretory dysfunction is defined as an abnormal 24 h GH secretory pattern from decreased number of GH pulses and decreased amplitude of the GH pulses. These children have the clinical picture of GH deficiency but normal GH stimulation testing. This diagnosis requires multiple GH sampling.
- Obesity reduces GH response, and labs need BMI-specific normal levels of response [21]. This is to prevent the false positive diagnosis of GH deficiency

that otherwise may occur. Despite a poor GH stimulation response, age-related serum IGF-1 usually remains normal in obesity. In obesity a poor GH response to stimulation with a low serum IGF-1 suggests true GH deficiency. Because there are neither widely accepted age- nor BMI-related ranges for any of the GH stimulation tests, normal ranges apply to the overall population.

- Genetic studies may be performed if thought to be a factor. Some genetic defects in transcription are found to be present with idiopathic GH deficiency, or with more extensive pituitary hypofunction.
- If three or more pituitary hormones are deficient, the probability of GH deficiency is over 95%, making stimulation testing unnecessary [22]. With pituitary disease, GH deficiency is the most common deficiency that is found.
- Hypoglycemia (induced by insulin) and injected glucagon act on the hypothalamus while GHRH acts directly on the pituitary. Hypothalamic/stalk disease causing GH deficiency may be missed if GHRH is used, as the pituitary response to direct acting stimuli may be normal.

KEY POINTS

GH deficiency in children causes short stature.

- GH deficiency in adults results in changes in body composition and a constellation of vague symptoms that affect the quality of life.
- Diagnosis of GH deficiency is made by the failure to respond to 1 or 2 GH stimulation test(s).
- Serum IGF-1 is not sensitive for detecting GH deficiency in adults, but is more so in children. It is therefore inappropriate to use IGF-1 as a screening test.
- There are recognizable factors that may affect the IGF-1 results.
- GH deficiency may be isolated, but more likely may result from pituitary disease that may involve one or more other hormones, at presentation or during long-term follow-up.
- IGF-1 levels must be interpreted according to age- and sex-matched reference ranges.

7.3.4 LABORATORY CONSIDERATIONS TO ASSESS GROWTH HORMONE-ASSOCIATED DISEASE

7.3.4.1 Growth hormone assays

- These hormones are secreted in a pulsatile fashion; therefore one single analysis may not provide appropriate information to endocrinologists for proper management of the patient. Thus it is recommended that at least three specimens be collected on different days at the same time and interpreted together.
- Less sensitive immunoradiometric assay (IRMA) methods have largely given way to newer immunometric assays, which utilize monoclonal anti-GH antibodies: competitive immunoassays involve patient GH competing for antibody binding with reagent GH; or two-site noncompetitive immunoassays employing a secondary binding antibody at a different site on the GH molecule. Fluorescent or chemiluminescent labeling, or an enzymatic reaction, is used for detection and quantitation.

- GH exists in vivo as a collection of monomeric, dimeric, and oligomeric isoforms. Posttranslational modification of monomeric forms (20−22 kDa) leads to further molecular diversity. Most GH consists of a 22 kDa monomeric form produced in the pituitary (GH-N or GH1); however, around 5%−10% of circulating GH consists of an isoform of GH-N that has undergone deletion of an internal 15 amino acid sequence, yielding a 20 kDa molecule which may dimerize. The 22 kDa variant is more biologically active than the 20 kDa GH. Some assays selectively measure this 20 kDa variant.
- Measurement of the 20kDa GH variant may have utility in detection of GH doping in sports: doping with exogenous recombinant GH (the 22 kDa isoform) will cause suppression of endogenous GH (a portion of which is the 20 kDa variant). Therefore decreased levels of 20 kDa GH may provide evidence to support GH doping; in fact the ratio of recombinant GH to "natural" isoforms of GH is used by some organizations to monitor doping (measurement of IGF and IGFBP may also used in GH doping detection) [23]. Finally a further variant of GH, GH-V (or GH2) is produced by the placenta. Oligomeric GH has decreased biologic activity and decreased renal clearance owing to its larger molecular size.
- About half of the GH in circulation is bound to growth hormone binding protein (GHBP) (the latter derived from the extracellular component of the GH receptor). In vitro, GHBP will dissociate from GH; this increases affinity of the anti-GH antibodies for GH, in the analytical reaction. Therefore immunoassays are reflective of total GH concentration. Methods for measuring the free (unbound) GH exist but are not widely used as the clinical utility for this is not known.
- Serum is the best specimen for GH measurement although plasma with EDTA or heparin anticoagulant may also be used. Specimens are generally stable up to 8 h at room temperature; storage at 2−8°C, or freezing to at least −20°C, may be performed depending on the delay in testing from time of collection.
- There is variation in normal ranges among various commercially available immunoassay methods; some methods correlate well with each other, while other immunoassay methods do not. Standardization and comparison of GH immunoassays are necessary to ensure consistent interpretation of results, but is hampered by the heterogeneity of GH (with its isoforms), interference from matrix components such as GHBP, and availability of different reference preparations for method calibration [24,25]. Current methods detect concentrations as low as 0.030 ng/mL (0.030 µg/L). It is important that each laboratory establish its own reference range based upon its own patient population.

7.3.4.2 Insulin-like growth factor-1 assays

- Serum IGF-1 levels are stable and correlate with the mean baseline 24 h secretion of GH in a log-linear manner, but this does plateau, reflecting the maximum hepatic ability to respond to GH. The correlation between GH response to glucose load and serum IGF-1 is generally good, but not always the case. Serum IGF-1 is normally elevated during puberty. Age-related normal ranges are needed.

- The biological half-life of IGF-1 (somatomedin-C) is much longer than that of GH when complexed with IGFBP-3 (12 h), but only 10 min when not complexed.
- IGF-1 concentrations differ significantly between children and adults; therefore separate reference intervals for the two groups need to be used.
- Chemiluminescent immunoassays have little to no cross-reactivity with IGF-2 and have replaced the previously used ^{125}I radioimmunoassays (RIAs).
- IGF-binding proteins interfere with the assay performance; therefore cryoprecipitation, acid-ethanol precipitation, gel filtration, reverse chromatography, or C-18 column extraction should be performed. Assays using extraction techniques are better able to differentiate between age-matched controls and GH-deficient patients.
- Most procedures use recombinant IGF-1 as calibration material as opposed to hormone isolated from human serum that was previously used.
- Although within-laboratory precision in current immunoassay platforms is excellent (CV <3% for low levels, <5% for high levels), agreement among the various immunoassay platforms is not good due to interference due to binding proteins and other factors, hence concentrations are not transferable among assays [26,27]. Some centers have therefore advocated that IGF-I concentrations be expressed in terms of a standard deviation score according to a matched normal population of the same age and gender [27].
- Liquid chromatography/mass spectrometry (LC-MS) is used instead of immunoassay methodology by some laboratories.
- Depending on the methodology, samples submitted for IGF-1 measurement should be submitted as either serum or plasma (EDTA or heparin) within 60 min; specimens frozen at $-20°C$ may be stored for up to 30 days. Some methods use serum collected on filter paper or dried whole blood.

7.3.4.3 IGFBP-3

- IGFBP-3 (somatomedin-C binding protein) is produced by the liver as a 264-amino acid polypeptide molecule. When not complexed with IGF-1, its half-life is 30−90 min; when complexed with IGF-1, the half-life is significantly extended to 12 h.
- IGFBP-3 is often measured by enzyme-labeled chemiluminescent immunometric assay, but may be measured by RIA; reagent polyclonal rabbit anti-IGFBP-3 is used. There is very little cross-reactivity by this method (<0.2%) [28]. Enzyme-linked immunosorbent assay (ELISA) kits are also commercially available. Levels as low as 0.006 mg/L (0.006 μg/mL) may be detected by chemiluminescent immunometric assay [29].
- Since levels of IGF-1 and IGFBP-3 fluctuate only very little, compared to GH, they are a good indicator of GH production and tissue effect.
- Patients, particularly those with autoimmune diseases or exposure to animals or animal serum products may have heterophilic antibodies in their serum that

may potentially interfere with reagent immunoglobulins in the IGF-1 and IGFBP-3 assays producing results that may be erroneously higher or lower. Therefore results must be taken into context with the clinical impression.

7.3.4.4 Other insulin-like growth factor hormones

- IGF-2 is a 67-amino acid polypeptide thought to be secreted by the liver. Similar to IGF-1, it resembles insulin in that it has two chains (A and B) connected by disulfide bonds.
- IGF-2 is important in gestation as a fetal growth-promoting hormone, with mitogenic, growth-regulating and insulin-like function. It is also clinically relevant in that it may be produced by islet cell tumors, causing hypoglycemia when produced in excess.
- Another role of IGF-2 is as a cohormone with FSH and LH, in that it promotes progesterone secretion after ovulation, during the luteal phase of the menstrual cycle. In this setting it is synthesized and secreted by ovarian theca cells in an autocrine manner as well as in a paracrine manner on ovarian granulosa cells. There is no known clinical relevance to IGF-2 in this setting.
- It is important to be aware that IGF-2 may cross-react in some IGF-1 laboratory assays (see Section 7.3.4.2).

7.3.5 COMMON MISUSE OR ERRORS IN BIOCHEMICAL TESTS OF GROWTH HORMONE DISORDERS

- *Failure to recognize the low value of random GH levels.* Random serum GH cannot be used to diagnose GH hypersecretion. GH secretion is pulsatile, so suppression and stimulation tests are needed to diagnose GH excess or deficiency.
- *Reliance upon IGF-1 instead of dynamic stimulatory testing to diagnose GH deficiency.* Serum IGF-1 cannot be used to diagnose GH deficiency in adults because of overlap between normal- and GH-deficient patients, and IGF-1 may be reduced by disorders other than GH deficiency.
- *Failure to use age- and sex-specific reference ranges.* Serum GH and IGF-1 fall with age, after the third decade, even though the response to provocative testing remains normal. The age-related decline in GH secretion is not true GH deficiency. Present recommendation is not to medically replace GH unless underlying disease is present to account for this and stimulation testing is abnormal.
- *Failure to consider repeat testing for uncertain GH-related diagnoses.* Stimulation tests for GH have a high false positive rate, so some guidelines recommend two different stimulation tests to be abnormal before therapy is initiated.
- *Inappropriate screening for GH deficiency without a reasonable clinical context.* GH stimulation testing for GH deficiency should not be done merely because of the presence of nonspecific symptoms, but rather with symptoms and the presence or high suspicion of underlying pituitary disease.

7.3.6 EXAMPLE WITH DISCUSSION

A 29-year-old woman was sent by her dentist. He performed dental X-rays and noticed the sella was enlarged. She was asymptomatic regarding headaches or visual changes. Pituitary function appeared to be normal as she had regular periods. She was clinically euthyroid with a slightly enlarged normally palpable thyroid gland, had no symptoms of cortisol deficiency, and was tanned from a recent holiday. She did admit to increased sweating, but had no growth of hands or feet, nor any noticed change in her appearance. Looking at her driver's license, there was no overt facial change over the past 3 years.

Question 1: What testing would you do for pituitary function and possible hormonal oversecretion?

Answer: First the abnormal sella needs to be confirmed. MRI demonstrated a 1.2 cm sellar mass with minimal extrasellar extension, away from the optic chiasm.

   Normal pituitary function (absence of hyposecretion) was found with normal early morning serum cortisol, and serum-free thyroxine.

   In the absence of Cushingoid features a screen for Cushing's disease is not needed at this time.

   Serum prolactin was normal as expected with her normal menses.

Question 2: She has no overt features of acromegaly, would you screen for this?

Answer: Especially relatively early in the course of this chronic disease, there may be few clinical signs of GH excess. Most experts and guidelines do recommend a screening test be done. Her serum IGF-1 was elevated at 0.440 nmol/L (0.057 ng/mL) (age- and sex-matched reference range 0.098−0.330 nmol/L or 0.0128−0.0432 ng/mL).

Question 3: Would you confirm GH excess with a glucose suppression test?

Answer: Not if there was an obvious clinical picture of acromegaly. Without this clinical picture a suppression test may be confirmatory. It was done and the serum GH did not fall below 2.2 µg/L (2.2 ng/mL), which is abnormal since a glucose load should suppress serum GH to <0.4 µg/L (0.4 ng/mL).

Question 4: She underwent transsphenoidal resection of the mass, which immunostained as a GH-secreting pituitary adenoma. How should she be assessed to see if she was cured or not?

Answer: A serum GH drawn within 24 h of surgery may be low, and thus highly predictive of complete resection [30]. Otherwise the window of assessment should be left for at least a month as serum IGF-1 may take some time to fall into the range of the new steady state. After 6 weeks the serum IGF-1 could be measured, or the glucose suppression test repeated. There are examples of postoperative normal GH suppression, but serum IGF-1 remains elevated. Most centers do both, whereas many just follow with IGF-1.

Question 5: If she does appear to be cured, does she need any further follow-up?

Answer: Recurrence rates of 15% are generally observed in those previously thought to be cured. Annual serum IGF-1 measurement is recommended following adenoma resection, perhaps less often after 10 years.

7.4 ADRENOCORTICOTROPHIC HORMONE EXCESS AND DEFICIENCY

Refer to Adrenal Endocrinology chapter (see Chapter 5: Adrenal disorders) for a more extensive discussion.

7.4.1 ADRENOCORTICOTROPHIC HORMONE EXCESS

7.4.1.1 Clinical picture

Owing to the fact that ACTH leads to the production of cortisol, any syndrome of ACTH excess includes features of cortisol excess, as well as symptoms or signs of the underlying cause (such as pituitary tumor). Cortisol excess of any cause is called Cushing's syndrome, but if from an ACTH-secreting pituitary adenoma, it is Cushing's disease. Clinical symptoms and signs overlap with many other disease states, so a high index of clinical suspicion is required.

- Adipose changes result in a trend to central obesity, with relative limb sparing. Adipose deposits such as interscapular (buffalo hump) and supraclavicular fat pad overlap with simple obesity.
- Round (moon) face and facial plethora are characteristic but not specific.
- Hypercortisolism-induced protein catabolism is more specific and presents with thin skin, bruising, red striae, proximal muscle weakness, and osteoporosis/fractures.
- Syndromes mimicking Cushing's syndrome (such as polycystic ovarian syndrome) do not have signs of protein catabolism.
- Women usually have menstrual irregularity and hirsutism/acne. Men may have decreased libido.
- There may be psychiatric/depressive changes.
- Hypertension and diabetes are more common.

7.4.1.2 Causes of adrenocorticotrophic hormone excess

Once iatrogenic causes (exogenous glucocorticoid use) have been excluded (from many sources) the most common cause of Cushing's syndrome is an ACTH-secreting pituitary adenoma. It needs to be differentiated from ectopic ACTH secretion (ACTH production by a nonpituitary tumor), and primary adrenal hypersecretory states: adrenal adenoma, adrenal carcinoma, and rarely adrenal hyperplasia, pigmented nodular adrenal disease, and McCune–Albright syndrome.

7.4.1.3 Diagnostic approach

With pathologic ACTH excess, there is excessive cortisol secretion that is not responsive to normal physiological control. The first step is to document this excess; then, to determine the cause of the cortisol excess.

- Excessive cortisol is measured by the 24 h urine-free cortisol (and creatinine). This measures the excessive amount of free cortisol not bound to cortisol-binding globulin. It is important that the upper limit of normal be sensitive to function as a screening test. A markedly elevated 24 h urine cortisol (i.e., more than 500% above the upper reference limit) has very high specificity for true pathologic hypercortisolism.
- Abnormal physiological control of cortisol (autonomous secretion) is determined by the overnight 1 mg dexamethasone suppression test (DST), demonstrating failure to physiologically suppress. Abnormal diurnal rhythm as a distortion of normal physiology is demonstrated by the midnight salivary cortisol measurement, being abnormally elevated. Again, it is important to set a range for sensitivity with a set cut-off point to discover almost all cases, and also to be aware of the levels for which there is high specificity so as not to wrongly diagnose otherwise normal individuals.
- Any of the above three are recommended as screening tests for Cushing's syndrome. If the test is clearly abnormal and agrees with the clinical picture, the diagnosis is very probable. If the screening test is not definitive or not in agreement with the clinical picture, more than one test may need to be performed. Rarely the dexamethasone-CRH test is used for difficult cases.
- False positive screening tests may result from stressed states, depression, later pregnancy, and various medications that may interfere with each screening test. Details are left to Chapter 5, Adrenal disorders. Examples of medications that give false positive DST are oral estrogen, or other drugs that accelerate dexamethasone metabolism such as carbamazepine, phenobarbital, primidone, and others which induce the hepatic cytochrome CYP3A4. Some individuals are slow metabolizers of dexamethasone which can result in a false negative test.
- The rare situation of cyclic cortisol secretion has its own diagnostic issues (beyond the scope of this chapter).

Once the presence of Cushing's syndrome is established the underlying cause needs to be determined. The initial test is serum ACTH measurement to separate ACTH-independent causes (adrenal, exogenous) from ACTH-dependent causes (pituitary adenoma or ectopic secretion). The focus in this chapter is ACTH-dependent disease. If Cushing's syndrome is present and serum ACTH is not suppressed, then the probability is that a pituitary adenoma is the most likely cause, especially in women.

MRI of the pituitary may show a definite adenoma >5 mm in size. This may be sufficient for the diagnosis, but some centers confirm with a CRH stimulation test. If imaging is inconclusive, a screen for ectopic source can occur, such as lung CT. If normal, then inferior petrosal vein catheterization is used to separate pituitary source of ACTH secretion (there is a central pituitary to peripheral vein gradient) from nonpituitary (ectopic) source. This test is operator dependent.

7.4.2 ADRENOCORTICOTROPHIC HORMONE DEFICIENCY

7.4.2.1 Clinical picture

ACTH deficiency results in decreased adrenal cortisol secretion, as well as decreased androgens, but adrenal mineralocorticoid secretion remains close to normal as it is mainly under control of the renin−angiotensin system and potassium concentration.

- Clinical presentation may be of acute hypotensive crisis, especially under times of stress, and thus ACTH deficiency is a potential life-threatening state with hypotensive shock and hyponatremia, nausea, and vomiting, and perhaps abdominal pain.
- This is one cause of the increased mortality seen in hypopituitarism [4].
- Usually the clinical picture of ACTH deficiency is of slow onset of chronic symptoms such as fatigue, nausea, anorexia and weight loss, arthralgias, weakness, dizziness, and hypoglycemia.
- Failure to tan may be a new symptom, or even a noticed loss of pigmentation, resulting from the decrease in melanocyte-stimulating hormone (a by-product of pituitary ACTH production).
- Women may lose pubic and axillary hair, especially if ovarian function is not intact (from concurrent gonadotroph failure).
- A tendency to low blood pressure with a postural fall may occur. Serum electrolytes may show hyponatremia due to increased AVP, but do not show hyperkalemia, as aldosterone secretion is still adequate.
- There may be anemia, eosinophilia, and lymphocytosis. TSH may be elevated and normalizes with cortisol replacement.
- Associated pituitary hormone deficiency (in addition to ACTH) may add symptoms or modify symptoms of ACTH deficiency.
- There may be symptoms of the underlying pituitary disease, such a sellar mass.
- Rare presentations include muscle atrophy, cholestatic jaundice, and pericardial effusion.
- Associated autoimmune diseases may be present if there is autoimmune pituitary insufficiency.

7.4.2.2 Causes of adrenocorticotropic hormone deficiency

The most common cause of ACTH insufficiency is iatrogenic, due to exogenous administration of glucocorticoids (which suppresses CRH/ACTH), and depending on dose and duration there may be CRH–ACTH hyposecretion that may become clinically obvious when the glucocorticoids are reduced or stopped, or major stress is superimposed and the needed increased cortisol secretion is not produced.

Decreased secretion of hypothalamic CRH–pituitary ACTH is called secondary adrenal insufficiency. It may be an isolated problem or associated with other pituitary hormone hyposecretion. All pathological causes are possible, most commonly a mass in or around the sella. Other causes include infectious, infiltrative, and inflammatory lesions; head trauma or subarachnoid hemorrhage, pituitary infarction, and pituitary hemorrhage (apoplexy); and iatrogenic causes such as long-term effects of sella radiation.

All of these are very likely to have more extensive pituitary secretory defects, but ACTH deficiency may occur alone, such as with autoimmune hypophysitis or the medication ipilimumab.

Isolated ACTH deficiency may occur, in the absence of structural hypothalamic-pituitary disease. Brain trauma and autoimmune causes may give isolated ACTH deficiency without demonstrable structural abnormalities.

Isolated ACTH deficiency from birth or in childhood is usually due to genetic mutation in factors that control pituitary development or function [31].

7.4.2.3 Diagnostic approach

Morning serum cortisol measurement is the simplest first step. Using most common cortisol immunoassays, if over 500 nmol/L (18 μg/dL) then deficient ACTH secretion is excluded, and if <100 nmol/L (3.6 μg/dL), cortisol deficiency is diagnosed. Intermediate levels of cortisol need further testing, using judgement. More recent cortisol assays have lowered the cut-off level for defining a normal response.

If the morning cortisol is low, a concomitant, inappropriately nonelevated serum ACTH confirms the central/pituitary origin.

The gold standard test of adrenal reserve is IIH. The serum glucose should fall below 2.7 mmol/L (49 mg/dL) with symptoms of hypoglycemia. This test should not be done in elderly patients, patients with documented or suspected coronary artery disease, history of seizures, or children younger than age 2. The cortisol response generally gives good separation between normal and ACTH insufficiency. It needs to be done by trained personnel and under medical supervision.

Alternatives to IIH are:

- Glucagon stimulation test, but the cortisol response is not as robust so the definition of normal response is lowered [32].
- Metyrapone test, which is limited by virtue of metyrapone not being available in many centers or countries. Metyrapone inhibits the conversion of 11-deoxycortisol (compound S) to cortisol, which should stimulate ACTH

secretion to bring the cortisol back to normal, while the serum compound S markedly increases because of the block in its conversion to cortisol. Medical supervision is needed. Interpretation is dependent upon the rise in 11-deoxycortisol after metyrapone [33].

- ACTH stimulation test. Synthetic [1−25] ACTH (Cosyntropin) is injected intravenously and the serum cortisol response assessed one-half hour later (there are variations in this protocol regarding dosage and timing). With chronic ACTH deficiency for at least 3 months, there is atrophy of the zona fasciculata of the adrenal gland as ACTH is a trophic factor. Stimulation with injected ACTH may thus result in subnormal cortisol secretion. There is controversy in whether the adrenal stimulation dose of 250 μg ACTH is as sensitive as the ACTH stimulation dose of 1 μg to detect ACTH insufficiency. Consensus appears that there is little difference in response between these two doses. This test strongly correlates with the IIH results. ACTH stimulation test results may not be valid if there is recent onset of ACTH deficiency or if only mild ACTH deficiency.
- Borderline results should have the test repeated since it may then be normal on repeat. If inconclusive, IIH or other test may have to be done.

To separate pituitary from hypothalamic disease the CRH test for ACTH response has been tried. The rare situation where this may be needed, the lack of validity demonstrated for this test, and the cost of CRH all make this impractical, and is mentioned only for physiological interest.

KEY POINTS

ACTH deficiency and ACTH excess produce adrenal insufficiency and Cushing's disease, respectively.

- ACTH levels may be useful in both cortisol excess and deficiency, to assist the clinician in determining whether the cortisol abnormality arises from the adrenal gland or from its control by the pituitary.
- Random ACTH levels are of no clinical value.

7.4.3 LABORATORY CONSIDERATIONS TO ASSESS DISEASE INVOLVING ADRENOCORTICOTROPIC HORMONE-CORTISOL AXIS

Refer to Chapter 5, Adrenal disorders, for discussion of laboratory aspects of cortisol measurement.

- Enzyme-linked immunosorbent assays (ELISA), chemiluminescent immunoassays (a variation of ELISA), competitive-binding RIAs, or IRMAs are used for quantitative measurement of ACTH.
- Current chemiluminescent immunoassays are precise (intraassay coefficient of variation lower than 10%) and can distinguish between low normal and

suppressed hormone secretion in the majority of patients; however, low ACTH values should be interpreted with caution, not only because of the intraindividual circadian and ultradian variation and the marked pulsatility of ACTH secretion, but also because the variability in measurement across different methodologies and platforms increases with low ACTH values (the interassay coefficient of variation ranges from 7−22%). Because of these factors, there is a lack of an international ACTH reference standard, and hence no interassay harmonization [34,35].

- RIAs typically involve competitive binding (with patient ACTH competing for antibody binding sites with radioactive-labeled reagent ACTH (most commonly ^{125}I)). Separation systems employ charcoal adsorption, second-antibody precipitation, or PEG. Some methodologies require preextraction of ACTH. The polyclonal antibodies involved in RIA methods bind to intact ACTH (amino acids 1−39), N-terminal ACTH fragments (amino acids 1−24), or ACTH precursors (including pro-ACTH and proopiomelanocortin POMC).

- IRMA typically utilizes a sandwich principle where two antibodies (either monoclonal/monoclonal or polyclonal/monoclonal combination) bind to different sites on the ACTH molecule and are very sensitive (with detection of ACTH concentrations as low as 0.22−0.88 pmol/L (1−4 pg/mL)). Radiolabeling is usually with ^{125}I.

- Although they perform similarly to other immunoassays, RIA and IRMA are no longer commonly performed in North America due to health hazards of using radioactive tracer substances (such as ^{125}I).

- ACTH is very labile, due to rapid degradation by plasma proteases during specimen freezing and thawing; moreover it is readily oxidized and adsorbs to glass surfaces. Therefore ideally blood samples should be collected into prechilled plastic EDTA tubes, placed immediately on ice, and centrifuged at 4°C. Protease inhibitors or antioxidants may also be used. If analysis is delayed, plasma should be frozen at −20°C or colder. Fibrin clots may interfere with assays, so after thawing frozen specimens, centrifugation should be done.

7.4.4 COMMON MISUSE OF LAB MEDICINE IN ADRENOCORTICOTROPIC HORMONE MEASURES

- *Failure to choose the correct cortisol measurement for the clinical situation.* Physicians sometimes mix up the cortisol screening tests. They may do a morning serum cortisol to screen for Cushing's syndrome, and a 24 h urine-free cortisol to screen for adrenal insufficiency. Not only is each test wrong in this context, but clinical decisions may be made based on the results of the wrong test.

- *Failure to interpret ACTH levels in the context of the concomitant cortisol measurements.* Random ACTH levels are therefore of no value on their own. A high ACTH level may be seen in conditions of cortisol deficiency or cortisol excess and thus should only be ordered after determining the cortisol status of the patient.

- *Misinterpretation of random cortisol measures in patients taking synthetic glucocorticoids.* Serum cortisol measurement may occasionally be of use to monitor the oral dose of hydrocortisol, but not if the glucocorticoid replacement is a synthetic such as prednisone or dexamethasone. If the serum cortisol is measured while taking these medications, the measured serum cortisol may result in the misdiagnosis of symptoms or in inappropriate dose adjustments of their replacement therapy.
- *Inappropriate clinical interpretation of ACTH levels prior to proving hypercortisolism in suspected Cushing's syndrome.* ACTH is not needed unless Cushing's syndrome has been diagnosed, to differentiate ACTH-dependent from ACTH-independent causes.

7.5 THYROID-STIMULATING HORMONE EXCESS AND DEFICIENCY FROM PITUITARY DISEASE

Refer to Thyroid Endocrinology chapter (see Chapter 2: Thyroid disorders) for a more extensive discussion.

7.5.1 THYROID-STIMULATING HORMONE—SECRETING ADENOMAS

7.5.1.1 Clinical picture

TSH-producing pituitary tumors are rare. These present with clinical hyperthyroidism and goiter, and biochemical hyperthyroidism but with inappropriately nonsuppressed serum TSH. The adenomas tend to be large, so there may be signs and symptoms of a sellar mass, and may have some degree of hypopituitarism involving other hormones.

7.5.1.2 Special situation

The serum TSH does not need to be elevated, just inappropriately normal in the presence of a hyperthyroid state. This is similar to the syndrome of thyroid hormone resistance, from which TSH-secreting adenoma must be differentiated. Pituitary MRI almost always shows the adenoma, which is usually large.

7.5.2 THYROID-STIMULATING HORMONE INSUFFICIENCY

7.5.2.1 Clinical picture

Central hypothyroidism due to deficiency of TSH (or deficiency of TRH) produces the same clinical picture of hypothyroidism that is seen with primary thyroid disease. The thyroid gland is atrophic and thus small and soft, as opposed to firm and possibly enlarged which occurs with primary thyroid disease.

7.5.2.2 Causes

The most common cause of suppressed TSH secretion with resulting hypothyroidism is iatrogenic. Long-term administration of thyroxine in an excessive dose, does suppress TRH—TSH secretion, and there may be a short or prolonged delay in recovery of secretion.

Developmental defects in the pituitary thyrotroph cells resulting in insufficient TSH secretion may involve transcription factors only for TSH, or other associated pituitary hormones (Pit-1, PROP1 transcription factors) resulting in childhood onset of hypopituitarism [31].

Any hypothalamic-pituitary pathological process (see hypopituitarism) may result in TSH deficiency and central hypothyroidism. The clinical picture will be modified by clinical features of the underlying cause, and of any concomitant pituitary hormone deficiency. TSH deficiency may be clinically silent and only found with laboratory screening.

7.5.2.3 Special situation

When giving thyroxine for primary hypothyroidism, the person usually feels better. If that person feels worse, then concomitant cortisol deficiency may be present due to associated primary autoimmune adrenal disease, or the hypothyroidism may simply be part of broader, unrecognized panhypopituitarism with ACTH deficiency. The increased metabolic changes from the administered thyroxine accentuate the relative cortisol insufficiency that is unable to meet the new demands. Thus clinical deterioration ensues with thyroxine administration.

7.5.2.4 Diagnostic approach

- To diagnose primary target gland hyposecretion, elevation of the pituitary trophic hormone may be used. For example, elevated serum TSH is used to diagnose primary thyroid hypofunction. Thus the pituitary reflects how the target gland is doing.
- In contrast, to diagnose pituitary hypofunction, measurement of the target gland hormone is used. Measurement of the target gland hormone is used to screen for pituitary hormone hyposecretion, and then confirmed by measuring the appropriate pituitary trophic hormone to demonstrate that it is not elevated (low or inappropriately normal). Thus secretion of the target gland reflects how the pituitary tropic hormone is doing.
- For example, if there is known or suspected pituitary disease, the first step is to measure serum-free thyroxine; if normal, then TSH secretion is adequate; if low, then measurement of TSH to show it is not elevated will then diagnose TSH deficiency (central or secondary hypothyroidism). This principle also applies to LH-testosterone, FSH-estradiol, and ACTH-cortisol.

7.5.2.5 Laboratory considerations to assess hypo- or hyperthyroid diseases

See Chapter 2, Thyroid disorders.

7.5.2.6 Common misuses of thyroid laboratory tests in pituitary disease

- Failure to remember that a "normal" TSH does not exclude all forms of hypothyroidism. If, in the presence of central hypothyroidism, the serum TSH is measured and found to be normal, then there may be an incorrect assumption that a hypothyroid state is not present and the diagnosis is missed.
- *Failure to remember that TSH measurement cannot be used to assess adequacy of thyroid hormone replacement in the presence of pituitary disease.* Patients with known central hypothyroidism being treated with thyroxine need their dose monitored by serum free T4, not TSH. Measuring the serum TSH is useful to follow replacement therapy in the patient with primary hypothyroidism, but not central hypothyroidism. Not only are the TSH results meaningless but unfortunately, inappropriate therapeutic decisions may be made based on the result.

7.5.2.7 Example with discussion

A 43-year-old woman had panhypopituitarism from a previous surgically treated nonsecreting pituitary macroadenoma. She was being replaced with thyroxine, hydrocortisone, and estrogen/progestin. Her primary healthcare provider measured her serum TSH and found it to be 0.1 mIU/L (normal 0.4−4.0 mIU/L). If the free thyroxine had been measured (as is needed in the presence of pituitary disease), it would have been in the normal range. Based on the erroneous assumption that the patient was overmedicated, the thyroxine dose was decreased by half. The serum TSH subsequently was repeated and found to be unchanged and still below the normal range. Her thyroxine dose was then discontinued out of concern that she was hyperthyroid (which is usually correct when primary thyroid disease is the concern). Over the next month she became quite symptomatic of hypothyroidism, still had a low serum TSH, so she was sent in for review. The 3-h trip was in winter, with poor visibility, and her vehicle collided with a moose. The patient ended up in intensive care for 3 weeks. Her thyroxine dose was restarted on arrival at the ICU.

7.6 GONADOTROPIN DEFICIENCY AND EXCESS

Refer to Reproductive Endocrinology chapter (Chapter 9: Neuroendocrine tumors) for a more extensive discussion.

7.6.1 "PRIMARY" VERSUS "SECONDARY" REVISITED

- See Section 7.1.3.9 for background discussion.
- Note that in reproductive endocrinology, the terms "primary" and "secondary" take on yet another meaning.
- "Primary" amenorrhea means that the female patient has never had an instance of menstruation. Note that this may be due to a congenital

("primary") disease and may or may not pertain to a "primary" ovarian disorder.

- In the male patient, "primary" hypogonadism is intended to convey that he has never shown signs of puberty.
- "Secondary" amenorrhea means that the female established ovulation/ menstruation at some point in her life but that such menstrual cycles have now ceased. Similarly in men, "secondary" hypogonadism means that he has at least started puberty spontaneously but has later become hypogonadal.
- This terminology is extremely important for the understanding of the differential diagnosis in reproductive disorders.

7.6.1.1 Gonadotropin deficiency

This may be primary (no pubertal development and primary amenorrhea/delayed puberty in men) or secondary (loss of previously present menses, or loss of previously normal testosterone in men).

Congenital causes include defects in gonadotropin cell formation and function (such as the transcription factor Pit-1), or defects in GnRH secretion or its receptor, or controlling factors such as Kisspeptin-1 [36].

Acquired gonadotropin deficiency may give primary and secondary amenorrhea, depending on time of onset (childhood vs. adulthood). In men, this presents as delayed puberty, or symptomatic loss of previously normal testosterone secretion. Organic causes involve any pathological process in the hypothalamic-pituitary axis. Hyperprolactinemia of any cause does suppress GnRH secretion.

Nonorganic causes affect the normal physiologic secretion of GnRH and are generally reversible. Any cause of poor nutrition may result in this, thus any comorbid disease process. The most common nonorganic causes are marked disorders of nutrition, eating disorders, very excessive exercise, or combinations of weight loss, exercise, and stress that have been shown to be additive [37]. Eating disorders are often the underlying cause, especially in younger women.

7.6.1.2 Gonadotropin excess

- Gonadotropin excess (largely laboratory finding) may be detected in either gender with gonadal failure.
- Pituitary adenomas may rarely secrete biologically active gonadotropins that result in a clinical syndrome of gonadal hyperstimulation. There are rare reports of women presenting with menstrual changes, pelvic pain, ovarian cysts, and ascites resulting from excessive FSH secretion from a functioning pituitary adenoma.
- The most common subtype of nonfunctioning pituitary adenomas is of gonadotroph cell origin. Pathological examination of the tumor finds local production of fragments of alpha and beta subunits of FSH and LH.
- Measurement of serum levels of alpha and beta subunits has been attempted to predict the pathological nature of such tumors, but have been poorly sensitive for this purpose. Because of the nonbiologically active production of these

gonadotropin fragments, these tumors present with mass symptoms, mild hyperprolactinemia, or as an incidental finding but without sex steroid excess; there may be sex steroid deficiency due to mass effect upon the normal surrounding pituitary.

7.6.1.3 Laboratory considerations to assess gonadotroph cell-associated diseases

See Reproductive Endocrine section (Chapter 8: Endocrinology and disorders of the reproductive system).

7.6.1.4 Misuse of laboratory tests of pituitary hormones

- *Failure to remember that pituitary hormones may only be interpreted with knowledge of the target organ.* It is often forgotten that pituitary hypofunction is tested by testing the target organ. A pituitary investigation of measuring ACTH, TSH, LH, FSH, and GH results in a waste of resources, the presence of disease may be missed, or further investigations/therapy/clinical decisions may be made on the basis of these tests. With pituitary disease the pituitary hormones are uninterpretable without the accompanying target gland hormone levels.
- *Failure to consider the effects of oral contraceptive pills (OCPs) upon reproductive hormone measurements.* A woman on the oral contraceptive for estrogen replacement may have the serum estradiol measured, which would be very low as the ovarian secretion of estradiol is suppressed, yet she is well estrogenized (as shown by the presence of menses) from the ethinyl estradiol in the oral contraceptive that is not measured by the serum estradiol assay. With the finding of a low serum estradiol, she may be considered to be hypoestrogenized despite the pharmacological ethinyl estradiol she is taking. This misinterpretation is further fueled because the serum LH and FSH would be measurably low because of the suppressive effect of the same oral contraceptive use.

7.6.1.5 Example with discussion

A 29-year-old woman moved to our center. She presented with amenorrhea for 2 years, fatigue, cold intolerance (it was summertime), loss of pubic and axillary hair, and generally not feeling well. Two years ago she was pregnant, underwent a second trimester therapeutic abortion, and had a postpartum bleed, and four units of blood were transfused. She did not lactate. One month later she was readmitted with meningitis, hospitalized for a month for treatment. One year later she moved to our city. On examination her blood pressure was 90/60 with no postural fall. She had cold hands, delayed relaxation of her deep tendon reflexes, and her thyroid gland was very small and soft. There was no axillary or pubic hair, no increased pigmentation, and no polyuria or polydipsia. She had normal visual field and cranial nerve examinations.

Question 1: Does this woman have hypopituitarism?

Answer: She likely has TSH deficiency due to hypothyroid signs and a small soft thyroid gland; likely ACTH deficiency with low blood pressure and loss of androgenic hair (both adrenals and gonads make androgens, so both will cease androgen production with ACTH and gonadotroph loss); in this context the amenorrhea is likely due to LH and FSH deficiency; most certainly with panhypopituitarism, she also has GH deficiency.

Question 2: How do we use the lab to assess pituitary function?

Answer: She was sent for an early morning serum cortisol (very low, 90 nmol/L (3.26 μg/dL)), serum-free thyroxine (low at 5.1 pmol/L (6.56 pg/mL)), TSH (inappropriately normal, 0.9 mIU/L), serum LH (0.5 IU/L), and FSH (1.0 IU/L) with a low serum estradiol (19 pmol/L (5.175 pg/mL)). Target gland hormone levels are low, and pituitary tropic hormones not elevated; thus the patient has hypopituitarism. ACTH did not need to be measured although if it had been, it would likely have also been low/normal.

Question 3: She is to have replacement hormone therapy. How does the lab monitor this?

Answer: Cortisol is given (hydrocortisone 15 mg a day in split doses) and clinically assessed. Thyroxine is given and serum-free thyroxine is kept in the upper half of the normal range. Estrogen is given as physiologic or pharmacologic, and clinically assessed. Estradiol supplementation could be monitored by serum estradiol, but usually this is not needed. Menstrual bleeding is a good index of sufficient estrogen biological activity. Pharmacologic replacement by the oral contraceptive requires no hormonal monitoring, and as mentioned, measurements of serum estradiol are not useful and may be confusing since the ethinyl estradiol of the OCP is not measured in the serum estradiol assay.

Question 4: What is the cause of her hypopituitarism, and does it make any difference?

Answer: Her sella was visualized by MRI and found to be normal.

She may have infarcted her pituitary with the postpartum bleed, or basal meningitis may have involved the diaphragma sella through which the pituitary stalk passes, with postinflammatory scarring obstructing stalk function. The damage would be low in the stalk as diabetes insipidus is not present.

Her serum prolactin was 27 nmol/L (normal 3−20 nmol/L) (0.62 μg/L; normal 0.07−0.46 μg/L); if pituitary infarction was present, prolactin would be expected to be low; if a suprasellar cause was present and disrupting dopaminergic suppression, it would be expected to be higher than normal. Thus mild elevation supports a suprasellar cause. This may be important as the natural history would be different between these two etiologies.

Question 5: *She was on physiologic (non-OCP) replacement therapy for 2 years, with continued hydrocortisone and thyroxine. She returned to the clinic with the complaint of nausea and vomiting for 1 month. What test would you do?*

Answer: Her serum beta HCG was quite elevated. With suprasellar scarring, she can spontaneously recover pituitary function by revascularizing the stalk. She did this, ovulated, and conceived. She has since been able to successfully discontinue all her pituitary replacement medications.

7.7 DIABETES INSIPIDUS (POSTERIOR PITUITARY DYSFUNCTION/ARGININE VASOPRESSIN DEFICIENCY)

7.7.1 CLINICAL PICTURE

There is chronic production of abnormally large volumes of dilute urine. The symptoms will be large volume of polyuria, nocturia, and compensatory polydipsia. Volume status and serum electrolytes remain normal unless there is interference with the compensatory polydipsia, such as coma, or concomitant disorders of thirst sensation.

Normal physiology: total body water and osmotic normalcy are maintained by ADH (also known as AVP) secretion from the hypothalamus, stored in the posterior pituitary, and released upon stimulation. Stimuli for AVP secretion are increasing serum osmolality and/or falling blood volume. AVP activates the renal vasopressin-2 receptor, increasing tubular water permeability by insertion of aquaporin-2 into the apical membrane. The resulting water reabsorption brings the rising serum osmolality down to normal, and/or contributes to restoring blood volume in association with the concomitant sodium retention resulting from the volume depletion. The sensation of thirst is also activated by the same stimuli, directing water-seeking behavior.

Polyuria is defined as

- urine volume >40 mL/kg/24 h
- >3 L/24 h
  Or, in children
- >100 mL/kg/24 h.

7.7.2 CAUSES

Once it is established that the polyuria is a water diuresis (dilute urine) and not a solute diuresis with isotonic urine (such as due to glycosuria), then there are several causes of a water diuresis.

Causes of large volume of dilute urine include:

1. *Central diabetes insipidus (DI)* from absolute or partial deficiency of AVP secretion.

2. *Nephrogenic DI* resulting from renal resistance to AVP. This may be congenital (AVP receptor or postreceptor mutations) or acquired. Acquired causes include several drugs such as lithium; hypokalemia or hypercalcemia; infiltrating renal disease such as sarcoidosis, multiple myeloma, or vascular disease such as sickle cell disease; or any chronic renal disease.

3. *Primary polydipsia* resulting in excessive water intake which then suppresses the serum osmolality and thus AVP secretion. This then results in a large volume of dilute urine, to compensate for the excessive water intake. It may be psychogenic in origin or dipsogenic resulting from defective thirst sensation. Dry mouth could be the initial factor in initiating drinking behavior. The dipsogenic causes have the same underlying central pathological causes as does central DI. Thus most still require an MRI to rule out central pathology. There may be a reset osmostat, so they may demonstrate normal responses to osmolar and volume stimuli, but at a lower set point [38].

   a. Central DI may be congenital (autosomal dominant, autosomal recessive, X-linked) or acquired. Because AVP is synthesized in the hypothalamus, the causative injury in acquired DI needs to be well above the level of the pituitary (high in the median eminence), even if the pathologic lesion originates within the pituitary gland.

   b. All pathological processes in the hypothalamus−stalk−pituitary area such as mass, trauma, or vascular disease may result in DI. Granulomatous infiltrative disease such as sarcoidosis seems especially prone to produce central or dipsogenic DI. Idiopathic DI is a diagnosis of exclusion, but some cases are immunological in origin [37]. Autoimmune destruction is directed to the stalk and posterior pituitary, with thickening of the stalk seen on MRI. This may clear after around 2 years, resulting in a subsequent normal MRI, but with persisting DI. With central DI of any cause, loss of the posterior bright spot is seen on unenhanced T1-weighted MRI [38]. This is a useful clinical sign.

7.7.3 SPECIAL SITUATIONS

DI may be transient or permanent after pituitary surgery. How high up the median eminence the surgery occurred is a factor. After pituitary surgery, a "triple response" has been observed with transient DI for a few days, remission with possible hyponatremia from unregulated AVP release for a few days, and then permanent DI [39].

7.7.4 RARE SITUATIONS

- Gestational DI results from increased placental AVP metabolism causing relative ADH deficiency due to enhanced AVP removal and inadequate compensatory secretion.

- Central DI may in some situations only be a partial deficiency of AVP. Such patients may be able to concentrate their urine above that of serum concentration with simple water deprivation, but not to maximal urine concentration. In the water deprivation test, they achieve a stable urine concentration above that of serum concentration in response to water deprivation, but further significantly increase their urine concentration after exogenous 1-desamino-8-D-arginine vasopressin (DDAVP), revealing relative AVP deficiency as the cause of being unable to maximally concentrate their urine. The water deprivation test must be strictly performed to diagnose this partial AVP-deficient state.

7.7.5 DIAGNOSTIC APPROACH

7.7.5.1 Simple screening tests

First the polyuria needs to be documented, then documented that it is dilute (which establishes a water diuresis as the cause of the polyuria). There may be a clinical clue as to the etiology based on drug history, known disease, or evidence of underlying disease.

Renal function, serum potassium, and calcium need to be measured and excluded as causes of nephrogenic DI.

Serum osmolality and serum sodium may suggest the etiology. Low normal values of each suggest water overload and primary polydipsia, whereas high normal serum sodium and osmolality with dilute urine suggest the inability to concentrate the urine due to AVP deficiency or resistance. With intact thirst, both serum sodium and osmolality are often found to be mid-normal and thus often not predictive.

7.7.5.2 Water deprivation tests

The water deprivation test has been the gold standard diagnostic test for DI since 1970. By restricting water intake and regularly monitoring urine osmolality until balance has occurred (for example, stimulating AVP secretion by increasing plasma osmolality and decreasing intravascular volume), and then assessing the urine concentration in response to an injection of synthetic AVP (DDAVP), the cause could be separated into primary polydipsia, nephrogenic DI, and partial or complete central DI. This test is the most important test used but has limitations. Overlap of responses by the pathophysiological states may occur.

- It may be modified by some degree of renal insufficiency.
- Chronic polyuria may result in decreased urinary concentration ability (solute washout) and may downregulate the AQP-2 water channels, and there may be enhanced response to AVP due to upregulation of V2R receptors.
- Some patients with nephrogenic DI may be only partially resistant to AVP which may be overcome by the administration of DDAVP, simulating central DI.
- A prospective study found a diagnostic accuracy of 41% in patients with primary polydipsia [41].

The addition of measuring serum AVP to this test has improved this test performance with some borderline cases, but the overall diagnostic accuracy still remains unknown, especially separating partial central DI from primary polydipsia and partial nephrogenic DI [42].

7.7.5.3 Additional laboratory measures during a water deprivation test

7.7.5.3.1 Plasma arginine vasopressin levels

Measuring AVP and correlating with serum osmolality are useful if there is relatively low AVP for the level of serum osmolality.

- AVP can be measured but must be interpreted regarding the serum osmolality at the same time. Thus it is not to be done under random conditions but rather under stimulated conditions, such as the water deprivation test or hypertonic saline infusion.
- A stimulation test for AVP secretion may be done with a hypertonic infusion of saline [43]. This is useful for central DI if the stimulated AVP is low despite serum hyperosmolality, and suggestive of noncentral DI if the stimulated AVP rises appropriately with hyperosmolality. It thus correlates with the water deprivation test in most cases. The water deprivation test and stimulated serum AVP may not always agree in cases of partial central DI and primary polydipsia [44].

7.7.5.3.2 Copeptin levels

Copeptin has been investigated as a surrogate marker of AVP, the C-terminal end of the AVP prohormone called copeptin, as it may be more stable than ADH and may give better separation of the diagnostic groups [43]. It may be used along with baseline serum osmolality in a nomogram to separate groups, or its response to water deprivation or saline infusion. More data for comparative purposes are needed, but it may prove to be an easier test to measure and be more accurate in the diagnostic process (see "Laboratory aspects" in Chapter 5, Adrenal disorders.). Indeed, recent studies have found promise in its use in managing sepsis as copeptin levels were found to distinguish normal controls from septic patients, and septic patients from septic shock patients [44]. It may also hold promise as a cardiology biomarker in conjunction with troponin for acute myocardial infarction rule out [45,46].

7.7.5.3.3 MRI findings

The loss of the posterior pituitary bright spot on MRI as previously mentioned does suggest central DI. However, early in the course of central DI it may be present and can be lost in otherwise older normal subjects. If thickening of the stalk is also seen, then idiopathic (immunological) central DI is likely present. Infiltrating granulomatous lesions also may give this appearance.

KEY POINTS

AVP deficiency presents with large volume of dilute urine resulting in urinary frequency and nocturia, with compensatory increased thirst and water intake.

- Acquired causes are usually hypothalamic-pituitary in origin.
- It must be differentiated from polyuria due to solute diuresis, primary polydipsia due to increased water intake (i.e., psychogenic or dipsogenic), and renal resistance to AVP action.

(Continued)

7.7.6 LABORATORY CONSIDERATIONS TO ASSESS POSTERIOR PITUITARY/ARGININE VASOPRESSIN DISEASE

Serum osmolality is reliably measured by freezing point depression. This is a direct measurement; osmolality may also be estimated from serum sodium, glucose, and urea:

$$mOsmol/kg = 2 \times [Na^+] + \frac{glucose\ in\ mg/dL}{18} + \frac{BUN\ in\ mg/dL}{2.8}$$

This equation is useful; however, it underestimates true osmolality by 5−10 mOsmol/kg due to contributions to overall osmolality by remaining constituents in serum. Laboratories may provide a reference interval accounting for that difference, or may use a different reference interval with an added average factor (typically 8−9 mOsmol/kg) to make calculated osmolality a closer estimate of measured osmolality. The following equations may also be used:

$$mOsmol/kg = 1.86 \times [Na^+] + Glu\ (mmol/L) + urea\ (mmol/L) + 9$$

or

$$mOsmol/kg = 1.86 \times [Na^+] + Glu\ (mg/dL) + \frac{urea\ (mg/dL)}{2.8} + 9$$

Urine specific gravity (SG) may be used to estimate urine osmolality, in the absence of osmotically active substances in urine (glucose, protein, heme, radiocontrast dyes).

The correlation between urine osmolality and SG depends on the method used for SG (refractometer versus ionic). The refractometer method is affected by high urine protein concentrations or the presence of radiocontrast dyes, whereas the ionic method (dipstick) is not affected.

AVP is a nonapeptide (amino acids 20−28 of the prohormone, preprovasopressin). AVP is typically measured by noncompetitive RIA methods; however, due to the extremely low concentration of AVP (picomolar range), preassay extraction is usually necessary. Extraction techniques include ethanol, petroleum ether, acetone, or chromatographic isolation with C-18 columns. Besides RIA, enzyme immunoassays are available for ADH quantitation, but are not as widely used.

Similar to ACTH, AVP has poor stability. AVP is largely attached to platelets, and rapidly cleared. ADH should be collected into prechilled tubes with EDTA anticoagulant. Specimens should be delivered to the laboratory on ice and centrifuged at 4°C within 30 min of collection, then plasma separated and frozen until analyzed. These preanalytical requirements, and the lack of readily available and fast AVP assays, have limited clinical use of AVP measurement. Urine may also be tested for AVP concentration; 24-h urine specimens collected in 10 mL of 6 mol/L hydrochloric acid may be used.

Copeptin, as mentioned earlier, is a glycoprotein comprising the C-terminal end of preprovasopressin (amino acids 126−164) and is 5 kDa in size. It is stable at room temperature for at least 7 days, and for at least 14 days at 4°C; therefore it is a more stable surrogate than the active hormone AVP for measuring AVP release; also, unlike AVP, copeptin is advantageous in assays in that it is not significantly bound to platelets and has been found to correlate better with serum osmolality than AVP itself [47]. A single baseline copeptin level of >21.4 pmol/L differentiated nephrogenic DI from other etiologies of polyuria−polydipsia syndrome with a 100% sensitivity and specificity, rendering water deprivation testing unnecessary [44].

A chemiluminescence sandwich immunoassay using two polyclonal antibodies to amino acids 132−164 of preprovasopressin has improved correlation between copeptin and ADH. This method has been developed on an automated immunofluorescent assay platform (B-R-A-H-M-S Copeptin pro AVP, Thermo Scientific) and may be used with EDTA plasma, serum, or heparinized blood samples (sample volume 50 μL) with a detection limit of 0.69 pmol/L (functional assay sensitivity of 1.08 pmol/L (4.34 pg/mL)) and is linear to 500 pmol/L (2010 pg/mL) [48].

AVP reference ranges are determined, by convention, based on platelet-poor EDTA plasma from individuals fasting no longer than overnight. The rationale for this is because platelets contain significant amounts of AVP, therefore measured AVP concentrations may be significantly affected by the platelet concentration in the specimen; moreover various conditions affecting platelets may also affect AVP levels. AVP concentration in platelet-rich specimens may be up to 10 times higher than in platelet-poor specimens [49]. Reference range determination is problematic, however, as there is no standardization on the actual length of fasting prior to specimen collection. Moreover there is no correlation between AVP reference levels and the corresponding serum osmolality. Therefore the use of a reference range for AVP is of little value because AVP is generally ordered in conjunction with a water deprivation test, when the serum osmolality would be high. It would be more meaningful to have a laboratory-validated "true" AVP reference range where the serum osmolality was controlled during the validation process.

7.7.7 EXAMPLE WITH DISCUSSION

A 22-year-old medical student presented with sudden onset of polyuria and polydipsia over 4 weeks. Nocturia occurred four times a night, large volume in nature, and water drinking occurred at the same time. When he restricted water intake for 6 h, the high urine volume persisted and he developed excessive thirst. Fluid intake, mainly cold water, over 24 h was found to be 6 L. He was previously well, taking tetracycline for acne for the last week (medication left over from 6 months ago). His past health was normal, and there was no family history of note. He denied headache, double vision, or loss of peripheral vision. His clinical volume status was normal. His mother looked this up on the Internet and told him the polyuria was due to the tetracycline medication. Physical exam was normal.

Question 1: Can tetracycline cause polyuria?

Answer: Demeclocyline is the only tetracycline that can cause nephrogenic DI, but this was not what he was taking. Also the symptoms preceded the drug ingestion.

Question 2: What lab tests can help?

Answer: Urine specific gravity (SG). was tested and was 1.001, a dilute urine.
Serum sodium was 143 mmol/L, serum osmolality 298 mOsmol/kg, and urine osmolality 124 mOsmol/kg. Serum glucose was 5.1 mmol/L (92 mg/dL).
Normal serum calcium, albumin, potassium, and creatinine excluded overt causes of nephrogenic DI. Even though the dilute urine pointed away from solute diuresis, it was nevertheless comforting to find a normal serum glucose (further excluding glucosuria as a cause of solute diuresis).
In the setting of high normal serum osmolality the patient's urine should be very concentrated, and thus is inappropriately dilute for the serum hyperosmolar state that results in the stimulation of AVP release/action. In the absence of an overt renal cause the rapidity of onset and lack of clinically evident central disease (symptoms/signs) suggest central idiopathic DI, possibly immunological in origin.

Question 3: Is a water deprivation test needed, with or without ADH measurements?

Answer: This certainly can be done. By not allowing any fluid intake the urine concentration can be monitored until it is no longer increasing over 3 consecutive hours, then an injection of DDAVP is administered to see if the urine does then further concentrate (thus showing that the problem was the inability to produce AVP) or does not concentrate (thus showing the problem is resistance to AVP). In our opinion a water deprivation test is unnecessary in this case due to the baseline gross abnormalities in serum osmolality and urine osmolality.
Instead a trial of DDAVP was undertaken, as he was leaving the country in 2 days. He was administered 10 µg intranasal twice over 24 h. He reported that the excessive urine volume disappeared, the excessive thirst disappeared, and he slept through the night.

Question 4: He is clinically normal, so is that good enough to exclude central disease?

Answer: No, so visualization is needed. MRI demonstrated no mass or infiltrate in the pituitary or hypothalamus. The pituitary stalk was thickened, and the posterior pituitary bright spot was absent. This suggests idiopathic (immunological) DI. Because granulomatous infiltrates are variable in appearance, a chest X-ray was done and was normal.

REFERENCES

[1] Regal M, Paramo C, Sierra SM, et al. Prevalence and incidence of hypopituitarism in an adult Caucasian population in northwestern Spain. Clin Endocrinol 2001;55 (6):735−40.

[2] Daly AF, Tichomirowa MA, Beckers A. The epidemiology and genetics of pituitary adenomas. Best Prac Res Clin Endocrinol Metab 2009;23:543−4.

[3] Korbonits M, Storr H, Kumar AV. Familial pituitary adenomas—who should be tested for AIP mutations? Clin Endocrinol 2012;77:351−6.

[4] Burman P, Mattsson AF, Johannsson G, et al. Deaths among adult patients with hypopituitarism: hypocortisolism during acute stress, and de novo malignant brain tumors contribute to an increased mortality. J Clin Endocrinol Metab 2013;98:1466−75.

[5] Faje AT, Sullivan R, Lawrence D, et al. Ipilimumab-induced hypophysitis: a detailed longitudinal analysis in a large cohort of patients with metastatic melanoma. J Clin Endocrinol Metab 2014;99:4078−85.

[6] Thorner MO, Perryman RL, Cronin MJ, Rogol AD, Draznin M, Johanson A, Vale W, Horvath E, Kovacs K. Somatotroph hyperplasia. Successful treatment of acromegaly by removal of a pancreatic islet tumor secreting a growth hormone-releasing factor. J Clin Invest. 1982 Nov;70(3):965−77.

[7] Melmed S, Casanueva FF, Hoffman AR, et al. Diagnosis and treatment of hyperprolactinemia: an Endocrine Society clinical practice guideline. J Clin Endocrinol Metab 2011;96:273−88.

[8] Batrinos ML. Extensive personal experience. Validation of prolactin levels in menstrual disorders and prolactinomas. Hormones 2009;8:258−66.

[9] Corenblum B, Taylor JP. Idiopathic hyperprolactinemia may include a distinct entity with a natural history different from that of prolactin adenomas. Fertil Steril 1988;49:544−6.

[10] Newey PJ, Gorvin CM, Cleland SJ, et al. Mutant prolactin receptor and familial hyperprolactinemia. N Engl J Med 2013;369:2012−20.

[11] Ribeiro-Oliveira Jr A, Barkan A. The changing face of acromegaly—advances in diagnosis and treatment. Nat Rev Endocrinol 2012;8:605−11.

[12] Dekkers OM, Biermasz NR, Pereira AM, Romijn JA, Vandenbroucke JP. Mortality in acromegaly: a metaanalysis. J Clin Endocrinol Metab 2008;93:61−7.

[13] Holdaway IM, Bolland MJ, Gamble GD. A meta-analysis of the effect of lowering serum levels of GH and IGF-1 on mortality in acromegaly. Eur J Endocrinol 2008;159:89−95.

[14] Eskander ET, Bonert V. Acromegaly as a complication of growth hormone therapy. AACE Clin Case Rep 2015;1:e282−94.

[15] Parkin JM. Incidence of growth hormone deficiency. Arch Dis Children 1974;4:904−5.

[16] Clemmons DR. The diagnosis and treatment of growth hormone deficiency in adults. Curr Opin Endocrinol Diab Obes 2010;17:377−83.

[17] Van Bunderen CC, van Nieuwpoort IC, Arwert LI, et al. Does growth hormone replacement therapy reduce mortality in adults with growth hormone deficiency? Data from the Dutch National Registry of Growth Hormone Treatment in Adults. J Clin Endocrinol Metab 2011;96:3151−9.

[18] Koltowska-Haffstrom M, Mattson AF, Shalet SM. Assessment of quality of life in adult patients with GH deficiency: KIMS contribution to clinical practice and pharmacoecomonic evaluations. Eur J Endocrinol 2009;161(Suppl. 1):S51−64.

[19] Scacchi M, Pincelli AI, Cavagnini F. Growth hormone in obesity. Int J Obes Relat Metab Disord 1999;23:260−71.

[20] Biller BMK, Samuels MH, Zagar A, et al. Sensitivity and specificity of six tests for the diagnosis of adult GH deficiency. J Clin Endocrinol Metab 2002;87:2067−79.

[21] Tzanela M, Zianni D, Bilariki K, et al. The effect of body mass index on the diagnosis of GH deficiency in patients at risk due to a pituitary insult European. J Endocrinol 2010;162:29−35.

[22] Hartman ML, Crowe BJ, Biller BMK, et al. Which patients do not require a growth hormone (GH) stimulation test for the diagnosis of adult GH deficiency? J Clin Endocrinol Metab 2002;8:477−85.

[23] Hanley JA, Saarela O, Stephens DA, Thalabard J-C. hGH isoform differential immunoassays applied to blood samples from athletes: decision limits for anti-doping testing. Growth Horm IGF Res 2014;24:205−15.

[24] Amed S, Devlin E, Hamilton J. Variation in growth hormone immunoassays in clinical practice in Canada. Horm Res 2008;69:290−4.

[25] Bidlingmaier M, Strasburger CJ. Growth hormone assays: current methodologies and their limitations. Pituitary 2007;10:115−19.

[26] Granada ML, Ulied A, Casanueva F, et al. Serum IGF-I measured by four different immunoassays in patients with adult GH deficiency or acromegaly and in a control population. Clin Endocrinol 2008;68:942−50.

[27] Gomez-Gomez C, Iglesias EM, Barallat J, et al. Lack of transferability between two automated immunoassays for serum IGF-I measurement. Clin Lab 2014;60(11):1859−64.

[28] Blum WF, Ranke MB, Kietzmann K, Gauggel E, Zeisel HJ, Bierich JR. A specific radioimmunoassay for the growth hormone (GH)-dependent somatomedin-binding protein: its use for diagnosis of GH deficiency. J Clin Endocrinol Metab 1990;70:1292−8.

[29] Owen WE, Roberts WL. Performance characteristics of the IMMULITE 2000 insulin-like growth factor binding protein-3 assay. Clin Chim Acta 2005;353:141−5.

[30] Dutta P, Korbonits M, Sachdeva N, et al. Can immediate postoperative random growth hormone levels predict long-term cure in patients with acromegaly? Neurol India 2016;64:252−8.

[31] Valette-Kasic S, Pulichino A-M, Gueydan M, et al. Congenital isolated adrenocorticotropin deficiency: an underestimated cause of neonatal death, explained by TPIT gene mutations. J Clin Endocrinol Metab 2005;90:1323−31.

[32] Hamrahian AH, Yuen KCJ, Gordon MB, et al. Revised GH and cortisol cut-points for the glucagon stimulation test in the evaluation of GH and hypothalamic-pituitary-adrenal axes in adults: results from a prospective randomized multicenter study. Pituitary 2016;19:332−41.

[33] Fiad TM, Kirby JM, Cunningham SK, et al. The overnight single-dose metyrapone test is a simple and reliable index of the hypothalamic-pituitary-adrenal axis. Clin Endocrinol 1994;40:603−9.

[34]   Giraldi FP, Saccani A, Cavagnini Fthe Study Group on the Hypothalamo-Pituitary-Adrenal Axis of the Italian Society of Endocrinology. Assessment of ACTH assay variability: a multicenter study. Eur J Endocrinol 2011;164:505−12.

[35]   Giraldi FP, Ambrogio AG. Variability in laboratory parameters used for management of Cushing's syndrome. Endocrine 2015;50(3):580−9.

[36]   Wahab F, Quinton R, Seminara SB. The kisspeptin signaling pathway and its role in human isolated GnRH deficiency. Mol Cell Endocrinol 2011;346:29−36.

[37]   Williams NI, Berga SL, Cameron JL. Synergism between psychosocial and metabolic stressors: impact on reproductive function in cynomolgus monkeys. Am J Physiol Endocrinol Metab 2007;293:E270−6.

[38]   Fenske W, Allolio B. Current state and future perspectives in the diagnosis of diabetes insipidus: a clinical review. J Clin Endocrinol Metab 2012;97:3426−37.

[39]   Tien R, Kucharczyk J, Kucharczyk W. MR imaging of the brain in patients with diabetes insipidus. AJNR 1991;12:533−42.

[40]   Schreckinger M, Szerlip N, Mittal S. Diabetes insipidus following resection of pituitary tumors. Clin Neurol Neurosurg 2013;115:121−6.

[41]   Fenske W, Quinkler M, Lorenz D, et al. Copeptin in the differential diagnosis of the polydipsia-polyuria syndrome—revisiting the direct and indirect water deprivation tests. J Clin Endocrinol Metab 2011;96:1506−15.

[42]   Milles JJ, Spruce B, Baylis PH. A comparison of diagnostic methods to differentiate diabetes insipidus from primary polyuria: a review of 21 patients. Acta Endocrinol 1983;104:410−16.

[43]   Morgenthaler NG, Struck J, Alonso C, Bergmann A. Assay for the measurement of copeptin, a stable peptide derived from the precursor of vasopressin. Clin Chem 2006;52:112−19.

[44]   Battista S, Audisio U, Galluzzo C, et al. Assessment of diagnostic and prognostic role of copeptin in the clinical setting of sepsis. Biomed Res Int 2016;2016:3624730.

[45]   Maisel A, Mueller C, Neath S-X, et al. Copeptin helps in the early detection of patients with acute myocardial infarction. JACC 2013;62:150−60.

[46]   Moeckel M, Searle J, Hamm C, et al. Early discharge using single cardiac troponin and copeptin testing in patients with suspected acute coronary syndrome (ACS): a randomized, controlled clinical process study. Eur Heart J 2015;36(6):369−76.

[47]   Balanescu S, Kopp P, Gaskill MB, et al. Correlation of plasma copeptin and vasopressin concentrations in hypo-, iso-, and hyperosmolar states. J Clin Endocrinol Metab 2011;96(4):1046−52.

[48]   Robertson GL. Antidiuretic hormone. Normal and disordered function. Endocrinol Metab Clin North Am 2001;30(3):671−94.

[49]   Timper K, Fenske W, Kuehn F, et al. Diagnostic accuracy of copeptin in the differential diagnosis of the polyuria-polydipsia syndrome: a prospective multicenter study. J Clin Endocrinol Metab 2015;100(6):2268−74.

FURTHER READING

Schoenmakers N, Alatzoglou KS, Chatterjee VK, et al. Recent advances in central congenital hypothyroidism. J Endocrinol 2015;227:R51−71.

Endocrinology and disorders of the reproductive system

8

Bernard Corenblum, MD[1] and Jessica Boyd, PhD[2]

[1]*Professor, Department of Medicine, Division of Endocrinology, University of Calgary, Calgary, Alberta, Canada*

[2]*Assistant Professor, Department of Pathology and Laboratory Medicine, University of Calgary, and Clinical Biochemist Calgary Laboratory Services, Calgary, Alberta, Canada*

CHAPTER OUTLINE

8.1 FEMALE REPRODUCTIVE ENDOCRINOLOGY

8.1.1 BASIC PHYSIOLOGY OF HYPOTHALAMUS—PITUITARY—OVARIAN AXIS

8.1.1.1 Puberty

Pubertal development occurs with changes of estrogen-driven breast development (thelarche), androgen-driven axillary and pubic hair growth (adrenarche), and finally the first menstrual period (menarche). Abnormal delay is if there is no thelarche by age 13.5, adrenarche by age 14, or menarche by age 16 (average onset is 12.8 years). There is a consideration to lower the normal age of menarche to 15. These all require estrogen production from the ovaries, and androgen production from the ovaries and adrenal glands. To have the first menstrual period, there

Endocrine Biomarkers. DOI: http://dx.doi.org/10.1016/B978-0-12-803412-5.00008-2

must be activity involving higher brain centers, the hypothalamus, the pituitary, ovarian follicles, and an intact uterus and anatomically normal outflow tract.

The initial control of the onset of puberty remains unknown. At the onset of pubertal development, there is activation of kisspeptin neurons in the periventricular and arcuate nuclei of the hypothalamus that act on the G-protein coupled receptor 54 (GPR54) receptor on the gonadotropin-releasing hormone (GnRH) neurons in the median preoptic area of the hypothalamus. This augments GnRH secretion in a pulsatile manner, increasing it in amplitude from the prepubertal state. The pituitary gonadotrope response requires this pulsatile secretion of GnRH in an adequate amplitude and frequency to elicit a normal luteinizing hormone (LH)/follicular-stimulating hormone (FSH) response.

8.1.1.2 Menstrual cycle

- The same gonadotrope cell of the pituitary secretes both LH and FSH. The relative amounts of each vary throughout the normal menstrual cycle and are modified by the frequency of GnRH pulses, feedback action of estradiol potentiating LH over FSH, and by various ovarian peptides such as inhibin. Lack of estrogen promotes FSH secretion, and the relative presence of adequate estrogen promotes LH secretion.
- The ovarian follicle consists of the oocyte ("egg") surrounded by a layer of granulosa cells. There is a constant growth of follicles, which develop by acquiring increased numbers of granulosa cells for over 6 months. Most of these follicles become apoptotic and cease to further develop.
- On any given day, there are a few developing follicles that are sufficiently mature at that time to respond to a rise of FSH secretion and thus continue to develop.
- If this signal is received, there is further granulosa cell proliferation and the acquisition of a new layer of cells (theca) to surround the granulosa cells.
- Although multiple follicles are initially recruited in each ovary, only one (the dominant follicle) will be selected to fully develop, thus resulting in a single ovulation, while the others then regress. This follicular growth is under the control of FSH/LH and intraovarian factors.
- LH acts on the theca cells to produce androgens, mainly androstenedione and testosterone.
- Androgens have local actions in that they diffuse from the theca cell layer to the granulosa cell layer, where they are converted by the aromatase enzyme to estrogen, the main estrogenic hormone secreted by the ovary.
- FSH acts on the granulosa cells to have many actions, one of which is to stimulate aromatase so that androgens may be aromatized to estrogens, which are subsequently secreted for a systemic effect. The major role for estrogen in this context is to act on the lining of the uterus, the endometrium, to induce a proliferative response, that may be subsequently changed by progesterone to a secretory response that will accept implantation of the fertilized oocyte (embryo).

- One problem is that the same estradiol needed for the endometrial response will also systemically suppress FSH secretion and thus halt further development of the follicle and continued estradiol secretion. Thus a complex interaction system is in place to prevent this. In the first few days, theca cell androgen secretion is somewhat held back by granulosa cell factors such as activin, acting in a paracrine action. At the same time, FSH acts on the granulosa cell to make multiple factors, resulting in induction of aromatase activity, upregulation of FSH receptors, induction of LH receptors (for the first time), and production of multiple other peptides to act in an autocrine manner on the same granulosa cell, and paracrine action on its layer of theca cells. Peptides such as inhibin are now produced by the granulosa cell in response to FSH and now can promote LH-induced androgen secretion from the theca cell, which are then aromatized to estradiol in the granulosa cells, and then systemically secreted. Thus there is a switch from inhibiting theca cell androgen secretion to then promoting it, as the dominant follicle is selected.

- Of the cohort of follicles that are recruited by the rise of FSH (antiatretic action) on Day 1 of the cycle, one will be biologically the most responsive to the actions of FSH. It will be the first to induce more FSH receptors, new LH receptors, and switch from the theca cell-inhibiting peptides to the theca cell-stimulating peptides. In this way, by about Day 5 of the cycle, the follicle that is destined to continue to grow, become dominant, and take part in ovulation, now begins to secrete more estradiol. This estradiol relatively suppresses FSH secretion, and thus inhibits further development of the remaining cohort of follicles, who biologically are not as developed as the now dominant follicle. This now results in the continued growth of only the single follicle and results in a single ovulation. The growing follicle from the preantral phase (Day 1) to the ovulatory phase (Day 14) is mainly responsible for the cyclic secretion of estrogen in the first 2 weeks.

- Ovulation results from increased LH pulsatility in both frequency and amplitude, resulting in a cumulative surge of LH that induces an inflammatory change in the dominant follicle wall, and the oocyte is extruded. Whether there is also a concomitant surge in GnRH secretion is suggested by the change in LH pulse frequency, but still not established. The major site of enhanced LH secretion is from estradiol potentiating the action of GnRH on the gonadotrope. Again, estradiol changes the gonadotrope from an FSH-secreting cell to an LH-secreting cell as does the increased (presumed) frequency of GnRH pulsation. Just prior to ovulation the preceding estradiol surge (signaling the presence of a dominant follicle) begins to fall, and the serum progesterone begins to rise, suggesting a change in steroidogenesis and the beginning of the corpus luteum.

- Following ovulation, progesterone begins to be produced from the remaining cells (corpus luteum). This changes the endometrial lining from the proliferative phase to the secretory phase. If a pregnancy is not achieved, then after about 2 weeks the corpus luteum stops making progesterone, the levels fall, and the absence of progesterone results in a menstrual bleed. At this time the fall of sex steroids and inhibin from the corpus luteum act to enhance new FSH secretion, and the cycle begins again.

- The fluctuations of hormones through the menstrual cycle reflect this: FSH rises for about 5 days (initiating follicle recruitment), then falls (reflecting estrogen secretion from the selected follicle); LH is steady until the preovulatory surge. Early in the cycle, with estrogen levels the lowest, FSH is higher than LH. But as estradiol is progressively made, FSH falls, and thus LH is higher than FSH. This very much reflects the feedback of estradiol on the gonadotrope. As the follicle becomes dominant and close to ovulation, the estradiol levels surge reflecting the mass of granulosa cells, which then induces the LH surge and ovulation, now that a dominant follicle has signaled its readiness for ovulation by the amount of estrogen being produced. FSH and LH are variable and generally low during the luteal phase, with progesterone acting along with estradiol in the feedback control on GnRH/LH/FSH.
- There are multiple other locally acting factors. For example, C-type natriuretic peptide is a follicle-stimulating factor. Recent work has shown a role for the local Hippo signaling pathway. There are several oocyte-derived factors that promote granulosa cell growth and development. Anti-Mullerian hormone (AMH) used to be called Mullerian-inhibiting factor for its role in male sexual differentiation. It is also produced by granulosa cells of growing follicles and is measurable in blood. It has a role in follicular growth and development.
- Because AMH is made in developing follicles, it may be used as a measurement of the ovarian reserve and thus ovarian aging. It is suggested to be superior to serum FSH and inhibin as a marker for aging and fecundity. It is increased in disease states that have greater numbers of developing follicles, such as polycystic ovary syndrome (PCOS), and may be used to make this diagnosis. This is becoming more routine in clinical practice.

8.1.1.3 Example of the menstrual cycle physiology in clinical use

A woman with hypothalamic amenorrhea was given ovulation induction with exogenous pulsatile GnRH. Serum estradiol on Day 1 was 49 pmol/L, and Day 5 was 205 pmol/L, indicating normal growth of the follicles, and likely selection of the dominant follicle. Day 10 had a serum estradiol of 710 pmol/L, and ultrasound demonstrated a dominant follicle 14 mm in size. Day 13 the serum estradiol was 290 pmol/L and concern was raised that a "crash" had occurred and the follicle was lost. But the estradiol does fall just before ovulation, so the pulsatile GnRH was continued, and the serum progesterone was taken the next day. It was 21 nmol/L, indicating ovulation had occurred, and this was in fact a normal cycle. She did conceive and delivered a healthy infant. Early discontinuation of the GnRH would have been a mistake.

KEY POINTS

The menstrual cycle depends on a careful balance of endocrine and paracrine/autocrine effects of steroid and peptide factors.

- The ratio of FSH to LH reflects the estrogen status of that woman at that time. (FSH > LH is a low estrogen state; LH > FSH is a normal to high estrogen state).
- A very, very high LH to FSH is an ovulatory surge.
- Interpretation of serum levels of LH, FSH, estradiol, and progesterone all depend on the stage of the cycle. Random measurements will be difficult or impossible to interpret.
- In a woman with amenorrhea, FSH higher than LH indicated a hypoestrogenized state, and LH higher that FSH indicates a well-estrogenized state.
- In a hypoestrogenized state, high FSH indicates ovarian insufficiency, whereas normal or low FSH indicates a central cause (it is inappropriate for the low estrogen state). This interpretation is physiology in action.

8.1.2 CLINICAL CONDITIONS

8.1.2.1 Amenorrhea/oligomenorrhea/anovulation

8.1.2.1.1 Clinical picture

The menstrual cycle is a vital sign in reproductive aged women. In the nonpregnant population, amenorrhea (no menses for 6 months) occurs in about 4%, and oligomenorrhea (9 or fewer menses a year) in about 10%. This could be of recent onset or longstanding. The identification of the underlying cause and pathophysiological disorder is important, establishing if there is estrogen deficiency (with issues of bone health and possibly cardiovascular health), or unopposed estrogen (with concerns of menstrual chaos, endometrial hyperplasia, and risk of endometrial carcinoma). Patient concerns such as anxiety, prolonged, or heavy menses (dysfunctional bleeding due to anovulation), or infertility need to be addressed. There may be associated problems, such as hirsutism (androgen excess). Depending on the underlying endocrine disorder, there may be comorbid disorders such as diabetes, autoimmune diseases, or pituitary tumors.

Amenorrhea may be primary (never had a spontaneous menstrual period) or secondary with loss of previously established menses. The same causes may underlie both presentations.

8.1.2.1.2 Causes

The causes of amenorrhea can be classified according to high gonadotropin and low/normal gonadotropin, presence of sexual characteristics or absence of sexual characteristics. We recommend to separate causes into five clinical groups that are easily definable, with a differential diagnosis of causes within each of these groups. Over 95% of cases will fall into one of these groups. Pregnancy is always excluded first.

1. *Outflow tract obstruction.* There could be a congenital problem such as agenesis of the Mullerian system (female embryologic structures) for various reasons. This would cause primary amenorrhea in a woman with otherwise normal pubertal development. Acquired outflow tract problems are almost always discovered on history taking. The hormonal changes of the menstrual cycle are intact, so a simple test would be to do a serum progesterone; if normal then repeat in 2 weeks. If the luteal phase progesterone is documented but no menses, then a disorder of uterine lining/patency to outflow tract is diagnosed. This would be further assessed by gynecology or pelvic ultrasonography.

2. *Ovarian failure or insufficiency.* About 1% of women before age 40 develop amenorrhea, hypoestrogenism, and elevated serum FSH (0.1% < age 30, 0.01% < age 20) [1]. The only way to make this diagnosis (or rule it out) is with the serum FSH (which will be high). At age 50 it would be a natural menopause (with 2SD around the age of menopause being 40−60 years). Before age 40 is premature. The natural history of premature ovarian insufficiency is variable with 25% returning to menses, and 5% still achieving conception without assisted reproduction therapy. Long-term complications of hypoestrogenism exist, which include osteoporosis and fracture, and possibly increased and earlier cardiovascular disease. Many underlying causes exist such as iatrogenic, immunological, chromosomal (such as Fragile X), and specific gene mutations. The latter may underlie familial tendencies. If thought to be immunological, associated immunological disorders may exist at presentation or be more likely to occur in the future. The most common would be autoimmune thyroid disease.

3. *Polycystic Ovarian Syndrome (PCOS).* This is the most common endocrine disorder, in fact of any disorder, in reproductive aged women. It occurs in 6%−10% of women. It is defined clinically by the variable presence of irregular cycles (chronic estrogenized anovulation), hyperandrogenism as shown clinically (hirsutism, and to a lesser degree, acne and scalp alopecia) or biochemically, polycystic ovaries on ultrasound, and the exclusion of other causes that may give estrogenized anovulation and/or hirsutism.

   Examples of disorders that must be excluded clinically or biochemically are Cushing's syndrome, hypothyroidism, nonclassical congenital adrenal hyperplasia (NCCAH), androgen-secreting tumors, and simple obesity. A normal serum FSH and prolactin are needed to rule out ovarian insufficiency and hyperprolactinemia. This is an estrogenized state, so the serum LH is usually higher that the serum FSH although modified by obesity and menstrual status. The history of the symptoms is important. Chronic symptoms that are slowly progressive reflect a benign process, whereas ominous, rapidly progressive causes such as the rare androgen-secreting tumors, would be very unlikely in this situation.

   There is no unified hypothesis for this clinical syndrome, and multiple areas of dysfunction in the entire course of follicular growth and secretion have been shown. Not surprisingly the most consistent finding is of increased androgen secretion from the theca cells, from what may be a disorder of regulation or an inherent secretory disorder in the theca cells. Insulin binds to its receptor on the theca cells, and if supraphysiological (due to insulin resistance/obesity elsewhere), then it may augment LH action on androgen synthesis. Thus states of insulin resistance may aggravate the clinical features of PCOS. Weight gain is a good example of an environmental factor (obesity) via enhanced insulin secretion that can bring out the genetic tendency to PCOS so that it may clinically manifest for the first time after weight gain, or by weight gain aggravating already existing symptoms of PCOS. This also shows the potential for symptoms to reverse to some degree in response to weight loss. Impaired folliculogenesis and steroidogenesis thus result from several extraovarian and intraovarian factors, as well as yet to be defined genetic factors.

The clinical problems reflect the presenting symptoms, and may be or may not be a concern to the woman, but may be of separate concern to the physician. These include hirsutism/acne/alopecia, anovulatory bleeding which is erratic/unpredictable/heavy, and infertility. Medical issues include a risk of endometrial hyperplasia and carcinoma, and increased risk of depression, sleep apnea, and hepatic steatosis [2]. There is twice the risk of insulin resistance with associated metabolic syndrome, and 2−10 times the risk of type 2 diabetes compared to women of similar body mass index (BMI) [2]. Obesity incidence is twice that of the background population (60%), with geographic/racial differences. Obesity, especially abdominal, will aggravate all these concerns. On the other hand, many of the long-term negative outcomes of PCOS may reflect obesity itself. There is increased risk of PCOS and/or metabolic syndrome in first degree relatives, the latter including males. PCOS is found in 40% of sisters and mothers [2].

4. *Hyperprolactinemia.* Prolactin does have a physiological inhibition on pulsatile GnRH secretion, possibly via kisspeptin-1. Mild elevation disrupts GnRH secretion somewhat and results in estrogenized anovulation. More marked elevation disrupts the GnRH secretion enough to result in hypoestrogenized amenorrhea. Thus prolactin needs to be measured in all oligomenorrheic or amenorrheic cases. Regardless of clinical presentation, such as the presence or absence of galactorrhea, the only way to diagnose hyperprolactinemia is by its measurement, and this is also the only way to rule it out.

The cause of the hyperprolactinemia needs to be determined as hyperprolactinemia is a biochemical measurement, not a diagnosis. Mild elevation needs to be repeated two more times as often it is normal on repeat. Not repeating and starting further detailed investigation/therapy is a common clinical error.

If truly high, the degree may determine the cause. Prolactin concentrations over 100 μg/L usually is associated with a prolactin-secreting pituitary tumor.

In general the higher the serum prolactin, the larger the tumor. Unrelated incidental findings in the pituitary found on MRI need to be considered, so the diagnosis of a microprolactinoma is often presumptive.

An increased prolactin under 100 μg/L may still be associated with a small pituitary tumor, but other causes need to be excluded. These include drug ingestion, chest wall irritation, primary hypothyroidism, and kidney and liver failure. If none of these are present, the MRI separates the entities by being normal (idiopathic hyperprolactinemia, by exclusion), small tumor (presumed microprolactinoma), and large nonsecreting tumor. The latter mildly increases the serum prolactin by its position inhibiting full dopaminergic transmission to the pituitary, resulting in disconnection hyperprolactinemia (see Chapter 7, Pituitary disorders), whereas a macroprolactinoma of a large size would have a much higher serum prolactin. Of course, exceptions occur.

Elevated serum prolactin without clinical symptoms suggests the presence of macroprolactinemia (a laboratory assay artifact), which is clinically insignificant. Mild hyperprolactinemia may not have an identifiable underlying cause, and there may be no symptoms, so this may be otherwise normal.

Common errors in laboratory investigation of hyperprolactinemia (also see Chapter 7*, Pituitary disorders)*

a. *Serum prolactin should only be measured in the presence of symptoms (menstrual change, galactorrhea, decreased libido) or the finding of a sellar mass. Measuring it at other times, such as unexplained infertility, risks inappropriate diagnosis, and treatment by finding an elevated level that is not clinically significant.*

b. *Not repeating if mild hyperprolactinemia is found.*

c. *Treating hyperprolactinemia without diagnosing the underlying cause. The treatment is first of the underlying cause, and natural history does differ depending on the cause. For example, dopamine agonists would shrink the size of a macroprolactinoma, but not likely for a nonsecreting macroadenoma.*

5. *Hypothalamic–pituitary disease or dysfunction.* There may be organic disease along this axis. It often, but not always, may have clinical clues such as known underlying disease (e.g., sarcoidosis, head trauma), clinical evidence of a mass or infiltrate, clinical evidence of more extensive pituitary hormone hyposecretion, or clinical evidence of increased hormonal secretion. If there are no clinical clues to the presence of organic disease, then to do or not to do an MRI is a clinical decision.

Clinical pearl: in nonorganic hypothalamic amenorrhea, the serum prolactin is usually low normal. If it is high normal or above normal, then a central mass effect from an underlying organic cause is strongly suspected.

Much more commonly there is a disruption of normal pulsatile secretion of GnRH resulting in insufficient pituitary secretion of LH and FSH to drive the menstrual cycle. It is physiologically a return to the set point of the prepubertal/ early pubertal state. This may be a prepubertal problem presenting with delayed puberty and primary amenorrhea (never had any menses), or acquired as secondary amenorrhea in a woman who has previously completed pubertal development and established some menstrual function. She is usually hypoestrogenized as shown by serum estradiol or lack of response (menses) to a 5-day course of oral progesterone. Serum prolactin is low normal. Serum FSH and LH are normal, which are inappropriately normal for the hypoestrogenized state. To reflect the decreased effect of estradiol on the gonadotrope of the pituitary, serum FSH is usually greater than LH.

The three major causes of nonorganic hypothalamic amenorrhea (also called functional, idiopathic) are stress, excessive exercise, and impaired nutrition. Some of these may be extreme, such as athletic amenorrhea or eating disorders. These three factors are also additive in their effect [3]. There are some women that appear to be more responsive to these factors than other women, indicating an individual susceptibility. In some, there may be a genetic basis that enhances susceptibility to these factors. Lifestyle changes (weight gain, decreased exercise) and cognitive behavioral therapy have been shown to help reverse this hypothalamic amenorrheic state [4].

Idiopathic hypothalamic amenorrhea is a diagnosis by exclusion. The natural history is that if weight is maintained or increased, then 75% return to normal menses within 5 years.

The clinical problems may reflect her concern, such as infertility. Medical concerns, besides identifying the underlying cause, include the short-term and long-term effects of hypoestrogenization.

Major concern: an underlying eating disorder may or may not be clinically obvious, but needs to be considered. The associated psychopathology would be important.

Major concern: any malnourished state may give this clinical picture. The underlying causes of decreased appetite and weight loss are extensive. A careful evaluation for the presence of underlying systemic disease must occur. If no evidence of systemic disease is obvious, this still needs to be considered with continued follow-up.

8.1.2.1.3 Iatrogenic causes

- Ovarian failure may result from oophorectomy, radiation, or chemotherapy.
- Many medications may produce symptomatic hyperprolactinemia, especially psychoactive and antinausea agents.
- Anticonvulsants such as valproate have been associated with PCOS [5].
- Some physicians forget that the oral contraceptive use may result in amenorrhea and yet yields uninterpretable laboratory tests for menstrual cycle evaluation.

KEY POINTS

After pregnancy is excluded the five pathophysiological groups are: outflow tract obstruction, ovarian insufficiency, PCOS, hyperprolactinemia, and hypothalamic—pituitary.

- Clinical evaluation often indicates which of these is present.
- Basic investigation includes a pregnancy test, serum FSH, serum prolactin, and progesterone challenge (or serum estradiol) to establish estrogenization status.
- Each diagnostic group may be identified: outflow tract (history), ovarian insufficiency (elevated FSH), hyperprolactinemia (elevated prolactin), PCOS (hyperandrogenism, normal FSH and prolactin, normal estrogen state) and hypothalamic—pituitary (normal FSH and prolactin, low estrogen state). Some overlap may exist in some cases.

8.1.2.2 Diagnostic approach to amenorrhea

- First, rule out pregnancy.
- Exclude primary hypothyroidism by serum thyroid-stimulating hormone (TSH).
- Is an irregularly menstruating woman ovulating? Do a serum progesterone about 1 week before expected menses. This timing may be unpredictable, so may have to be performed weekly a few times, until menses occur. This would not be practical if there is severe oligomenorrhea of 3 months or more. Ovulation detection kits have similar limitations and are much more expensive. Basal body temperature charts are unreliable.

   With marked oligomenorrhea, even if ovulation may occasionally occur, it would be infrequent enough to be treated as if it did not occur.
- What is the cause of the oligo—amenorrhea? Clinical assessment is important and does rule out outflow tract obstruction. All women need a serum FSH and prolactin to either diagnose or rule out ovarian insufficiency and hyperprolactinemia.

   Hyperprolactinemia requires its own investigation as to the underlying cause.
- Ovarian insufficiency may require screening for other autoimmune disease, such as Hashimoto thyroiditis and Graves' disease, and although rare, Addison's disease (adrenal insufficiency) should be considered. Antiadrenal antibodies may be somewhat predictive of future adrenal failure.
- If FSH and prolactin are normal, then the next step is to separate PCOS from hypothalamic amenorrhea.
- PCOS may be supported by the well-estrogenized state (LH > FSH, normal serum estradiol, positive response (menses) to progestin administration), and evidence of hyperandrogenism, clinically, or biochemically (serum testosterone, assessment of unbound testosterone), with or without ovarian ultrasound showing polycystic changes. Before PCOS is diagnosed, Cushing's syndrome may need exclusion clinically, or by screening tests of 1 mg dexamethasone suppression, 24-h urinary free cortisol, or midnight salivary

cortisol. Adult onset or NCCAH may need to be screened for by an early morning serum 17-hydroxyprogesterone (see Chapter 5, Adrenal disorders). Rapid onset of hyperandrogenic may indicate a testosterone-secreting ovarian tumor or androgen-secreting adrenal tumor be screened by a serum testosterone and serum dihydroepiandrosterone-sulfate (DHEA-S).

Some clinicians routinely do all of these tests, regardless of the clinical presentation.

- Hypothalamic amenorrhea is usually poorly estrogenized (LH < FSH, low serum estradiol, negative response to progestin administration), and no hyperandrogenism. The ovarian ultrasound is not helpful and may be confusing.
- In the presence of hypothalamic—pituitary dysfunction, it may be necessary to assess other pituitary hormone secretion with an early morning serum cortisol and free thyroxine. Organic central disease may have to be excluded by MRI of the pituitary—hypothalamus. The serum prolactin may be mildly elevated in this situation.

Clinical overlap: these diagnostic categories are usually easy to separate by the above approach in over 95% of cases. There is recognized overlap in some cases.

- Some women with hypothalamic amenorrhea may recover and appear to have PCOS. The combination of hypothalamic amenorrhea and PCOS has been observed and now described [6]. Thus a woman with hypothalamic amenorrhea may not upon recovery get return of regular menses.
- Some women with oligomenorrhea may have features of both PCOS and hypothalamic oligomenorrhea, and the overlap may be difficult to separate them and clearly make a diagnosis.
- Mild hyperprolactinemia may give estrogenized oligomenorrhea similar to PCOS, and PCOS may be associated with mild hyperprolactinemia. Separating mild hyperprolactinemia as the primary problem from PCOS may be difficult. Laboratory testing is inconclusive.

8.1.2.3 Diagnostic definitions and misuse of laboratory tests in menstrual disorders

- *Incorrect timing of and insufficient measurement frequency of serum progesterone in attempts to document ovulation.* Serum progesterone to indicate the presence of ovulation or anovulation needs to be drawn within 12 days (but not within 2 days) before the subsequent period; not 3 weeks after the previous period if menses are more than 4 weeks apart. This is one test that may need to be repeated, and ovulation may vary from cycle to cycle; reproducibility is important.
- Assumption that regular menstrual cycles are synonymous with ovulation. Up to 10% of women with infertility and PCOS have somewhat regular cycles yet still have anovulation; the regular cycles being regular anovulatory withdrawal

bleeds. Thus even with regular cycles a Day 21–23 serum progesterone is needed to ensure that the regular cycles are indeed ovulatory.

- Failure to investigate other causes of irregular menses besides PCOS. PCOS has several different diagnostic categories. The laboratory needs to ensure a normal serum FSH and prolactin (mildly high prolactin may be accepted), then the clinical diagnosis is a variable group of irregular anovulatory cycles, hyperandrogenism (can be clinical or biochemical), polycystic ovaries on ultrasound, and ruling out other similar disorders (laboratory again). There may be mild hyperprolactinemia or mild elevation of serum 17-hydroxyprogesterone with the presence of PCOS. The latter should not be measured during the luteal phase.

- Diagnosis of PCOS without clinical or laboratory evidence. Polycystic ovaries found on ultrasound do not diagnose PCOS. This may be found in otherwise normal women. PCOS requires the clinical or laboratory picture to use polycystic ovaries on ultrasound as evidence of PCOS.

- Over or under-investigation of associated metabolic risks that accompany PCOS. If PCOS is present, laboratory investigation of diabetes, cardiovascular risk factors, and liver function is needed. This can be overdone if supposed risk factors are measured that are not established to be useful nor necessarily need to be treated. There is no role for insulin measurement.

- Failure to consider that some menstrual disorders may fluctuate in both clinical and laboratory presentation. The natural history may fluctuate, such as spontaneous resolution of premature ovarian failure and thus laboratory abnormalities such as high FSH may change. Even established hyperprolactinemia may normalize, especially after a pregnancy. PCOS may reverse after weight loss. Hypothalamic amenorrhea may reverse after weight gain.

- Premature conclusions or imaging tests ordered based on a single hormone measurement. Laboratory abnormalities such as elevated prolactin and FSH need to be repeated and may require further consultation with a Clinical Chemist.

- Laboratory investigation of menstrual disorders while the patient is taking contraceptive estrogens. Laboratory tests done on the oral contraceptive cannot be interpreted (other than prolactin); they may be attempted to be measured on the 7th day off the pills.

- Random reproductive hormone measures ordered without reference to the specific phase of the menstrual cycle. Knowledge of the hormonal changes of the menstrual cycle is needed to interpret baseline laboratory measurements of FSH, LH, estradiol, and progesterone. This is best assessed by correlating the results with the subsequent menstrual period, to determine where in the follicular phase, ovulatory phase, or luteal phase the sample was taken. For example a high LH (and lower FSH) is an ovulatory surge, not a menopausal reading. A minimally high serum estradiol and concomitantly nonsuppressed serum progesterone may be normal if it is preovulatory. Follicular phase levels of FSH and LH and their relative ratio will vary from early to mid to late follicular phase, and may be rather low in the luteal phase.

8.1.2.4 Emerging test of ovarian function: Serum anti-Mullerian hormone [7]

- AMH is produced by the granulosa cells of small growing follicles and continues to be produced from the selected follicles into the menstrual cycle up to the antral stage, then production declines with selection of the dominant follicle. It may have a role in the physiological production of the dominant follicle.
- Because it is produced from the multiple small follicles in the ovarian pool, it may be used as a measurement of ovarian reserve. It may be predictive of response to assisted conception, and of the approaching menopausal state.
- PCOS has increased numbers of growing follicles, and AMH measurement has a potential role in the diagnosis of PCOS by being elevated above the normal range.
- Its physiological role in normal ovarian function, the normal menstrual cycle, and ovarian health and disease is still being investigated. Its role in oocyte growth and differentiation, and on follicular steroidogenesis is still being defined.

Example: A 34-year-old woman had secondary hypothalamic amenorrhea and infertility. Serum FSH, LH, and estradiol were low. Ovulation induction with pulsatile GnRH and then gonadotropins (FSH and LH) failed to produce and response in steroidogenesis (rise of estradiol) nor folliculogenesis (follicle growth). Serum AMH was found to be very low, indicating poor ovarian reserve of follicles, in a woman who could not show this by having a high FSH, because this could not occur in the state of hypothalamic amenorrhea.

8.1.2.5 Case example with discussion

A 27-year-old woman presented with infertility. She and her partner had been trying to conceive for 2 years.

Since her first menstrual period at age 15, she has been irregular, 2–3 months apart. At age 17 there was an onset of dark coarse hair on her chin, between her breasts, and below her umbilicus. Previous teenage acne has not disappeared. Her BMI at age 16 was about 25 and is now 28. She denies symptoms of a pituitary mass, breast discharge, hot flushes, or thyroid dysfunction.

She has no risk factors on history to increase the risk of having tubal obstruction.

Family history finds her mother and maternal aunt with type 2 diabetes, father has had angina since age 50. No family history of hirsutism or irregular menses.

Investigation:

- Partner was sent for semen analysis which was normal.
- She had a negative pregnancy test.

Question 1: What initial laboratory tests should be ordered?

Answer: Her serum FSH was 5 IU/L (reference intervals: follicular phase 2−10 IU/L; luteal phase 1−9 IU/L; mid-cycle peak 3−33 IU/L), LH was 9 IU/L (reference intervals: follicular phase 1−13 IU/L; luteal phase 1−17 IU/L; mid-cycle peak 8−76 IU/L), prolactin 17 μg/L (reference interval: 0−25 μg/L). This rules out ovarian failure and hyperprolactinemia. The LH is higher than the FSH, so she is estrogenized.

Question 2: With a working diagnosis of PCOS, what additional conditions should be considered both clinically and by laboratory testing?

Answer:
- Because of her being overweight, an overnight dexamethasone suppression test found her morning cortisol to be 29 nmol/L (1 μg/dL). This rules out Cushing's syndrome.
- On a different day, her early morning serum 17-hydroxyprogesterone was 4 nmol/L (13.33 ng/mL) which makes nonclassical (adult onset) congenital adrenal hyperplasia very unlikely.

Question 3: How might you determine if she is having any ovulatory cycles?

Answer: Serum progesterone was measured after 7 weeks from the previous period, weekly for 3 weeks, until the subsequent period did occur. The serum progesterone taken 7 days before the next period was 1.2 nmol/L (0.4 ng/mL) (reference interval: luteal phase 5.0−76.0 nmol/L or 1.6−24 ng/mL). Thus she is anovulatory.

Question 4: To what extent should you use laboratory tests to investigate the hirsutism?

Answer: It is debatable if serum androgens need to be measured (despite guidelines). Because of the 10-year history of slow onset and progression of hirsutism, a benign underlying cause is most certain. If measured, the serum testosterone is high normal, serum sex hormone binding globulin (SHBG) low, and calculated free androgen index (FAI) high. If measured, serum DHEA-S is normal, androstenedione is 25% above normal (which really does not need to be measured).

Question 5: Should ovarian U/S be ordered to confirm "PCOS"?

Answer: It is debatable if ovarian ultrasound is needed as the diagnosis is already made *without* it. If done, it is compatible with polycystic ovaries (so it changes nothing, even if it was normal). If negative, it does not change the clinical diagnosis.

Question 6: Given the associated metabolic risks with PCOS, what additional tests should be considered?

Answer: Associated laboratory tests: fasting serum glucose (especially if contemplating pregnancy), lipid profile, liver function, and TSH all normal. Rubella titer in the immunized range.

Question 7: What treatment approach may be considered?

Answer: She has been taking folate supplementation all along.

Weight loss should help many of the manifestations and should be encouraged. From a practical standpoint, her menses were quite irregular even at her lower and normal BMI.

Nevertheless if time is an option, lifestyle recommendations are useful, especially with her diabetes risk.

She does not wish to delay ovulation induction. Clomiphene citrate 50 mg a day for 5 days was given. Ovulation occurs about a week after discontinuation of the clomiphene, so serum progesterone 16 days after stopping the clomiphene was in the anovulatory range at 1.4 nmol/L. Menses did occur, and clomiphene citrate was given at 100 mg a day for 5 days; 16 days after stopping the clomiphene her serum progesterone was in the luteal phase at 34 nmol/L (11 ng/mL) (reference interval: luteal phase 5.0−76.0 nmol/L or 1.6−24 ng/mL). This indicates that ovulation almost certainly has occurred. Menses did occur, so she did not conceive. The same dose of clomiphene citrate was given again, 16 days later the serum progesterone was 39 nmol/L (12 ng/mL). Menses did not occur, and 10 days later the serum β human chorionic gonadotropin (ßhCG) was positive. A healthy birth occurred 9 months later.

This shows how the same serum progesterone used to determine that ovulation is not occurring was then used to indicate when the ovulatory defect had been corrected. If that was the only reason why pregnancy had not occurred, then by correcting the ovulation defect, this couple now had a close to normal rate of conception.

KEY POINTS

The menstrual cycle is a dynamic cyclic event, and hormonal measurements used to assess the cycle must be interpreted as to when they were taken. They are a snapshot of a dynamic process.

- Disruption of the menstrual cycle results in oligo/amenorrhea. Investigation needs to be problem-oriented and interpreted in context. For example, if well estrogenized then a normal serum FSH is appropriate, but if hypoestrogenized the same level of FSH is inappropriate and indicates a central cause.
- Diagnosis of the underlying pathophysiological entity then dictates a new set of problems that arise from that diagnostic category, with appropriate further investigations.
- Investigation is problem specific. The appropriate test is that which is used to answer the clinical question being considered.

8.1.3 ANDROGEN EXCESS IN WOMEN

8.1.3.1 Clinical picture

Androgen excess is a clinical presentation, not a biochemical one. The most closely associated clinical sign of hyperandrogenism is the presence of hirsutism, lesser correlated with acne, and least correlated with scalp alopecia. In some women and populations with minimal hair response to androgens, hirsutism is a poor clinical sign of hyperandrogenism. Marked androgen excess results in signs of virilization such as deepening of the voice, clitoromegaly, increased muscle mass, breast atrophy, and amenorrhea.

Hirsutism is the most common focus, and its complaint is dependent on societal norms.

Hirsutism is an increased amount of androgen-mediated hair (terminal hair, coarse and pigmented) in the male pattern distribution. This requires both objective and subjective input. Hirsutism is terminal hair on the face, upper lip, neck, chest, lower abdomen, inner thighs, and lower back. The hair becomes coarse, grows faster, continues to spread over the area, and is pigmented in an appropriate manner to her background pigmentation. It is the biological response of the hair follicle to androgen action (intracellular dihydrotestosterone (DHT)). It depends on the amount of androgen, and the androgen receptor in the target tissue. The latter is why there is racial differences in response, and variations within any given racial group. The wide variation in hair distribution among normal adult men, despite similar androgens, will be reflected by a similar variation in hair response among women. Hirsutism may be found in the presence of apparently normal serum androgen levels. Androgens also increase sebum production and thus acne.

Hirsutism as defined is common (5%–15% of non-Asians and 2% of Asians) [8]. Onset may be at the time of puberty, or any time after, including after menopause. The course is usually rather stable, slow in onset and in progression. More rapid onset and progression, and onset in later years all suggest a new and possible ominous cause.

The degree of hirsutism can be clinically semi-quantified by various scoring methods, such as the Ferriman–Gallwey score, which may also be used to follow responses to therapy.

It is important to look at hair only in the androgen distribution; increased hair in the nonandrogen distribution is called hypertrichosis and is not androgen mediated. In the Caucasian North America population, Ferriman–Gallwey index is 6, 8, or 10 in 8%, 5%, and 2% of the population [8]. It is limited by intra- and interobserver variation. It is considered high if 8 or more in non-Asians, and 3 or more in Asians. Experienced clinicians usually do not use these tools and ultimately patient perception often drives investigation and/or management.

Increased skin thickness may be seen in women and may be present in those who do not manifest hirsutism in response to increased androgen production.

Androgens are synthesized in both the adrenals and ovaries, peripherally converted from weaker androgens such as DHEA and DHEA-S, then intracellularly converted to DHT, which then acts on the androgen receptor. The DHT receptor has variable degrees of sensitivity due to polymorphisms in the number of glutamine repeats in exon 1 [9]. Fewer repeats result in increased androgen receptor activation, and this form is more commonly seen in women with hirsutism, acne, and androgenic alopecia.

Furthermore, binding proteins in the blood may modify androgen delivery to the target tissues. The circulating level of androgens correlates poorly with the severity of hirsutism/acne. Also, measurement of androgens in the blood may not correlate well with the actual biological activity in the skin. They may not reflect the intracellular production of DHT.

In the clinical state where hyperandrogenism is present in a woman, there are multiple possible causes. It is important to establish the cause by clinical evaluation and any further investigation that is focused and based on the individual case. Needless over investigation is discouraged.

8.1.3.2 Causes of androgen excess

The cause is identified by clinical features, relevant laboratory assays, and occasionally by genetic analysis. There are several major causes.

8.1.3.2.1 Benign causes

- Even in tertiary referral centers, benign causes account for more than 98% of cases. By far the most common cause is PCOS.

     Variations that do not make the PCOS diagnosis are idiopathic hyperandrogenism (high androgens, normal ovulatory cycles, normal ovarian ultrasound) and idiopathic hirsutism (normal androgens, normal cycles, normal ultrasound). The source of the elevated androgens is almost always the ovary, occasionally with an adrenal contribution.
- Idiopathic hirsutism has normal androgens, cycles, and ovarian ultrasound. This may result from enhanced intracellular conversion of testosterone to DHT, increased androgen receptor sensitivity, or the mild defects in androgen secretion may not be detected by the sensitivity of the laboratory test, or a clear normal range is not established for the population in question.
- NCCAH—A benign adrenal cause is an enzyme defect in steroidogenesis (e.g., 21-hydroxylase deficiency from loss of function mutations in the CYP21A2 gene). The resulting increase in compensatory ACTH stimulation and excessive androgen synthesis depends on where the enzyme defect is, and degree of deficiency of normal action of the enzyme. The spectrum of possible gene mutations and resultant enzyme activity can influence age of onset and symptoms. Onset at birth (1 in 20,000) can present with salt wasting and ambiguous genitalia in females (enzyme activity estimated at <1%), in childhood with premature adrenarche and growth spurt followed by short stature (enzyme activity 2%−10%), or postpubertal onset that may have

simple hirsutism or associated menstrual irregularity that resembles PCOS (1 in 1000) (enzyme activity 11%−75%). In the hirsutism population, adult onset (nonclassical) 21-hydroxylase CAH occurs about 3%. Certain ethnic groups have a higher incidence, such as Latinos, Pacific Northwest aboriginals, Indian, Middle Eastern, and eastern European Ashkenazi Jews. Half of these affected women have polycystic ovary morphology on ultrasound.

• Other mutations involving 11β-hydroxylase (CYP11B1) and 3β-hydroxysteroid dehydrogenase (HSD3B2) are rare.
• Insulin resistance.

Insulin resistance can bring out or aggravate hirsutism. Insulin binds to its own receptor in the theca cell of the ovary and can potentiate LH stimulation of androgen synthesis. Insulin also acts on the liver to decrease SHBG production, lowering the total serum testosterone but then increasing the free serum testosterone.

In fact a low serum SHBG is a marker of insulin resistance with hyperinsulinemia.

Insulin acts over its spectrum of elevation. Thus a primary problem with hyperinsulinism or acquired from weight gain may be a factor. The clinical clue to marked insulin resistance may be the presence of acanthosis nigricans in the skin. The presence of insulin resistance increases the risk of components of metabolic syndrome and type 2 diabetes.

8.1.3.2.2 Rare causes

Rare ominous causes of hyperandrogenism include ovarian tumors (large arrhenoblastoma, small lipoid, and hilus cell tumors) and adrenal tumors (large carcinomas, small adenomas). Ovarian tumors mainly secrete testosterone, whereas DHEA-S is a marker for an adrenal source. Androstenedione may come from both.

KEY POINTS

Most mild or slow onset cases of hirsutism have a benign etiology

• Rapid or severe forms of hirsutism may be due to a malignant cause.
• Regular menses with mild hirsutism is usually a sign of idiopathic androgen excess and almost never due to a serious underlying cause.
• Irregular menses with obesity and hirsutism are almost always due to PCOS but a small list of other causes should still be considered and excluded before PCOS is diagnosed.

8.1.3.3 Special situations

• A rare but severe form of ovarian hyperandrogenism is hyperthecosis. This consists of nests of luteinized theca cells scattered throughout the ovarian stroma. There is much greater production of androgens that what occurs with PCOS.

- Severe insulin resistance, such as with lipodystrophy syndromes, is associated with marked hyperandrogenism with resulting virilization.
- Some medications may produce hirsutism (such as steroids, hCG, some anticonvulsants such as valproate) or hypertrichosis (such as some anticonvulsants, diazoxide).
- Gestational hirsutism may occur during a pregnancy, from a luteoma or theca-lutein cyst.
- Cushing's syndrome has cortisol excess and causes other than most adrenal adenomas have concomitant increased androgen secretion. Cushing's syndrome is sufficiently rare that screening is not done as a routine, but rather with clinical evaluation.
- Glucocorticoid resistance syndrome (Chrousos' syndrome) has enhanced androgen production due to the compensatory increased ACTH stimulation of the adrenal. Despite high cortisol levels that are nonsuppressible to dexamethasone, the person does not look Cushingoid.

8.1.3.4 Diagnostic approach

There needs to be identification and then quantification of the hirsutism, then determination of the cause. Some women may complain of terminal hair in the androgen areas, but if the amount is not excessive compared to the racially matched population, then no further investigation is needed, just a discussion of therapeutic options, because it is still androgen mediated.

- Clinical evaluation with history (including family history) and physical exam are the most important diagnostic tool.
- A slow onset and progression of hirsutism so strongly suggest a benign cause, that an extensive investigation may not change the clinical approach and thus not be cost-effective. This is especially true if the menses remain regular over the course.
- In part, standard androgen measurements may have poor accuracy and lack specificity. For example a hirsute woman found to have a normal total serum testosterone may be misdiagnosed as thought not having an androgen-mediated problem. Androgens commonly measured are serum testosterone (better with a measurement of the unbound fraction), androstenedione (controversial), and DHEA-S (usually not needed).
- The presence of suspected PCOS dictates some clinicians to check ovarian morphology, and some to measure serum AMH. If PCOS is present, screening for cardiovascular risk factors, and liver function is needed. In addition a careful screen for type 2 diabetes at this time, and repeatedly in the long term, is recommended, especially before a planned pregnancy.
- The role of obesity may be indirectly assessed by the serum SHBG, which will reflect insulin resistance, which decreases the SHBG.
- Suspected Cushing's syndrome is screened by the 24-h urine-free cortisol, 1 mg overnight dexamethasone suppression test, or midnight salivary cortisol.

- Adult onset congenital adrenal hyperplasia is screened by the morning 17-hydroxy progesterone; if borderline then its response to an injection of ACTH may be measured. These tests need to be done in the follicular phase, not in the luteal phase with its elevated serum progesterone. Genetic analysis for gene mutation may be needed to be definitive.
- Rapid onset and progression of hirsutism, especially with some virilization, dictate a serum testosterone to screen for a testosterone-secreting ovarian tumor, and serum DHEA-S to screen for an androgen-secreting adrenal tumor (likely a carcinoma). If elevated, such as over twice normal, imaging studies are initiated. However, in these tumors, there is a spectrum of elevation that runs from the benign levels to the ominous tumor levels. The laboratory test is no substitute for clinical evaluation and must be interpreted in that context. More ominous tumors are of rapid clinical onset, but some more benign tumors may run a slow progressive course and an index of suspicion is always needed.

  Again, clinical judgement is needed. For example, there may be recent onset of hirsutism because an oral contraceptive was recently discontinued and return of ovarian activity resulted in the new onset of hirsutism.
- It is controversial, but there are advocates for an assessment of the hormonal profile in all cases, while others support a more selective approach for any investigation [10]. The most basic test is a serum testosterone, most commonly by double antibody radioimmunoassay or immunometric assay, hopefully with the interassay coefficient of variation <10%, with a carefully determined local normal range from a local healthy nonhirsute population. Improvements in techniques such as by liquid chromatography combined with mass spectrometry make these measurements potentially more clinically relevant. As well, the range of steroids measured may be extended when using LC–MS. For all assays, it is critical to establish normal ranges in healthy regularly menstruating nonhirsute women in the target population
- In the past, stimulation tests and suppression tests for androgen secretion had been used to assess the source of secretion and the risk of autonomous secretion, but these are of minimal clinical use and have generally been replaced by baseline androgen assays.

8.1.4 LABORATORY CONSIDERATIONS TO ASSESS FEMALE REPRODUCTIVE ENDOCRINOLOGY

8.1.4.1 Preanalytical considerations

8.1.4.1.1 Biological variation affects when specimens should be collected

As with other hormones, reproductive hormones concentrations are affected by several factors including patient sex, patient age (prepubertal, pubertal, adult, menopause), diurnal variation, menstrual cycle, and trimester of pregnancy. These variations will dictate when it is appropriate to collect the specimen.

8.1.4.1.2 Serum/plasma is the specimen of choice

The specimen of choice for most reproductive hormone testing is serum (tubes with no additives or serum separator tube) or plasma (EDTA, heparin). It is always important to check the specimen requirements for your local laboratory as different testing methodologies may have different requirements.

8.1.4.2 Analytical considerations

8.1.4.2.1 Immunoassay is the most common method for measuring reproductive hormones

Currently, most assays for reproductive hormones are performed by immunoassays, which use an antibody to recognize the analyte of interest. Also, immunoassays can be run on automated analyzer which allows for multiplexing of tests and faster turnaround times. However, immunoassays suffer from cross-reactivity that can impact their specificity and can be expensive.

8.1.4.2.2 Immunoassay interferences

As with any immunoassay, tests for reproductive hormones can be subject to interference by several different mechanisms. A brief description is provided later. For a more extensive discussion on immunoassays, please see Chapter 1, Endocrine laboratory testing: Excellence, errors, and the need for collaboration between clinicians and chemists.

- Heterophile antibodies: Endogenously produced antibodies that can produce false negative or false positive results.
- Hook effect: High concentrations of analyte causing a false negative result.
- Assay specificity: Selection of antibodies for use in immunoassays can be challenging considering the many different isoforms (products of a single gene with different posttranslational modifications) available for most reproductive hormones, particularly hCG, FSH, and LH. In addition, antibody epitopes are not well standardized (except for hCG) which has two practical implications for interpretation of these laboratory results. First the absolute value obtained from methods developed by different manufacturers can differ from each other, so that results cannot be directly compared. Thus it is preferable, for a patient to be monitored over time by the same method and technology from the same manufacturer. Secondly the degree of interference from hormones and drugs that share structure homology to the analyte of interest will vary between assay manufacturers. Often laboratories state the common interferences with their assay in an interpretive comment attached to laboratory results. However, consultation with the Clinical Chemist about potential sources of interference is recommended when investigating a result that does correlate well with the patient's clinical presentation.
- Biotin: Also known as vitamin B7, biotin has recently become a popular supplement to improve hair, skin, and nail health. Biotin, because of its strong affinity for the protein streptavidin, is a common component of immunoassays

used in clinical chemistry laboratories. Consumption of large doses of biotin supplements can interfere with immunoassays that use biotin. Thus most of the current immunoassays, including those used to measure reproductive hormones, can be affected when patients are taking highdoses of biotin. The direction of interference depends on the principle of the immunoassay used. In general, for competitive assays biotin causes falsely high results, whereas for noncompetitive immunoassays biotin causes falsely low results.

8.1.4.2.3 Liquid chromatography–tandem mass spectrometry for analysis of reproductive hormones is becoming more common due to increased specificity over immunoassays

Liquid chromatography–tandem mass spectrometry (LC–MS/MS) has superior specificity compared to immunoassays and is also able to measure several analytes simultaneously. However, it has several shortcomings that have so far prevented this technology from becoming the first line test for reproductive hormones. First, LC–MS/MS has a high initial cost to purchase the instrument which is prohibitive for many laboratories. Second, it is a very sophisticated technology that requires experienced technologists to run and interpret the data. Finally, there are currently few options for front end automation of LC–MS/MS systems meaning that sample preparation is still performed manually by laboratory staff which is time consuming and labor intensive. In general, it takes at least 4 h to prepare and analyze a sample by LC–MS/MS.

8.1.4.2.4 Liquid chromatography–tandem mass spectrometry interferences

LC–MS/MS is now considered the gold standard for measurement of many reproductive hormones. Currently the vast majority of LC–MS/MS assays for reproductive hormones available at clinical laboratories are laboratory-developed tests (LDTs). The term LDT refers to any assay that has been developed in house by the laboratory instead of purchased from a manufacturer. Although many laboratories may measure the same analyte by LC–MS/MS, there may be differences in the chromatography (separation), the MS conditions used, and the interferences in each assay. Therefore the LC–MS/MS results from different laboratories may not be directly comparable.

In addition, although LC–MS/MS analysis is much more specific than immunoassays, they are not immune to interferences. One documented interference is from collection of specimens for steroid analysis in gel separator tubes. Some gels contain a compound with similar molecular weight to steroids and can interfere with hormone measurement by some LC–MS/MS methods [11].

8.1.4.2.5 Technical requirements for assays measuring reproductive hormones

There are several important points that should be considered for measuring steroid hormones as in many cases more than one hormone needs to be measured. Also, these hormones have different blood concentrations that can make their measurement more challenging. Thus the method of choice should have: (1) a wide linear measuring range so, hormones with very high and very low blood levels can be measured simultaneously, (2) acceptable specificity to ensure measuring active hormone not the metabolites or other steroid hormones with similar structure, (3) acceptable precision to measure clinically significant changes in hormone concentrations over time. Not all assays meet these requirements (particularly with respect to low end sensitivity), which has spurred development of better immunoassays as well as LC−MS/MS assays which are much more specific.

8.1.4.2.6 Assays for reproductive hormones are not standardized

With the exception of hCG, assays for measuring reproductive hormones are not standardized. Practically, this means that the numeric values produced from assays from different manufacturers can be very different, even if the same sample is tested. This makes tracking results from patients who are tested at different laboratories extremely difficult as changes in the result may be due to the change in assay rather than a physiological change. To avoid this issue, patients should be tested at the same laboratory so that the results are from the same assay.

Assay standardization is a complex process that requires research into the analyte and its isoforms, agreement on nomenclature and antibody epitopes, development of primary reference materials and establishment of a reference laboratory network. These tools can then be used by assay manufacturers to standardize their assay back to the primary reference material and continually monitor the value assignment using the reference laboratory network. Assay standardization can take many years to complete, but has been shown to be very successful (such as in the case of cholesterol measurement).

Many working groups have been formed to improve assay standardization for the reproductive hormones. This includes the Centers for Disease Control (CDC) Hormone Standardization Program (HoSt) which is working on estradiol and testosterone measurements. The World Health Organization and the International Federation of Clinical Chemistry (IFCC) have developed international standards for FSH, LH, hCG, and prolactin.

8.1.4.2.7 Protein binding

Steroid hormones, being lipophilic, are carried in the blood by proteins. For the sex steroids, SHBG and albumin are the major binding proteins. This means that these hormones are present in three forms: free, SHBG bound, and albumin bound. SHBG has high affinity, but low capacity for sex steroids whereas albumin has high capacity but low affinity. The lower affinity binding by albumin means these steroids are more available to interact with receptors.

For routine assessment of reproductive steroid hormones, usually the total hormone concentration is measured. Free testosterone may be ordered in select cases (further discussion is provided in Section 8.2.3).

8.1.4.3 Postanalytical considerations

8.1.4.3.1 Reference intervals should be specific for the assay and the population served by the laboratory

Reference intervals are a critical component for interpretation of laboratory results. Because of the assay standardization issues discussed earlier, reference intervals for reproductive hormones need to be developed for the specific assay being used by the laboratory. It is also extremely important that the laboratory uses samples from the local population to establish the reference intervals.

8.1.4.3.2 Trimester or menstrual phase specific reference intervals are often appended as an interpretive comment

Age- and/or sex-specific reference intervals are available for most analytes. As age and sex information is routinely collected by the laboratory on all patients, it can be used to automatically append the correct reference interval to laboratory results. However, automatic application of reference intervals specific for phase of the menstrual cycle or trimester of pregnancy is more difficult as these data are not usually collected by the laboratory. In these cases, reference intervals are appended as an interpretive comment to the laboratory result.

8.1.4.4 Discussion of specific analytes

- *Human chorionic gonadotropin*
  - hCG is primarily measured to detect pregnancy. However, it is also used in the diagnosis and follow-up of ectopic and molar pregnancies, as a tumor marker for several gestational trophoblastic and germ cell cancers, and for prenatal screening marker for Downs syndrome.
  - In early pregnancy, serum hCG measurements predictably double every 48–72 h. This doubling time, along with ultrasound, can be used to confirm pregnancy. hCG continues to rise until weeks 12–14. In ectopic pregnancy, hCG doubling time is much longer and should be considered if serial serum hCG measurements do not double by 72 h.
  - Structurally hCG is a glycoprotein dimer composed alpha and beta subunits. The alpha subunit is shared with three other hormones (FSH, LH, and TSH) whereas the beta subunit is specific for hCG. Most laboratory testing refers to this analyte as beta hCG (βhCG) to indicate that the assay detects the beta subunit and can differentiate hCG from TSH, LH, and FSH. hCG is present in several forms including the main active form, intact hCG (which has both alpha and beta subunits) and many forms that have either reduced or no biological activity (eg. free beta hCG, free alpha hCG, nicked hCG, nicked free beta hCG, and the beta core fragment). Because of the many forms of hCG, considerable work has been performed to ensure that assays detect the clinically significant forms [12].
  - hCG also has several posttranslational modifications that can change over the course of a pregnancy or from a tumor. In early pregnancy the

hyperglycosylated form of hCG (HhCG) predominates. However, in the second and third trimesters the normally glycosylated hCG form takes over. These modifications can affect detection in certain immunoassays and may explain why a urine pregnancy test may not be positive in late pregnancy. HhCG does play a role as a marker of implantation and has been investigated as a marker for failed and ectopic pregnancies [13]. In trophoblastic disease, HhCG can be used as a tumor marker to detect recurrence [13]. HhCG testing is performed at reference laboratories.

- In 1994, the International Federation of Clinical Chemistry formed an hCG working group with the goal of standardizing hCG assays. This group has standardized the characterization and nomenclature of the different hCG isoforms, standardized antibody epitopes for immunoassays, and developed hCG primary reference standards [12,14,15].

- ßhCG can be tested from either urine or serum. Urine tests are usually performed using a point of care slide in the laboratory, in medical clinics, or as a home pregnancy test. This testing is qualitative and usually has a higher detection limit than for serum tests. ßhCG testing is performed using sandwich immunoassay, using two antibodies to recognize ßhCG. Depending on the specificity of these antibodies, different hCG isoforms are detected which also varies by manufacturer [16]. Urine ßhCG tests are susceptible to the hook effect which can occur in late pregnancy or in patients with molar pregnancy or trophoblastic disease where hCG concentrations can be much higher than what is seen in pregnancy. Serum ßhCG tests are typically performed on a main chemistry analyzer in the laboratory and also use sandwich immunoassay methodology. These tests are quantitative and usually have a detection limit of 5 IU/L. hCG doubling time can be determined from serial serum hCG measurements to aid in diagnosis of ectopic pregnancy. Differences in detection limits between serum and urine assays can lead to a serum test being positive 1−3 days before the urine test. In general, serum testing is preferred over urine testing for hCG. However, a urine test is invaluable for detecting pregnancy in situations where time is of the essence (e.g., female patients going for radiological studies, administration of drugs, etc.).

- In perimenopausal and menopausal women, ßhCG can sometimes be detected at low concentrations just above the pregnancy cut-off. However, the hCG is not due to pregnancy but is produced along with FSH and LH from the pituitary due to the lack of negative feedback control from estradiol. Although FSH and LH are the main products of the lack of negative feedback control, a small amount of hCG is also produced because of its structural similarity. To identify these cases, serial measurements of hCG can be performed to confirm that the values are not increasing. FSH or LH can also be measured, with a high result indicative of pituitary production of hCG [17].

- *FSH/LH*
  - FSH and LH are commonly measured to assess proper functioning and response of the hypothalamus–pituitary–gonadal axis in both females and males.
  - Structurally, FSH and LH are both glycoprotein dimers with an alpha and beta subunit. The alpha subunit is shared between FSH, LH, TSH, and hCG while the beta subunit confers specificity. As with hCG, circulating FSH and LH have many different isoforms and degrees of glycosylation [18]. Glycosylation of LH and FSH changes their biological activity, and the pituitary plays a role in changing the glycosylation status during the course of the menstrual cycle [19]. LH and FSH isoforms are not as well standardized or characterized as hCG.
  - Typically, plasma or serum is used to measure FSH and LH. Timing of specimen collection is important as FSH and LH levels change by age, sex, and menstrual cycle stage.
  - Measurement of FSH and LH is commonly performed using sandwich immunoassay, which uses two antibodies to detect FSH or LH, with at least one antibody recognizing the beta subunit of each hormone. Owing to FSH and LH isoforms, as well as the assay standardization issues discussed earlier, the value obtained for FSH and LH from different instrument platforms may not be the same. However, recent studies comparing the analytical performance of various chemistry analyzers for measurement of FSH and LH show relatively good agreement between the methods [20].
  - In addition to erroneous FSH and LH results due to human anti-mouse antibodies (HAMA) and hook effect, cases of macro-FSH (attachment of an immunoglobulin to FSH) have also been reported [21]. Macro-FSH is cleared much more slowly from the body than FSH, and so causes prolonged, elevated levels of FSH that do not fit the clinical picture.
- *Estradiol*
  - As a steroid hormone, estradiol has a cholesterol backbone. Because of its lipophilic nature, 97% of circulating estradiol is protein bound either by albumin or SHBG. Albumin binds steroid hormones, like estradiol, with high capacity but low affinity, whereas SHBG binds them with high affinity and low capacity.
  - Factors that change binding protein concentrations will affect the total concentration of estrogen in the blood. For example, SHBG is increased by estrogens (and so are naturally higher in women than in men), pregnancy, hyperthyroidism, and certain drugs (e.g., phenytoin). Thus estradiol is increased in these conditions.
  - Estradiol concentrations also change by age, gender, and during the menstrual cycle.
  - Estradiol is measured in serum or plasma.
  - Currently available assays for estradiol can measure a wide range of concentrations; however, some are not sensitive enough to be clinically useful

in patient groups with very low estradiol concentrations (postmenopausal women, breast cancer patients undergoing antiaromatase therapy, children, men) [22,23]. For these patients, samples may need to be sent to reference laboratories that have assays (either immunoassay or mass spectrometry) developed specifically for measurement of low estradiol concentrations.

- Estradiol is measured using immunoassay as a total measurement (i.e., both free and protein bound forms). Some assays may experience up to 10% cross-reactivity with other estrogens (estrone, estriol) or metabolites, which will falsely elevate the estradiol result. Cross-reactivity with some drugs, including mometasone, has also been reported.
- Estradiol immunoassays are not standardized. In addition, results from immunoassays are usually very different compared to LC−MS/MS [22]. This is likely due to increased interference from cross-reactivity in immunoassays. The CDC has recently begun a harmonization project to develop primary reference materials for estradiol.

- *Prolactin*
  - Prolactin is most commonly measured in patients undergoing workup for infertility, amenorrhea, and/or galactorrhea. In women, one of the more common reasons for ordering prolactin is to exclude hyperprolactinemia in the investigation of PCOS.
  - Prolactin can be elevated due to many causes. In pregnancy, prolactin is produced in preparation for breastfeeding. Similarly, chest wall stimulation (usually from a suckling infant) promotes prolactin production. In hypothyroidism, increased concentrations of thyroid-releasing hormone promote prolactin release. Drugs can increase prolactin levels through several different mechanisms. The major pathway is via disruption of the dopamine inhibitory signal that is used to suppress prolactin release from the anterior pituitary [24]. A second, but less well understood, pathway involves increased production of serotonin. Antipsychotics are the most common cause of drug-related hyperprolactinemia [24,25]; however, other drugs such as morphine, estrogens, verapamil, tricyclic antidepressants, selective serotonin reuptake inhibitors, monoamine oxidase inhibitors, and H2 antagonists may also elevate prolactin. Natural products, supplements, herbal teas, or other natural products may contain ingredients such as milkweed, which can increase prolactin secretion. Prolactinomas ectopically produce prolactin. Macroprolactin (discussed in more detail later) is an immunoglobulin conjugated form of prolactin, which is cleared slowly from the body and can produce prolonged positive results in laboratory testing [26].
  - Prolactin is subject to diurnal variation, with a peak early in the day and nadir while sleeping.
  - Prolactin is present in blood in three forms: monomer (active), dimer (inactive), and macroprolactin (inactive). The prolactin monomer represents 80% of all circulating prolactin and has a serum half-life of

25−40 min. Macroprolactin is formed when prolactin is combined with an immunoglobulin and has been suggested to be more common in women with autoimmune diseases or who are pregnant. Macroprolactin has a much longer half-life than the other two forms and so can present as a persistently elevated serum prolactin result, but without the symptoms of hyperprolactinemia.

- In cases where macroprolactin is suspected the laboratory can perform a polyethylene glycol (PEG) precipitation. PEG will precipitate macroprolactin, leaving behind the biologically active and relevant prolactin monomer. The specimen can then be remeasured for prolactin and the difference in values pre- and postprecipitation compared. A postprecipitation prolactin value ≤ 30% of the preprecipitation value has been suggested as an appropriate cut-off to identify the presence of macroprolactin [27]. Electrophoresis is the definitive method to confirm macroprolactin and may be used in difficult cases where PEG precipitation is inconclusive [28].
- Prolactin is measured using immunoassay. Although an international standard is available, prolactin measurements are not well standardized between manufacturers. The differences arise due to different reactivities to the various forms of prolactin described earlier. All assays react well with the biologically active prolactin monomer but have differing reactivities to big prolactin and macroprolactin [29].

- *Anti-mullerian hormone*
  - AMH has been proposed for evaluation of ovarian reserve [30] and in evaluation of PCOS [31].
  - AMH is a glycosylated homodimer in the transforming growth factor beta family.
  - As it is a relatively new marker, there are few assays for measuring AMH. Much of the early work on AMH was done on one of two commercial immunoassays that used antibodies targeting different parts of the AMH molecule. Because of these differences in specificities the absolute values obtained from the assays were often different which has made setting cutoffs for clinical use of AMH challenging [32]. AMH assays are now available for automated chemistry analyzers, some of which have been standardized back to older assays to provide equivalent results.

KEY POINTS

Immunoassays are the main assay for measuring reproductive hormones. They are subject to several interferences that may cause false positive or false negative results.

- Mass spectrometry assays for many reproductive hormones are now available at reference laboratories.
- Patients being serially monitored should use the same laboratory for measurement of reproductive hormones. This will allow for the results to be directly comparable.
- The Clinical Chemist should be consulted regarding laboratory results that are not consistent with the clinical picture.

8.1.4.5 Misuse of laboratory tests in investigation of female androgen excess

- *Excessive importance given to "normal" androgen levels in mild androgen excess states.* A hirsute woman can have a normal serum testosterone (and DHEA-S). This may reflect the relative inaccuracy of standard tests, or enhanced biological effects at the intracellular level. A common cause is low SHBG. If testosterone is repeated and adjusted for low SHBG, the free androgen level will often now be elevated. The mistake is the physician may consider the "normal" laboratory test means "no problem," rather than relying upon the patient presentation. Androgen-mediated hair that is bothersome to the woman requires androgen lowering and/or blocking therapy. Treat the patient, not the laboratory test.

- *Failure to recognize that PCOS can have mild increases in various androgens.* Women with PCOS may have mild elevation of serum prolactin, 17-hydroxyprogesterone, and DHEA-S. These are often overinterpreted and thus the over investigation begins.

- *Attributing androgen excess to high prolactin.* An elevated serum prolactin does not cause hirsutism.

- *Excessive repeated testosterone measurements by multiple means.* There are several methods to measure non-SHBG testosterone. There is no reason to do all variations of these measurements in the same woman.

- *Inappropriate isolated use of androgen levels to gauge "severity" of clinical disease.* Hirsutism will result from androgen concentration and intracellular sensitivity/genetics. There is a poor correlation with the measured laboratory tests and the degree of hirsutism.

- *Use of poorly defined female androgen reference ranges.* The normal range of androgens is determined by sampling a group of regular cycling women. But up to 10% may have some degree of hirsutism or PCOS, expanding the normal range of the measurement. The normal controls would best have the presence of hirsutism or anovulation excluded.

KEY POINTS

Hyperandrogenism presents differently in children than in adults. Girls may show early growth spurt, early onset of adrenarche (axillary and pubic hair), body odor, and behavioral changes. Adult women show a variable spectrum of hirsutism, acne, and androgen alopecia.

- Hyperandrogenism is common in adult women, but clinical presentation depends on personal assessment modified by cultural norms.
- A very high percentage in adult women are of benign causation, most commonly PCOS.
- The rare ominous cause of hyperandrogenism is usually different enough in presentation that it appears distinct from the much more common benign causes.
- Clinical evaluation is the most important diagnostic tool.
- Laboratory evaluation is used to refine the clinical impression and diagnose rare ominous causes. It must be used in conjunction with the clinical impression, and not by itself.

8.1.4.6 Case examples with discussion

8.1.4.6.1 Case example 1

A 54-year-old woman, menopausal for 2 years, presents with a 6-month history of rapid onset and progression of hirsutism. She was otherwise well and taking no medications. Clinical exam was normal, no abdominal or pelvic mass was felt, no clinical signs of Cushing's syndrome.

> *Question 1*: *What tests should be ordered at this stage?*
>
> *Answer*: Serum testosterone was over twice normal, 7.2 nmol/L (207 ng/dL) (reference interval: 0.5–2.0 nmol/L or 14–58 ng/dL). Serum DHEA-S was normal. This is worrisome because both the brief onset and the marked elevation in testosterone. However, abdominal and pelvic CT scan was normal.
>
> More careful history taking found that her partner has been treated for low libido for 9 months. He used testosterone cream before bedtime and did not wear pajamas. It did stimulate libido. Transfer of the cream to the patient was assumed. Her partner was asked to use pajamas to cover the cream, and in 1 month her serum testosterone had fallen back to the postmenopausal level.

8.1.4.6.2 Case example 2

A 39-year-old Caucasian woman presented with hirsutism and amenorrhea. Prior to 8 months ago, she had no increased hair in the midline and had regular monthly menses. Since then there has been a rapid onset of dark coarse hair above her lip, on the chin and neck, around the areolae and between the breasts, a line above the navel, and a thick line below the navel, inner thighs, and small area of the back. Teenage acne reappeared. Increased greasy skin and body odor were noted. Amenorrhea has occurred for the past 5 months.

There was no family history of similar problems, she took no medications, including over the counter or illicit supplements. Review of symptoms was normal. Blood pressure was normal.

Physical exam found the dark coarse hair in the androgen distribution, skinfold thickness increased, no acanthosis nigricans. She did not have any Cushingoid nor acromegalic features. Abdominal exam thought to be normal. Pelvic exam was refused as she was virginal. There was some temporal balding. Voice seemed deep for a woman.

> *Question 1*: *Is this just a usual case of hirsutism or are there other concerns?*
>
> *Answer*: The previously normal woman with rapid and late onset of clinical hyperandrogenism suggests an acquired process. The early signs of virilization suggest a marked increase in androgen secretion.

Question 2: *What laboratory tests, if any, should be done?*

Answer: Serum total testosterone was three times above normal, serum DHEA-S was four times above normal. These suggest a major androgen secretory process, adrenal in origin.

Question 3: *What should be done now?*

Answer: Imaging of the abdomen by CT scan showed a 6-cm mass in the left adrenal gland.

Question 4: Final diagnosis?

Answer: Adrenal carcinoma.

8.2 MALE REPRODUCTIVE ENDOCRINOLOGY

8.2.1 PHYSIOLOGY OF THE HYPOTHALAMUS—PITUITARY—TESTICULAR AXIS

- Testosterone production is needed to develop and maintain the male secondary sexual characteristics, and acts on target tissues to maintain body composition, cardiovascular function, cognitive function, etc. Adequate intratesticular testosterone is needed to initiate and maintain spermatogenesis and fertility. This situation requires 50 times the concentration of testosterone than what is in peripheral blood.
- With pubertal development, higher brain centers induce adequate hypothalamic pulsatile secretion of GnRH which acts on the gonadotrope of the pituitary to secrete LH and FSH, which then acts on testicular receptors. LH acts on the Leydig cells to initiate steroidogenesis for testosterone production and secretion. Secreted testosterone is bound to proteins in the circulation, mainly SHBG, and to a lesser extent albumin. The latter may be available to tissues, along with unbound "free" testosterone. Testosterone may be further aromatized to estradiol, or converted to DHT by the intracellular enzyme 5α-reductase. The net biological activity is the combined action of all three steroids. For example, bone effect is via estradiol, whereas virilizing (masculinizing) effects are via DHT.
- Feedback to the hypothalamus—pituitary is complex and involves all three steroids, but mainly testosterone. Sertoli cells produce two proteins, inhibin and activin, that have local effects on spermatogenesis and also selectively feedback to inhibit and stimulate FSH secretion (thus their names). There are multiple paracrine factors within the testes that have as of yet undefined roles in spermatogenesis and sperm function. Local production of a binding protein helps keep intratesticular testosterone in the concentration needed for spermatogenesis.

- There is a diurnal rhythm in normal testosterone secretion, highest upon awakening and then is falls throughout the day. This rhythm may normally disappear in the elderly.
- With increased aging the general male population demonstrates that SHBG production tends to increase while total testosterone decreases. The net effect is a gradual reduction in unbound "free" testosterone. With multiple levels of action, it is uncertain if these measurable changes in the circulation translate into changes in tissue effects or intratesticular levels. There may be a role of increased comorbid disease in the aging population augmenting these changes. Replacement therapy in this situation has not been definitive [33,34].

KEY POINTS

Puberty and subsequent adult testicular function depend on the brain, hypothalamus and pituitary gland—stimulating testicular steroidogenesis, and gametogenesis.

- A steady-state classical feedback loop results from hypothalamus—pituitary and testicular interaction.
- Testicular function may decrease over the span of adult to elderly life.
- Physiological/clinical changes reflect both the normal and decreased action of testosterone.
- Laboratory measurements and their interpretation need to take into account the secretion and transport of testosterone, as well as the variable phenotype in biological response.

8.2.2 CLINICAL CONDITIONS

8.2.2.1 Hypogonadism

- Hypogonadism at birth presents with ambiguous genitalia or micropenis.
- At puberty, hypogonadism presents with delayed or incomplete puberty, with failure to develop full secondary sexual characteristics.
- An adult who has otherwise normally virilized demonstrates a different clinical picture. Decreased spermatogenesis may result in infertility. Decreased testosterone production results in many or no symptoms, depending on its degree. Borderline low values probably result in no symptoms and show a poor clinical response to replacement therapy. More marked deficiency is associated with decreased libido, decreased spontaneous erections, loss of sexual hair, decreased muscle mass and strength, loss of bone mass and increased fragility fracture and height loss, increased body fat, hot flushes, shrinking testes, gynecomastia, and various mood changes including depression. With true testosterone deficiency, most of these symptoms and altered body composition do improve with testosterone replacement therapy.
- The slow decline in testosterone levels with aging has been associated with aging changes such as fatigue, erectile dysfunction, and change in body composition (less bone and muscle, more fat). Although they are in association, a cause and effect relationship has not been demonstrated, and

replacement testosterone therapy has been inconclusive regarding benefits. Thus these aging symptoms are likely not due to a falling testosterone. It is possible the age-related change may be secondary to the increase in comorbid disease that occurs in the aging population, and the somewhat low testosterone being secondary to the comorbid chronic diseases. When men reporting good health are studied, there is little decrease of testosterone with aging. In this context, serum testosterone acts as a vital sign [35].

8.2.2.2 Causes of male hypogonadism

Acquired hypogonadism in the adult male will result from central deficiency of GnRH-LH/FSH secretion, or from testicular disease. This results in infertility from deceased spermatogenesis and/or the clinical picture of decreased testosterone secretion. The underlying causative disease process may add to the clinical picture.

8.2.2.2.1 Central deficiency of gonadotropin-releasing hormone—luteinizing hormone/follicular-stimulating hormone secretion

Central disease may be from underlying organic disease in the hypothalamus—pituitary. This could be mass, infiltrate, or any other pathological process. There may be associated mass symptoms or of other pituitary hormone deficiency. Hyperprolactinemia from any cause may inhibit GnRH production. Isolated defect in gonadotrope secretion may result from iron overload and deposition, such as hemochromatosis. Nonorganic suppression of GnRH secretion may result from marked systemic disease, nutritional deficiency, extreme exercise without adequate nutritional support, or eating disorders. Serum testosterone may fall during acute or subacute illness, due to central effects. This includes severe emotional stress. It is best not to measure testosterone at the time of acute illness, especially during hospital admission. Rather, wait until recovery.

Testicular disease may result from trauma, torsion, iatrogenic (chemotherapy, radiation), or congenital disease such as Klinefelter syndrome. There are many causes of isolated defects in spermatogenesis, with normal testosterone production.

Delayed puberty presents with the same range of central and testicular causes. Isolated defects in GnRH production or in the GnRH receptor would also present this way. A syndrome of congenital GnRH deficiency with anosmia is Kallmann syndrome and represents fetal problems of migration of GnRH neurons and olfactory neurons. Congenital defects in GnRH/GnRH receptor need to be differentiated from simple delayed puberty which eventually evolves normally.

8.2.2.3 Special situations

- One of the most common causes of suppression of the hypothalamus—pituitary control of testosterone secretion is the use of anabolic steroids. This could be supraphysiological such as for body building, or simple long-term testosterone replacement therapy. There is a recovery time after discontinuation with clinical testosterone deficiency and poor spermatogenesis until recovery may occur.

- Medications may affect hypothalamus—pituitary—testicular function. For example, chronic opioid use can suppress GnRH-LH secretion. This is especially the case with methadone, due to its long duration of action. Another example is the GnRH agonists that are used to inhibit testosterone secretion for diseases such as prostate cancer. Glucocorticoids can inhibit gonadotrope function.

- Obesity in its extreme results in measurable low total and free testosterone but inappropriately normal serum LH. There may be several mechanisms, such as decreased SHBG, but also increased peripheral aromatization of testosterone to estradiol which feeds back to suppress LH secretion.

- The aging male has been associated with decreasing serum testosterone. Cross-sectional longitudinal epidemiological studies have repeatedly shown this [14]. This has been associated with nonspecific symptoms such as fatigue, irritability, loss of vigor, and sexual symptoms such as decreased libido and erectile dysfunction. The abnormalities are more common with older age, increased waist size, and association of comorbid chronic disease such as chronic obstructive pulmonary disorder, diabetes, cardiovascular disease, and malignancy. Studies in men reporting good health do not show this age-related change [35]. Thus the fall in testosterone may reflect the response to comorbid chronic disease, as well as the changes with aging and changes that affect the total testosterone assay, such as changes in SHBG. The fall in measured testosterone is not associated with any increase in serum LH. Studies of testosterone replacement in this population have been inconsistent regarding outcomes. Present recommendations are that testosterone replacement not be offered to men with this clinical picture, unless there is known disease affecting the pituitary or testes, and there are clearly abnormal testosterone levels. Insensitive laboratory testing or poorly defined normal ranges contribute to the inappropriate use of testosterone replacement therapy. There is also not a defined lower range at which clinical symptoms may occur. It is likely that different symptoms and physiological changes occur at different testosterone levels, and vary amongst individuals.

- For the appropriately diagnosed older man with definitely low serum testosterone, there are emerging data that beneficial clinical outcomes may be achieved [33,34]. The long-term risk of therapy has yet to be clearly evaluated and remains controversial.

- Chronic illness is associated with a fall in total serum testosterone. The greater the degree of the illness, the lower the average serum testosterone. If treated, such as obstructive sleep apnea, type 2 diabetes, or obesity, then the serum testosterone does tend to rise toward normal. The lowest quartile of serum testosterone is associated with a higher mortality rate, especially with cardiovascular disease. This likely reflects the lower testosterone being related to the comorbid disease, making it a vital sign or of predictive outcome.

Testosterone replacement therapy has not been shown to have an effect on these outcomes and is not recommended.

KEY POINTS

Serum testosterone functions as a vital sign in men.

- Testosterone should only be measured if symptoms or clinical situations suggest it may be deficient.
- Many factors affect its interpretation, such as time of day, protein binding, associated comorbid disease, acute disease, and drugs.
- The correctly selected patient for testosterone replacement therapy usually demonstrates a good clinical response; but this is not seen in patients not appropriately identified.
- The fall of testosterone levels with aging has not had risk-benefit of replacement testosterone adequately determined.

8.2.2.4 Pediatric considerations

- Hypogonadism should be suspected in a normal male who is born with ambiguous genitalia. This demonstrates decreased action of DHT. This results from a defect in androgen production or action. Gonadal defects may occur from LH or hCG receptor defects, Leydig cell hypoplasia, or testicular agenesis or regression. Defects in testosterone biosynthesis may occur through each enzymatic step, and conversion from testosterone to DHT requires the 5α-reductase enzyme. Androgen receptor defects will result in the spectrum from normal female phenotype (without axillary or pubic hair, and no menses), to partially virilized appearing females, to poorly virilized appearing males.

- Delayed puberty is defined as no onset of secondary sexual characteristics by age 14 in an otherwise normal boy, or more than 5 years from onset of puberty to its completion. Causes are either central from inadequate GnRH/LH/FSH secretion or peripheral with inadequate testicular response to now increased LH/FSH secretion.

- Central causes must be differentiated from simple delayed puberty from constitutional growth delay. Gonadotrope defects are congenital or acquired. They may result from a congenital transcription factor defect, lack of stimulation of gonadotropes due to GnRH deficiency/GnRH receptor defect, or hypothalamic–pituitary disease such as craniopharyngioma. Poor growth and delayed puberty may result from chronic disease and malnutrition. Hypothyroidism may present in this manner.

- Testicular causes are also congenital or acquired. Congenital includes gonadal dysgenesis from chromosomal abnormalities, Klinefelter syndrome, disorders of testosterone biosynthesis, anorchia, and myotonic dystrophy. Acquired testicular disease may be iatrogenic, trauma or torsion, mumps, or autoimmune.

8.2.2.5 Rare causes

- Defects in steroidogenesis may result in newborn or pubertal discovery of hypogonadal males.
- GnRH deficiency may result in a spectrum of FSH and LH secretion, with variable effects on virilization and spermatogenesis. One phenotype is a hypogonadal male with few secondary sexual characteristics but close to normal sperm count (so-called "fertile eunuch").
- There may be isolated genetic defects in production of the beta subunit for LH or FSH, resulting in virilization without spermatogenesis, or spermatogenesis without virilization.

8.2.2.6 Gynecomastia

- Gynecomastia is enlargement of the male breast, consisting of true ductal tissue and not just fat. It is in response to an abnormal ratio of testosterone: estrogen action at the breast. A fall in testosterone biological activity and/or an increase in estrogen biological activity will result in ductal enlargement. Typical complaint is of local tenderness or cosmetic appearance.
- It may occur from a normal physiological relatively high estrogen state in newborns and at puberty. Both regress with reestablishment of the normal hormonal milieu. Pubertal changes may not always regress and thus persist from that time. In an otherwise normal man who does acquire gynecomastia, increased estrogen states may result from tumors-secreting hCG (or rarely secreting estrogen), drugs that have estrogen action (including transferrable topical compounds from another person) or anabolic steroids that are aromatized to estrogen. Increased peripheral aromatization of androgen to estrogen may occur with obesity and cirrhosis. Hyperthyroidism combines increased peripheral aromatization and increased SHBG.
- Any male that has chronic malnutrition, which induces regression of the postpubertal secretion of GnRH to the prepubertal state, will generally return to normal pubertal GnRH secretion upon recovery, and may appear to go through the changes of a second puberty, with its expected gynecomastia. This is referred to as refeeding gynecomastia [36] and is otherwise a normal physiological change, in this clinical context. The history is suggestive.
- Decreased androgen action results from disease in the hypothalamus–pituitary–testicular axis, or drugs such as antiandrogens. Thus the serum LH will be high or inappropriately normal or low, depending on the cause, and is a clue as what the cause may be.
- Longstanding chronic gynecomastia almost always has a benign underlying cause, and if no cause is found then it is idiopathic by exclusion. A more recent onset and rapid progression is worrisome of an underlying disease of concern, such as an hCG-secreting tumor, usually testicular but may occur from any tumor.

8.2.2.7 Diagnostic approach

- *The investigation for hypogonadism in men should be clinically based.*

     Infertility requires a semen analysis in all. If there are severe defects in sperm count, measurement of serum testosterone and LH is needed to assess this part of testicular function. A serum FSH is measured and if high, with normal serum testosterone and LH, then there is lack of feedback of inhibin to FSH secretion and thus a severe isolated defect in spermatogenesis from testicular disease. Low serum testosterone and nonelevated LH and FSH suggest a central cause for the disorder of spermatogenesis although this could be seen with severe obesity, or some drugs.

     When the normal range of serum FSH is determined, this often includes men who may have disorders of spermatogenesis, and thus the reference range is higher than if done in men with proven normal spermatogenesis (may decrease by 25%). Thus even a high normal FSH may suggest a testicular cause.

- The diagnosis of testosterone deficiency should be made in men with appropriate symptoms and signs and unequivocal low serum testosterone. Investigation for deceased testosterone should be based on known or suspected hypothalamus–pituitary–testicular disease. Otherwise, it is based on the presence of symptoms and signs suggestive of testosterone deficiency. These include delayed puberty, loss of libido, decreased spontaneous erections, gynecomastia, shrinking or small testes, fragility fracture, or hot flushes. Nonspecific symptoms are common and by themselves are usually not associated with true testosterone deficiency. These include decreased energy, loss of motivation, depression, sleep disturbance, loss of strength, increased body fat, or unexplained anemia. It is with these latter symptoms that low testosterone is over diagnosed and over treated.

- Testosterone has a pulsatile and diurnal rhythm and is affected by acute and subacute illness, some medications, and chronic diseases. Changes in serum levels of SHBG may need to be considered.

- The initial test is total serum testosterone, measured early in the morning (or after awakening in night workers), with a reliable assay. If low, it should be repeated as up to 30% are normal on repeat [37]. The repeat test may need to be considered for compensation for changes in SHBG, especially with already borderline results.

- There is concern of interpretation of this morning testosterone interpretation because:
  - the reliability of direct automated immunoassays and immunometric assays, especially in the lower ranges
  - the unclear definition of the lower range at which clinical symptoms may occur
  - the realization that the serum testosterone does not fully reflect the total biological intracellular activity due to multiple levels of action and interaction

- A true free testosterone can be accurately measured by equilibrium dialysis, or calculated from total testosterone, SHBG, and albumin. Bioavailable testosterone is measured after ammonium sulfate precipitation. Free testosterone measured by analog methods is still affected by the same assay limitations and changes in SHBG concentrations.

  The recommended initial screening test is the morning total testosterone. If equivocal or low, then it is repeated using a measurement that will also assess bound and unbound testosterone. There are several ways to achieve this. In our center the FAI appears to perform best in correlating with the clinical impression. Other centers and authors may prefer a different test.
- If the serum testosterone is found to be repeatedly low, then measurement of serum LH is needed to separate central causes from testicular causes of low testosterone.
- If a testicular cause, a karyotype may be needed to look for Klinefelter syndrome.
- If central (serum LH not elevated), then a serum prolactin, measurement of iron binding capacity (iron overload and deposition), and MRI if no obvious cause is otherwise present, such as opioid use. If pituitary disease is found, such as a mass, then investigation of the remaining pituitary function with serum free thyroxine and early morning serum cortisol is needed. The decision to search for growth hormone deficiency is individualized.

8.2.3 LABORATORY CONSIDERATIONS TO ASSESS MALE REPRODUCTIVE DISORDERS

As discussed in Section 8.1.4, interpretation of male sex hormone test results requires an understanding of the preanalytical, analytical, and postanalytical aspects that affect laboratory measurements including hormonal rhythms, sampling, specimen of interest, analytical methodology, and interferences.

8.2.3.1 Test-specific discussion

- Testosterone:
  - Testosterone is a steroid hormone that is bound primarily by albumin or SHBG. SHBG binds testosterone with high affinity but low capacity whereas albumin binds testosterone with low affinity but high capacity. A small fraction ($\sim 2\%$) of testosterone is free from protein binding.
  - SHBG binds testosterone very tightly and does not allow it to interact with receptors. However, testosterone is bound less tightly to albumin and so can dissociate off, interact with receptors, and cause a biological effect. The ratio between SHBG and albumin bound testosterone is different between men and women, with men having 54% albumin bound and 44% SHBG bound and women having 20% albumin bound and 78% SHBG bound.

- Both serum and plasma (heparin and EDTA) can be used to measure testosterone. Mass spectrometry−based assays may experience interference from gel separator containing tubes, in which case samples should be collected without separating gel. If free testosterone, FAI, or bioavailable testosterone is being ordered, other tests may also need to be collected at the same time (e.g., SHBG, albumin).
- Testosterone can be measured or estimated in several different ways:

  Total testosterone is the most common test and includes both free and protein bound testosterone. Protein bound testosterone is displaced from the protein prior to measurement, which is usually done by immunoassay. The main antibody target is testosterone; however, cross-reactivity with DHT can occur and varies by manufacturer. Mass spectrometry−based assays are now available at reference laboratories.

  Free testosterone is a determination of the free form of testosterone and none of the protein bound form and can be done either by calculation or direct measurement. This may be useful when the total testosterone does not correlate with the clinical picture. In most routine clinical laboratories, free testosterone is determined using a calculation (calculated free testosterone or CFT), which takes into account the total testosterone, SHBG, and albumin concentrations. Several equations have been proposed for CFT [38]; however, this approach has not been standardized between laboratories. Direct measurement of free testosterone requires complicated analytical techniques including equilibrium dialysis or competitive radioimmunoassay. Because of the time-consuming nature of these methodologies, direct measurement of free testosterone is only available at reference laboratories and is only performed for complex cases.

  FAI is another way to assess free testosterone. It is calculated by dividing the total testosterone by the SHBG concentration.

  Bioavailable testosterone is the sum of free and albumin bound testosterone. It is most commonly determined using an algorithm that takes into account the total testosterone, albumin, and SHBG concentrations. Algorithms may differ between laboratories and will need to be adjusted in any physiological condition where albumin concentrations are altered. Bioavailable testosterone can also be directly measured using methods that selectively precipitate SHBG (and the testosterone bound to it), which is available at some reference laboratories.
- Free testosterone, FAI, and bioavailable testosterone calculations will be affected by the SHBG and albumin concentrations as well as the accuracy and precision of the methods used to measure them. SHBG concentrations increase with age, cirrhosis, hepatitis, hyperthyroidism, drugs (anticonvulsants, estrogens), and HIV. SHBG concentrations decrease in nephritic syndrome, hypothyroidism, drugs (glucocorticoids, androgenic steroids), diabetes mellitus, and moderate obesity [39].

Albumin concentrations are decreased in liver disease, nephrotic syndrome, malnutrition, hypothyroidism, and inflammation and is increased by drugs (estrogens, opioids, steroids), dehydration, and pregnancy.

• Diurnal variation has a significant impact on testosterone concentrations. Testosterone concentrations are highest in the morning and can decrease as much as 25% in the afternoon [40]. This diurnal change has also been observed in women. The amplitude of this variation decreases with age, with men older than the age of 70 experiencing a change between a.m. and p.m. samples of closer to 10%. Many laboratories have defined their reference intervals using samples collected in the morning, which should be taken into consideration when interpreting testosterone results from specimens collected in the afternoon. Morning samples should also be collected in the investigation of androgen deficiency, particularly in younger men. Similar patterns of diurnal variation are also observed for free testosterone and bioavailable testosterone.

• Testosterone concentrations naturally decline as men age.

• Some commercially available testosterone assays are not sensitive enough for use in patient populations with low testosterone concentrations (e.g., women, children, men with hypogonadism). In these cases a discussion with a Clinical Chemist is recommended to determine the capabilities of the local testosterone assay.

• At present, data from the College of American Pathologist (CAP) proficiency testing surveys suggest that most available immunoassays do not meet desired specifications for bias, imprecision, and total error (5.4%, 4.7%, 13.1%, respectively) using biological variation [41]. When these methods are compared to the gold standard (LC−MS/MS), some show significant bias. To address issues with bias the CDC Hormone Standardization Program has recently started working on standardization of testosterone measurements. This will include development of a primary reference standard to allow for traceability of measurement and development of commutable standard to improve interlaboratory comparison.

KEY POINTS

Total testosterone is the most commonly ordered test for evaluation of testosterone. However, several other tests are available that may be of use depending on the clinical presentation.

• Testosterone exhibits diurnal variation. Most laboratories use reference intervals based on samples collected in the morning.
• Testosterone measurements are not standardized between laboratories.
• Consultation with a clinical chemist is recommended if the clinical presentation does not match the laboratory results.

8.2.3.2 Misuse of laboratory tests in male reproductive endocrinology

8.2.3.2.1 Serum testosterone is a test that is very much abused in general medical practice

- There is a significant body of epidemiologic evidence available suggesting a decrease of serum testosterone with aging, and its reported frequency in the general population [42]. Because of this evidence, routine screening of testosterone levels in older men is not recommended, yet is often done. In those men with borderline or slightly low levels, adverse consequences of these findings have not been demonstrated and overall benefits of replacement therapy (and safety) have not been shown [37]. Some studies that demonstrate some effects of testosterone replacement therapy in older men are of questionable significance and do not address the issue of risk of therapy. A possible increased risk of cardiovascular disease is suspected [37].
- Failure to consider nonendocrine causes or comorbid disease as the reason for borderline low testosterone and/or symptoms. In a man with nonspecific symptoms, slightly low serum testosterone is often used to explain such symptoms, especially in men with comorbid disease. Objective reviews do not recommend replacement testosterone therapy in these men [39,43]. Industry use of media for direct marketing to the general population is a driving factor. Social media may also aggravate the problem.
- Failure to measure testosterone to investigate suspected hypogonadism. Database reviews find that some men are replaced with testosterone without measurement of serum testosterone [44].
- False diagnosis of hypogonadism based on a single or improperly timed measurement. A low serum testosterone is often not repeated, nor ensuring it be done early in the morning. Timing of sampling may be different in shift workers.
- Unawareness that acute or subacute illness or hospitalization is not the time to measure serum testosterone. If done, then await recovery to repeat it.
- Failure to investigate the cause of true hypogonadism. A true and persistently low serum testosterone should be further investigated by measuring LH and then whatever else is necessary to determine the etiology; sometimes serious diagnoses may emerge and low testosterone should not be assumed to be idiopathic.
- Failure to appreciate that a "normal" LH or FSH is not physiologically normal in the setting of true hypogonadism. Primary testicular failure causes an elevation in LH levels; thus if the testosterone is truly low and LH nonelevated, that indicates the problem is with the hypothalamic/pituitary function—which may need further investigation.
- Failure to recognize that exogenous testosterone use (current or recent) or opioid use may suppress LH and testosterone; initiation of testosterone therapy

may be inappropriate, cessation of the offending drug would be the preferred treatment.

- Use of random testosterone measures to monitor intermittent testosterone replacement dosing. When giving replacement therapy, the serum testosterone is often utilized to monitor response. Often, this test is not done at the appropriate time for the formulation being used. Reproducibility has been shown to be a problem and there is no clear correlation between "on-therapy" testosterone levels and clinical benefits.
- Possible inaccuracy of "normal" FSH references due to failure to exclude subjects with abnormal sperm counts. The normal range of FSH is determined from a group of normal men. But some of them likely do have oligospermia, raising the serum FSH in the "normal" range. It would be best to determine normal semen analysis when constructing the normal group. Because this is usually not done, the normal range of FSH may be set too high by 25% [45].

KEY POINTS

Male hypogonadism may present at birth with ambiguous genitalia, at puberty with delayed puberty, or as an acquired problem in a man with previously normal testosterone.

- Adult onset of low testosterone may be suspected in a man with known or suspected disease in the hypothalamus—pituitary—testicular axis.
- There may be organic disease in the hypothalamic—pituitary—testicular axis, or nonorganic disease such as the response to drugs such as opioids, or malnutrition.
- Serum testosterone has difficulties in measurement, and in interpretation as to how it represents the total biological activity of a complex biological system.
- Measurement of serum testosterone is often inappropriately done, its interpretation abused, and replacement therapy often inappropriately given.
- The diagnosis of true hypogonadism requires both definite symptoms and signs and an unequivocal low serum testosterone measurement by a reliable assay.
- The decrease of serum testosterone with aging men may be related to comorbid disease and it is presently not recommended to treat this, unless there is demonstrated disease in the hypothalamus—pituitary—testicular axis. Safety and outcomes have not been established.

8.2.3.3 Case example with discussion

A 36-year-old man presented with primary infertility. His partner has regular ovulatory cycles. He had normal pubertal development and was fully virilized. Overall health was previously excellent. He was a professional football player who admitted to having at least 10 concussions of which he was aware. He admitted to a 3-year history of decreased libido, loss of early morning erections, and some loss of overall strength and wellbeing. Testicular size was low normal at 15 cc.

Semen analysis demonstrated azoospermia.

Question 1: With the history of suspected hypogonadism, what laboratory tests should be done?

Answer: Early morning serum testosterone was 3.1 nmol/L (89 ng/dL) (reference interval: 8−25 nmol/L or 231−720 ng/dL). On repeat the result was 3.8 nmol/L (110 ng/dL).

Serum LH was 2.5 IU/L (reference interval: 2−12 IU/L), serum FSH was 5.3 IU/L (reference interval: 2−12 IU/L). Thus this represents a central cause of hypogonadism since the LH/FSH are "inappropriately normal" (would expect them to be high if primary testicular pathology).

Question 2: Now, with evidence of a hypothalamic or pituitary problem, what additional tests must be considered?

Answer: Pituitary testing found serum free thyroxine 6.5 pmol/L (0.5 ng/dL) (reference interval: 10.0−25.0 pmol/L or 0.8−1.9 ng/dL), early morning serum cortisol 565 nmol/L (20.5 µg/dL) (normal).

MRI of the pituitary was normal.

Insulin-induced hypoglycemia test found a poor growth hormone response, and peak response to adequate hypoglycemia of 0.9 µg/L (16 mg/dL). A normal is considered to be >10 µg/L (180 mg/dL).

Question 3: What is the most likely diagnosis?

Answer: Pituitary dysfunction from repeated traumatic brain injury was diagnosed.

Replacement thyroxine was started. Growth hormone replacement was not given at this time.

Question 4: How does knowledge of the diagnosis shape his gonadal-replacement therapy?

Answer: It is assumed that at one time he had normal testicular function based on postpubertal changes. Thus intratesticular testosterone and normal testicular size in the past did likely induce normal spermatogenesis. This spermatogenesis may be restored if intratesticular testosterone could be restored. Human chorionic gonadotropin, a long-acting stimulator of the LH receptor and testosterone synthesis, was started. Three times a week s.c. injections of hCG 1000 units were started. After 3 weeks the serum testosterone was found to be 15.7 nmol/L (452 ng/dL) (reference interval: 8−25 nmol/L or 231−720 ng/dL). If the serum testosterone was now normal, then likely the intratesticular testosterone was sufficient to induce spermatogenesis. Semen analysis after 4 months did demonstrate oligospermia, and after 6 months his partner conceived and went on to deliver a healthy boy.

He was then placed on chronic testosterone replacement therapy.

REFERENCES

[1] Torrealday S, Pal L. Premature menopause. Endocrinol Metab Clin North Am 2015;44:543−57. Available from: http://dx.doi.org/10.1016/j.ecl.2015.05.004.

[2] Dumesic DA, Oberfield SE, Stener-Victorin E, Marshall JC, Laven JS, Legro RS. Scientific statement on the diagnostic criteria, epidemiology, pathophysiology, and molecular genetics of polycystic ovary syndrome. Endocr Rev 2015;36:487−525. Available from: http://dx.doi.org/10.1210/er.2015-1018.

[3] Williams NI, Berga SL, Cameron JL. Synergism between psychosocial and metabolic stressors: impact on reproductive function in cynomolgus monkeys. Am J Physiol Endocrinol Metab 2007;293:E270−6. Available from: http://dx.doi.org/10.1152/ajpendo.00108.2007.

[4] Zhu J, Chan Y-M. Fertility issues for patients with hypogonadotropic causes of delayed puberty. Endocrinol Metab Clin North Am 2015;44:821−34. Available from: http://dx.doi.org/10.1016/j.ecl.2015.07.011.

[5] Verrotti A, D'Egidio C, Mohn A, Coppola G, Parisi P, Chiarelli F. Antiepileptic drugs, sex hormones, and PCOS. Epilepsia 2010;52:199−211. Available from: http://dx.doi.org/10.1111/j.1528-1167.2010.02897.x.

[6] Sum M, Warren MP. Hypothalamic amenorrhea in young women with underlying polycystic ovary syndrome. Fertil Steril 2009;92:2106−8. Available from: http://dx.doi.org/10.1016/j.fertnstert.2009.05.063.

[7] Anderson RA. What does anti-Müllerian hormone tell you about ovarian function? Clin Endocrinol (Oxf) 2012;77:652−5. Available from: http://dx.doi.org/10.1111/j.1365-2265.2012.04451.x.

[8] Pasquali R, Gambineri A. Treatment of hirsutism in the polycystic ovary syndrome. Eur J Endocrinol 2014;170:R75−90. Available from: http://dx.doi.org/10.1530/EJE-13-0585.

[9] Lai J-J, Chang P, Lai K-P, Chen L, Chang C. The role of androgen and androgen receptor in skin-related disorders. Arch Dermatol Res 2012;304:499−510. Available from: http://dx.doi.org/10.1007/s00403-012-1265-x.

[10] Hunter MH, Carek PJ. Evaluation and treatment of women with hirsutism. Am Fam Physician 2003;67:2565−72.

[11] Shi RZ, van Rossum HH, Bowen RAR. Serum testosterone quantitation by liquid chromatography-tandem mass spectrometry: interference from blood collection tubes. Clin Biochem 2012;45:1706−9. Available from: http://dx.doi.org/10.1016/j.clinbiochem.2012.08.008.

[12] Stenman U-H, Tiitinen A, Alfthan H, Valmu L. The classification, functions and clinical use of different isoforms of HCG. Hum Reprod Update 2006;12:769−84. Available from: http://dx.doi.org/10.1093/humupd/dml029.

[13] Cole LA, Khanlian SA. Hyperglycosylated hCG: a variant with separate biological functions to regular hCG. Mol Cell Endocrinol 2007;260−262:228−36. Available from: http://dx.doi.org/10.1016/j.mce.2006.03.047.

[14] Bristow A, Berger P, Bidart J-M, Birken S, Norman R, Stenman U-H, et al. IFCC Working Group on hCG, Establishment, value assignment, and characterization of new WHO reference reagents for six molecular forms of human chorionic gonadotropin. Clin Chem 2004;51:177−82. Available from: http://dx.doi.org/10.1373/clinchem.2004.038679.

[15] Sturgeon CM, Ellis AR. Standardization of FSH, LH and hCG—current position and future prospects. Mol Cell Endocrinol 2007;260–262:301–9. Available from: http://dx.doi.org/10.1016/j.mce.2006.09.004.

[16] Cole LA. Human chorionic gonadotropin tests. Expert Rev Mol Diagn 2009;9:721–47. Available from: http://dx.doi.org/10.1586/erm.09.51.

[17] Schmid BC, Reilly A, Oehler MK, Schmid BC, Reilly A, Oehler MK. Management of nonpregnant women with elevated human chorionic gonadotropin. Case Rep Obstet Gynecol 2013;2013:580709. Available from: http://dx.doi.org/10.1155/2013/580709.

[18] Rose MP, Gaines Das RE, Balen AH. Definition and measurement of follicle stimulating hormone. Endocr Rev 2000;21:5–22.

[19] Choi J, Smitz J. Luteinizing hormone and human chorionic gonadotropin: origins of difference. Mol Cell Endocrinol 2014;383:203–13. Available from: http://dx.doi.org/10.1016/j.mce.2013.12.009.

[20] Radicioni A, Lenzi A, Spaziani M, Anzuini A, Ruga G, Papi G, et al. A multicenter evaluation of immunoassays for follicle-stimulating hormone, luteinizing hormone and testosterone: concordance, imprecision and reference values. J Endocrinol Invest 2013;36:739–44. Available from: http://dx.doi.org/10.1007/BF03347112.

[21] Webster R, Fahie-Wilson M, Barker P, Chatterjee VK, Halsall DJ. Immunoglobulin interference in serum follicle-stimulating hormone assays: autoimmune and heterophilic antibody interference. Ann Clin Biochem 2010;47:386–9. Available from: http://dx.doi.org/10.1258/acb.2010.010044.

[22] Ketha H, Girtman A, Singh RJ. Estradiol assays—the path ahead. Steroids 2015;99:39–44. Available from: http://dx.doi.org/10.1016/j.steroids.2014.08.009.

[23] Rosner W, Hankinson SE, Sluss PM, Vesper HW, Wierman ME. Challenges to the measurement of estradiol: an endocrine society position statement. J Clin Endocrinol Metab 2013;98:1376–87. Available from: http://dx.doi.org/10.1210/jc.2012-3780.

[24] La Torre D, Falorni A. Pharmacological causes of hyperprolactinemia. Ther Clin Risk Manag 2007;3:929–51.

[25] Peuskens J, Pani L, Detraux J, De Hert M. The effects of novel and newly approved antipsychotics on serum prolactin levels: a comprehensive review. CNS Drugs 2014;28:421–53. Available from: http://dx.doi.org/10.1007/s40263-014-0157-3.

[26] Serri O, Chik CL, Ur E, Ezzat S. Diagnosis and management of hyperprolactinemia. CMAJ 2003;169:575–81.

[27] McCudden CR, Sharpless JL, Grenache DG. Comparison of multiple methods for identification of hyperprolactinemia in the presence of macroprolactin. Clin Chim Acta 2010;411:155–60. Available from: http://dx.doi.org/10.1016/j.cca.2009.10.020.

[28] Suliman AM, Smith TP, Gibney J, McKenna TJ. Frequent misdiagnosis and mismanagement of hyperprolactinemic patients before the introduction of macroprolactin screening: application of a new strict laboratory definition of macroprolactinemia. Clin Chem 2003;49:1504–9. Available from: http://dx.doi.org/10.1373/49.9.1504.

[29] Fahie-Wilson M, Smith TP. Determination of prolactin: the macroprolactin problem. Best Pract Res Clin Endocrinol Metab 2013;27:725–42. Available from: http://dx.doi.org/10.1016/j.beem.2013.07.002.

[30] Meczekalski B, Czyzyk A, Kunicki M, Podfigurna-Stopa A, Plociennik L, Jakiel G, et al. Fertility in women of late reproductive age: the role of serum anti-Müllerian

hormone (AMH) levels in its assessment. J Endocrinol Invest 2016;39:1259−65. Available from: http://dx.doi.org/10.1007/s40618-016-0497-6.

[31] Bhide P, Homburg R. Anti-Müllerian hormone and polycystic ovary syndrome. Best Pract Res Clin Obstet Gynaecol 2016;37:38−45. Available from: http://dx.doi.org/ 10.1016/j.bpobgyn.2016.03.004.

[32] Nelson SM, La Marca A. The journey from the old to the new AMH assay: how to avoid getting lost in the values. Reprod Biomed Online 2011;23:411−20. Available from: http://dx.doi.org/10.1016/j.rbmo.2011.06.011.

[33] Storer TW, Basaria S, Traustadottir T, Harman SM, Pencina K, Li Z, et al. Effects of testosterone supplementation for 3-years on muscle performance and physical function in older men. J Clin Endocrinol Metab 2016jc20162771. Available from: http:// dx.doi.org/10.1210/jc.2016-2771.

[34] Snyder PJ, Bhasin S, Cunningham GR, Matsumoto AM, Stephens-Shields AJ, Cauley JA, et al. Testosterone trials investigators, effects of testosterone treatment in older men. N Engl J Med 2016;374:611−24. Available from: http://dx.doi.org/ 10.1056/NEJMoa1506119.

[35] Sartorius G, Spasevska S, Idan A, Turner L, Forbes E, Zamojska A, et al. Serum testosterone, dihydrotestosterone and estradiol concentrations in older men self-reporting very good health: the healthy man study. Clin Endocrinol (Oxf) 2012;77:755−63. Available from: http://dx.doi.org/10.1111/j.1365-2265.2012.04432.x.

[36] Dickson G. Gynecomastia. Am Fam Physician 2012;85:716−22.

[37] Albert SG, Morley JE. Testosterone therapy, association with age, initiation and mode of therapy with cardiovascular events: a systematic review. Clin Endocrinol (Oxf) 2016;85:436−43. Available from: http://dx.doi.org/10.1111/cen.13084.

[38] Ho CK, Stoddart M, Walton M, Anderson RA, Beckett GJ, Ho C. Calculated free testosterone in men: comparison of four equations and with free androgen index. Ann Clin Biochem 2006;43:389−97.

[39] Bhasin S, Cunningham GR, Hayes FJ, Matsumoto AM, Snyder PJ, Swerdloff RS, et al. Testosterone therapy in men with androgen deficiency syndromes: an endocrine society clinical practice guideline. J Clin Endocrinol Metab 2010;95:2536−59. Available from: http://dx.doi.org/10.1210/jc.2009-2354.

[40] Brambilla DJ, Matsumoto AM, Araujo AB, McKinlay JB. The effect of diurnal variation on clinical measurement of serum testosterone and other sex hormone levels in men. J Clin Endocrinol Metab 2011;94:907−13. Available from: http://dx.doi.org/ 10.1210/jc.2008-1902.

[41] Ricós C, Alvarez V, Cava F, García-Lario JV, Hernández A, Jiménez CV, et al. Current databases on biological variation: pros, cons and progress. Scand J Clin Lab Invest 1999;59:491−500. Available from: http://dx.doi.org/10.1080/0036551995 0185229.

[42] Kaufman JM, Vermeulen A. The decline of androgen levels in elderly men and its clinical and therapeutic implications. Endocr Rev 2005;26:833−76. Available from: http://dx.doi.org/10.1210/er.2004-0013.

[43] Morales A, Bebb RA, Manjoo P, Assimakopoulos P, Axler J, Collier C, et al. Canadian men's health foundation multidisciplinary guidelines task force on testosterone deficiency, diagnosis and management of testosterone deficiency syndrome in men: clinical practice guideline. CMAJ 2015;187:1369−77. Available from: http:// dx.doi.org/10.1503/cmaj.150033.

[44] Silverman E. FDA lowdown on "low T" drugs: little evidence to support widespread use. Wall Str J 2014 WSJ.com.

[45] Sikaris K, McLachlan RI, Kazlauskas R, de Kretser D, Holden CA, Handelsman DJ. Reproductive hormone reference intervals for healthy fertile young men: evaluation of automated platform assays. J Clin Endocrinol Metab 2005;90:5928−36. Available from: http://dx.doi.org/10.1210/jc.2005-0962.

Neuroendocrine tumors

Otto P. Rorstad, MD

Professor Emeritus, Department of Medicine, Division of Endocrinology,
University of Calgary, Calgary, Alberta, Canada

CHAPTER OUTLINE

Endocrine Biomarkers. DOI: http://dx.doi.org/10.1016/B978-0-12-803412-5.00009-4

ABBREVIATIONS

5-HIAA	5-hydroxyindoleacetic acid
5-HT	5-hydroxytryptamine
ACTH	adrenocorticotropic hormone
AJCC	American Joint Committee on Cancer
CgA	chromogranin A
CRH	corticotropin-releasing hormone
EC	enterochromaffin
ECL	enterochromaffin like
ELISA	enzyme-linked immunosorbent assays
GHRH	growth hormone–releasing hormone
GI	gastrointestinal
LCNEC	large-cell neuroendocrine carcinoma
MAO	monoamine oxidase
MEN1	multiple endocrine neoplasia type 1
MR	magnetic resonance
NET	neuroendocrine tumor
NIPHS	noninsulinoma pancreatogenous hypoglycemia syndrome
NSE	neuron-specific enolase
PNET	pancreatic neuroendocrine tumor
PP	pancreatic polypeptide
RIA	radioimmunoassay
SCLC	small-cell lung carcinoma
SEER	surveillance, epidemiology, and end results
SIADH	syndrome of inappropriate antidiuretic hormone
SI-NET	small intestinal neuroendocrine tumor
ULN	upper limit of normal
VIP	vasoactive intestinal peptide
WHO	World Health Organization

9.1 OVERVIEW OF NEUROENDOCRINE TUMORS AND THEIR BIOMARKERS

Neuroendocrine tumors (NETs) are medically important because of the diversity of clinical syndromes they may produce and the variety of hormones that the tumors may synthesize and secrete [1−3]. Often NETs present diagnostic questions for which the expertise of the clinical chemist represents an essential requirement of the diagnostic investigation. As will be apparent subsequently, tumors of the pituitary and parathyroid glands, the adrenal medulla, and paraganglia can formally be considered as satisfying the biological criteria of NETs. However, these important tumors are more appropriately considered in detail in the chapters on pituitary, calcium, and adrenal gland disorders. The remaining NETs predominately originate in the intestinal tract, the endocrine pancreas, the lung and its respiratory tract, and numerous other anatomical locations throughout the body. Although these NETs are relatively uncommon, they represent unique diagnostic and therapeutic challenges for the clinician, anatomical pathologist, and clinical chemist [4].

9.1.1 ORGANIZATION AND BIOCHEMISTRY OF THE NEUROENDOCRINE SYSTEM

- The endocrine system can be considered as consisting of four cell types which each synthesize their own distinctive molecular classes of hormones:
  - thyroid hormones from the thyroid follicular cells,
  - glucocorticoids, mineralocorticoids, and sex steroid hormones produced by the adrenal cortex and sex steroid hormones synthesized by the ovary and testes,
  - biologically active amines synthesized by neuroendocrine cells, and
  - polypeptide hormones produced by neuroendocrine cells.
- The concept of a unifying nomenclature of neuroendocrine cells derives from the observation that cells identified in widespread locations throughout the body share the cellular capacity to synthesize, secrete, and take up amine and peptide neurotransmitters and hormones.
- In 1938 Feyrter discovered that neuroendocrine cells could exist in a widely disseminated manner in the intestinal tract and originated the concept of the "diffuse endocrine system" [5]. For example the digestive system contains at least 16 distinct neuroendocrine cells [3]. Pearse in 1968 observed that diffuse neuroendocrine cells had the capacity to metabolize and take up amines [6]. The term "amine precursor uptake and decarboxylation" summarized this system.
- Biologically active amines relevant to NETs include:
  - serotonin in intestinal NETs (also termed carcinoid tumors),
  - histamine in gastric and lung carcinoid tumors, and

- catecholamines (epinephrine and norepinephrine) in pheochromocytomas and paragangliomas.
- Subsequently since the 1970s a large number of distinct neuroendocrine cells, new peptide hormones, and receptors for peptide hormones on diffuse neuroendocrine cells have been identified. Polypeptide hormones range in size from the three amino acid thyrotropin-releasing hormone, to the large glycoprotein pituitary hormones, thyroid-stimulating hormone, follicle-stimulating hormone, luteinizing hormone, and the placental human chorionic gonadotropin comprised of over 200 amino acids in two polypeptide subunits and including substantial glycosylation.

KEY POINTS

NETs are characterized by the capacity to synthesize and secrete biological amines and/or polypeptide hormones. The most important tumoral biological amines are serotonin (intestinal NETs), histamine (gastric and pulmonary carcinoid tumors), and catecholamines (pheochromocytomas and paragangliomas). NETs may arise in a wide diversity of locations consistent with the cells of origin belonging to the "diffuse endocrine system."

9.1.2 TUMOR BIOLOGY, CLASSIFICATION AND GRADING

9.1.2.1 Tissue classification

NETs of the digestive tract are thought to arise from the embryonic endoderm [3]. A former theory that NETs originated from the embryonic neural crest has not been supported by subsequent studies of cellular embryogenesis and migration [3]. Williams and Sandler in 1963 introduced a classification of carcinoid tumors based on their origin in the embryological foregut, midgut, or hindgut [7]. The principal limitation of the embryological classification as applied comprehensively to NETs is that tumors with very different cell biology and clinical behavior are encompassed under the same embryological designation of origin. For example an aggressive pancreatic neuroendocrine tumor (PNET) and an indolent lung typical carcinoid tumor are both foregut tumors but are very different in their hormone production and clinical prognosis. Modern World Health Organization (WHO) nomenclature favors designation of the organ of origin of the NET, such as the pancreas or rectum, rather than the presumed embryological foregut, midgut, or hindgut origin [8].

NETs can arise from different anatomical organizations of tissues as in the following examples [9]:

- Endocrine organs such as the parathyroid glands and the adrenal medulla.
- Clusters of cells in an organ such as the islets of Langerhans (pancreas) and the bronchial (lung) neuroendocrine bodies.

- Isolated neuroendocrine cells in the intestinal lining such as the serotonin-producing enterochromaffin (EC) cell, bronchial neuroendocrine cells, and the C cells producing calcitonin in the cellular lining of the thyroid follicle.

9.1.2.2 Common components of secretory vesicles that define a "neuroendocrine" tumor

- A common characteristic of neuroendocrine cells and tumors is the presence of cellular components for packaging amines and peptide hormones in secretory vesicles [9]. Several molecular markers of the cellular secretory vesicles have been investigated in immunohistochemical studies to evaluate their validity in the histological diagnosis of NETs.
- The secretory vesicle markers that have seen the widest adoption for diagnosis are *chromogranin A (CgA)*, a component of large secretory vesicles, and *synaptophysin*, found in small secretory vesicles.

9.1.2.3 Histological features for classification and grading

- Histological tumor characteristics such as cellular and nuclear pleomorphism, nuclear-to-cytoplasmic ratio, areas of necrosis, lymphovascular invasion, and the number of mitotic figures are important adjunctive features to assist the pathologist in diagnosing and classifying a NET.
- Subsequent to diagnosis the pathologist will classify (or type) a NET as well or poorly differentiated and assign a *grade* to designate the inherent biological aggressiveness of the tumor.
- Important criteria for grading emphasize indicators of proliferation such as
  - A count of the number of mitoses per standardized microscopic dimensions.
  - The Ki67 score. This is derived as the percentage of tumor cells which are positive on immunohistochemical staining for the nuclear proliferation protein, Ki67. A monoclonal antibody (termed MIB-1) generated to the Ki67 protein is very widely used for Ki67 immunohistochemistry and facilitates multicenter and international standardization of tumor grading.
    - The Ki67 score has been correlated with indicators of prognosis and clinical outcome with higher Ki67 scores being associated with poorer outcome [10]. For the GI tract NETs and PNETs, presently three grade (G) levels are designated with increasing Ki67 scores, G1 (Ki67 0%–2%), G2 (Ki67 3%–20%), and G3 (Ki67 >20%) [8].
    - The current grading system used for lung, respiratory tract, and thymus NETs gives priority to the cellular mitotic count for tumor grading and uses different terminology for the grade than is used for the digestive system [11].
    - NETs may contain nonneuroendocrine components such as adenocarcinoma and be classified as mixed tumors. An important example is the intestinal adenoneuroendocrine carcinoma in which the adenocarcinoma component is usually higher grade and more aggressive [8].

> **KEY POINTS**
>
> Modern WHO nomenclature designates the specific organ of origin of a NET, such as small intestine or pancreas, rather than the historical foregut, midgut, and hindgut classification. Immunohistochemical demonstration of the secretory vesicle proteins CgA and synaptophysin is widely used in the pathological diagnosis of a NET. NETs are classified as well or poorly differentiated. Histological grade is usually expressed as three or four levels of increasing biological aggressiveness. Different criteria for grading are used for the digestive tract and intrathoracic NETs.

9.1.3 TUMOR STAGING

The *stage* of a NET describes its anatomical extent as opposed to the cellular features described by the *grade* and is intended to confer a prognosis for the tumor. Before 2010 the principal resource for classification and staging of NETs was the National Cancer Institute (USA) Surveillance, Epidemiology and End Results (SEER) database which included NETs arising from the tubular GI tract, the pancreas, lung, and the rare sites of thymus, hepatobiliary system, ovary, and testis [12]. Pheochromocytomas and paragangliomas were not included. The historical staging designations were

- Regional: tumor confined to the organ of origin.
- Localized: tumor extending from the organ of origin directly into surrounding organs or tissue or involving regional lymph nodes.
- Distant: tumor metastatic to distant sites (e.g., liver, lung, bone).

The SEER database has served as the principal resource for epidemiological research on NETs and contributed to inclusion of NETs for the first time in the 2010 TNM Cancer Staging Manual (7th ed.), prepared by the American Joint Committee on Cancer (AJCC)/International Union Against Cancer (IUAC) [13]. The T, N, and M designations are used to assign a stage ranging from I to IV, with the higher stages conferring a poorer prognosis.

- T: Size and extent of the primary tumor.
- N: Involvement of regional lymph nodes.
- M: Distant metastases.

NET sites included in the 2010 TNM Staging Manual (7th ed.) are the tubular GI tract, pancreas, lung, and Merkel cell carcinoma of skin [13]. Mixed adenoneuroendocrine carcinomas are staged as for the adenocarcinoma component.

> **KEY POINTS**
>
> Staging refers to the anatomical extent of a tumor. NETs of the tubular GI tract, pancreas, lung, and skin are presently incorporated in the 2010 AJCC Cancer Staging Manual (7th ed.).

9.1.4 DECIPHERING NOMENCLATURE: NEUROENDOCRINE TUMORS, CARCINOIDS, AND HORMONE-SECRETING "OMAS"

A transition in nomenclature describing tumors addressed in this chapter from carcinoids to NETs is well advanced based on the unifying scientific characteristics of NETs. The term *carcinoid* (carcinoma like) has historical priority by virtue of originating from Oberndorfer's pathological observations in 1907 [14]. He described small intestinal tumors that appeared well differentiated and in general had a better prognosis compared to intestinal adenocarcinomas. In spite of recognition in 1949 that a significant proportion of carcinoid tumors could pursue a malignant clinical course, the terminology of carcinoid persists in several versions of usage, both formal and informal [15].

9.1.4.1 The confusion of "carcinoid" and "carcinoid syndrome"

- At its most colloquial, "carcinoid" is used by some as synonymous with NET, most inappropriately for PNETs.
- The clinical term, "carcinoid syndrome," describing an association of diarrhea, skin flushing, and other symptoms due to hormone secretion from a particular NET is historically entrenched in medical usage.
- The original classical carcinoid syndrome is associated with hypersecretion of serotonin, also termed 5-hydroxytryptamine (5-HT), from the *EC intestinal NET*.
- The term "variant carcinoid syndrome" is preferred for tumors secreting histamine as the active agent causing the syndrome. These include gastric NETs (carcinoids) arising from the enterochromaffin-like (ECL) cell or lung carcinoid tumors.
- One may also encounter the term "variant carcinoid tumor" for NETs that lack aromatic-L-amino acid decarboxylase and are unable to synthesize serotonin but produce its precursor 5-hydroxytryptophan.

9.1.4.2 Gastric, intrathoracic, and other "carcinoids"

NETs of the stomach arising from the histamine-producing ECL cells are classified into three types that are commonly informally described as gastric carcinoid tumors. The 2010 WHO classification of tumors of the digestive system formally uses NET nomenclature for these tumors [8]. In contrast, the 2015 WHO classification of tumors of the lung, pleura, thymus, and heart formally designates well-differentiated NETs of the lung as typical carcinoids and atypical carcinoids [11]. Description of a NET as having foregut, midgut, or hindgut origin should be considered historical and not used in modern nomenclature or classification.

9.1.4.3 Neuroendocrine tumors and endocrine secretions

NETs can be functionally divided into two groups:

- Nonsyndromic NETs which do not secrete hormones and may also be referred to as nonfunctional.

- Syndromic NETs that secrete hormones causing a clinically evident syndrome, such as an insulinoma. The suffix "oma" should be used if there is a clinical syndrome caused by excessive secretion of that hormone, as shown by increased blood or urine levels of the hormone.
- If there is no clinical syndrome, yet the NET stains histologically for a hormone, it is appropriate to add the staining result to the pathology report (e.g., duodenal well-differentiated NET, grade G2 with immunohistochemically demonstrated somatostatin production) [8].

KEY POINTS

NET nomenclature is preferred for scientific clarity in most cases in contrast to historical carcinoid terminology. As an exception the terms typical and atypical carcinoid representing well-differentiated tumors of the lung and thymus are used for formal classification by the WHO. The clinical term "carcinoid syndrome" is historically established and unlikely to be changed. Ideally, in clinical usage, this term should be applied to the syndrome of serotonin excess. The term "variant carcinoid syndrome" has been used for symptoms due to secretion of histamine, usually from NETs of the stomach or lung. For NETs that cause a clinical syndrome due to secretion of a hormone the suffix "oma" can be added to the name of the hormone (e.g., insulinoma).

9.1.5 EPIDEMIOLOGY OF NEUROENDOCRINE TUMORS

9.1.5.1 Incidence, prevalence, and predictors of clinical outcome

- The SEER database has contributed the most important epidemiological information on NETs derived from 35,825 cases [12]. The incidence of NETs originating from all sites studied increased by approximately fourfold between 1973 (1.07 cases per 100,000) and 2004 (5.25 cases per 100,000). This compares to an approximately 16% increase in the incidence of all malignant neoplasms over the same period. A Swedish nationwide epidemiological survey of 5184 NETs from all sites found about a doubling of the incidence of these tumors between 1958 and 1998 [16]. The estimated prevalence of all NETs in 2004 was 35 per 100,000 population in the United States [12].
- The SEER database may have underestimated the incidence and prevalence of NETs. Benign appearing small NETs such as appendiceal NETs and gastric type I NETs probably were underreported.
- The cause of the increased incidence of all NETs over the past four decades may be related to improved classification and diagnostic detection of these tumors, particularly with advancements of diagnostic imaging and endoscopy of the GI tract and bronchial tree. In support of this hypothesis the greatest increase in incidence was observed in NETs arising from the lung, rectum, and small intestine. There is no convincing evidence of environmental factors playing a role.
- Univariate and multivariate analyses of predictors of clinical outcome revealed the following significant parameters: sex, age, race, period of diagnosis, anatomical site of the primary tumor, histological grade, and

disease state [12]. The anatomical site of the primary tumor was the most important predictor of outcome.

9.1.5.2 Survival in patients with neuroendocrine tumors

- Duration of survival was analyzed by comparing patients diagnosed during two periods of time, 1973−87 and 1988−2004 [12]. This dichotomy was chosen because the somatostatin receptor agonist drug, octreotide, was introduced in 1987. Systemic peptide receptor radionuclide therapy was rarely used in the United States during the second time period.
- There was no significant difference between the survival duration of patients diagnosed in either time period who had localized or regional disease.
- However, in patients with metastatic disease diagnosed during the more recent time period there was a statistically and clinically significant improved survival (hazard ratio 0.67).
- Hypotheses to explain this observation in the modern time period include prevention and treatment of the potentially fatal carcinoid crisis associated with hemodynamic collapse. Also, reducing the circulating level of serotonin by octreotide or lanreotide may have ameliorated the development and severity of fibrotic carcinoid heart disease. Finally, randomized controlled trials of octreotide in patients with metastatic midgut carcinoid tumors and lanreotide in advanced PNETs, intestinal, and site-unknown NETs have demonstrated a prolongation of progression-free survival [17,18]. These studies support the longstanding theory of somatostatin analogs having a cytostatic effect on NETs.

KEY POINTS

The incidence of NETs has risen approximately fourfold over the past four decades, probably because of improvements in pathological diagnosis, diagnostic imaging, and endoscopy. Patient sex, age, race, time period of diagnosis, anatomical site of the primary tumor, histological grade, and disease state are all predictors of outcome for NETs. Duration of survival in the recent time period is improved for patients with metastatic disease.

9.1.6 NEUROENDOCRINE TUMORS AND HEREDITARY TUMOR SYNDROMES

9.1.6.1 Hereditary tumor syndromes associated with expression of neuroendocrine tumors [1−3].

Multiple endocrine neoplasia type 1 (MEN1)
- PNETs
- bronchial carcinoids
- thymic carcinoids
- gastric type II NETs associated with gastrinomas

Von Hippel Lindau disease
- PNETs

Neurofibromatosis type 1
- PNETs and duodenal somatostatinomas

Tuberous sclerosis
- PNETs

9.1.6.2 Inheritance of small bowel neuroendocrine tumors

- Concerning tumors causing the carcinoid syndrome, case reports, and series have identified rare kindreds with hereditary EC cell small intestinal NETs (SI-NETs) believed to be inherited in an autosomal dominant manner and that do not appear to be associated with other hereditary endocrine tumor syndromes.
- In support of a hereditary version of SI-NETs, epidemiological analysis of a Swedish database of 5184 patients with NETs of all sites determined that a first degree relative of a patient with a NET had a 3.6-fold greater chance than unrelated individuals of having this tumor [16].
- Hereditary SI-NETs are rare compared to their sporadic counterparts. Research into the molecular pathogenesis of SI-NETs has focused on aberrations to chromosome 18, including loss of one copy of this chromosome [19]. Molecular genetic aberrations of a similar nature affecting chromosome 18 have been reported in 8 familial and 37 sporadic patients with SI-NETs [20]. In addition the clinical features of sporadic and hereditary SI-NETs were indistinguishable. The chromosome 18 changes were present in tumor tissue samples of 100% of sporadic and 38% of hereditary tumors. A common pathogenic mechanism of tumorigenesis has been hypothesized to involve a tumor suppressor gene on chromosome 18q. If correct, the reason why hereditary SI-NETs are so rare compared to their sporadic counterparts is a mystery.
- A recent molecular genetic analysis of sporadic SI-NET tissue and the human SI-NET cell line CNDT2·5 identified TCEB3C (Elongin A3) as a potential tumor suppressor gene which is imprinted on chromosome 18 and subject to epigenetic repression [21]. This finding is a promising advance in the elucidation of the pathogenesis of SI-NETs.

> **KEY POINTS**
>
> PNETs are the most frequent NET associated with hereditary endocrine tumor syndromes of which MEN1 is the most important and complex. Hereditary SI-NETs are rare and progress is being made in understanding of their molecular pathogenesis.

9.1.7 CIRCULATING TUMOR BIOMARKERS

An important component of evaluating the status of a patient with a NET during the course of diagnosis and therapy involves assessment of biological activity, extent of disease, evidence of recurrence, progression, and response to therapy by measurement of tumor biomarkers [22−24]. This term refers to molecules such as CgA that are secreted directly into the circulation by tumors or metabolites of hormones like 5-hydroxyindoleacetic acid (5-HIAA) that is derived from serotonin and is excreted in the urine. Tumor biomarkers can be organized into two groups: *General* and *organ specific* (often hormones associated with recognized syndromes).

9.1.7.1 General neuroendocrine tumor markers: Molecules that are secreted nonspecifically or generally from neuroendocrine tumors arising from multiple organs

These may be structurally associated with the general hormonal vesicular secretory apparatus or be localized in the cytoplasm such as the enzyme neuron-specific enolase (NSE). The greatest benefit of general tumor biomarkers is probably for surveillance of NETs for which there is no clinical syndrome or specific hormone secreted as is the case for three-quarters of PNETs [25,26]. CgA is the most important example of a general tumor marker that is secreted into the circulation upon fusion of the secretory granules with the plasma membrane [24]. Pancreastatin, pancreatic polypeptide (PP), and NSE are other examples of the general variety of tumor biomarkers [22,23]. CgA measurements are readily available to tertiary care medical centers whereas obtaining assays of the other general biomarkers may require identification of specialized independent laboratories or academic centers with an interest in NETs.

9.1.7.1.1 Chromogranin A

CgA is the most extensively used and studied *general* tumor marker for NETs [22−24]. It is a large 439 amino acid single chain polypeptide that is intrinsically part of dense core secretory vesicles [27]. The fully translated molecule is proteolytically cleaved into a large number of smaller peptide fragments among which pancreastatin has been used as an alternative NET tumor marker. CgA is also related to other members of a family of high molecular weight acidic proteins including chromogranin B. This latter peptide has been less studied as a NET tumor marker than CgA.

9.1.7.1.2 Clinical sensitivity and specificity of chromogranin A for surveillance of neuroendocrine tumors

Studies of detection of NETs have favored CgA as the preferred tumor biomarker for well-differentiated NETs [22−24]. Bajetta et al. [28] evaluated CgA with other tumor markers in 106 patients with NETs from various sites. CgA had a sensitivity of 86% and specificity of 68% for detection of NETs, in addition to having 83% positivity in patients with progressive disease. Another relatively large study compared CgA with NSE in terms of detection of NETs in 99 gastroenteropancreatic tumor, 19 medullary thyroid cancer, and 10 pheochromocytoma patients [29]. For CgA the sensitivity was 59% and the specificity 68%. NSE had inferior tumor marker characteristics including a sensitivity of 38%. A large CgA study of 211 patients with NETs found a sensitivity of 53% and specificity of 93% [30]. The variability of sensitivity and specificity may have related to the types of NETs included and the proportion of patients with small tumors [30]. Korse et al. [31] analyzed CgA assay properties using receiver−operator curve (ROC) characteristics in NET patients with histological grade 1 (242 patients), grade 2 (38 patients), and poorly differentiated neuroendocrine carcinomas (42 patients). The ROC areas under the curves were respectively 0.86, 0.91, and 0.90.

It is important to note that these studies have defined the sensitivity and specificity of CgA in patients with known NETs and not with respect to "screening" for potential NETs in an otherwise nonspecific population of patients. This has major implications for the interpretation of sensitivity and specificity of the test as used during follow-up of a NET patient versus initial diagnosis in a patient presenting with NET-related or suspicious symptomatology (see Section 9.1.7.1.4). It is likely that random CgA measurements in undifferentiated patient populations would show vastly lower specificity for true NETs.

9.1.7.1.3 Correlation of chromogranin A with disease burden and survival

CgA has been found to be an independent predictor of survival in a study of 301 patients with NETs from many diverse anatomical sites [32], 15 patients with gastric NETs [33], 39 patients with metastatic GI NETs [34], and 102 patients with gastroenteropancreatic NETs with metastases to the liver or regional lymph nodes [35]. CgA levels also correlate with the extent or anatomical burden of NETs from numerous sites [29,30]. Janson et al. [32] found that the circulating CgA level was higher in patients harboring five or more hepatic metastases compared to those with fewer liver metastases [32]. Arnold et al. [35] identified a correlation between blood CgA and the tumor burden of metastases in 102 patients with GI and PNETs. The CgA level was found to be significantly elevated in all 42 patients with nonsyndromic PNETs who had liver metastases at diagnosis [36]. In a surveillance study of 56 patients who had undergone radical surgery for midgut carcinoid tumors, the circulating CgA level was the earliest indicator of disease recurrence, being superior to urine 5-HIAA and diagnostic imaging modalities [37].

9.1.7.1.4 Challenges to interpretation of chromogranin A measurements

1. *Poorly differentiated NETs*

    Although CgA is a valuable tumor marker for well-differentiated NETs, a caution is noteworthy for poorly differentiated NETs [22]. These high-grade NETs may lose the capacity to synthesis CgA and be associated with a lower level of CgA than would be expected for the tumor burden. Also, CgA has greater utility as a tumor marker for NETs of the digestive system, particularly for SI-NETs, rather than for NETs arising from other sites [22,30].

2. *Multiple conditions causing false positive high CgA levels*

Circulating CgA is often measured to assist in the diagnosis of a NET in patients with a variety of symptoms such as diarrhea, flushing, or abdominal pain and patients in whom a tissue mass has been incidentally discovered on abdominal or pulmonary diagnostic imaging. Marrotta et al. [38] investigated the usefulness of circulating CgA for diagnosis in three groups of patients: 42 with NETs, 120 with non-NET cancers, and 100 with a benign nodular goiter and no evidence of a neoplastic condition. CgA levels in NET patients were not significantly different from patients in the other two groups. Receiver−operator analysis failed to find a cut-off level of CgA for the specific diagnosis of NETs. The probable reason for the poor performance of circulating CgA as a diagnostic test is the high number of situations in which CgA is subject to false positive elevation.

- The clinically often-used PPI and histamine receptor blocker drugs elevate CgA concomitantly with hypergastrinemia and stimulation of the gastric ECL cell. A controlled study comparing patients treated with a PPI drug for 6 months or longer to control patients who had never received antacid drugs found the mean CgA level was significantly elevated in treated (131 ng/mL) versus control patients (15 ng/mL) [39]. This study also demonstrated a higher mean level of blood gastrin in patients using a PPI drug (168 pg/mL) versus control patients (35 pg/mL) [39].
- Common comorbid conditions which can falsely elevate CgA are renal, heart and hepatic failure, atrophic gastritis, helicobacter pylori gastritis, hyperparathyroidism, and inflammatory bowel disease among others.
- Non-NET neoplasms which may raise CgA levels are malignancies of breast, lung, GI tract, genitourinary tract, head and neck, and hematological cancer [24,38].
- More detailed technical interferences such as a hook effect that is seen with some immunoassays and which could also give falsely low CgA results are described later.

A study of the utility of CgA and PP for diagnosis of a PNET was undertaken in 81 and 73 patients respectively with MEN1 [40]. The diagnostic accuracy outcome measures based on receiver−operator area under the curve and sensitivity and specificity of CgA and PP measurements for diagnosis of PNETs were low, raising doubt about their usefulness for tumor surveillance in MEN1.

3. *Assay variation and clinical interpretation*

A separate problem plaguing CgA measurements is the lack of standardized assays. Radioimmunoassay (RIA) and enzyme-linked immunosorbent assays (ELISA) methodologies are available and assays are targeted to different epitopes of the very large CgA polypeptide. In addition, there are a number of commercial and institution-specific assays of CgA reflecting the lack of widespread standardization. An oncology clinic may have to refamiliarize itself with changes to different CgA assays over time. The opinion of the author based on experience and supported by published data is that CgA should be utilized as a tumor marker for patients with known NETs, but not as a diagnostic test for those suspected of harboring a NET [38].

9.1.7.1.5 Pancreastatin

- Pancreastatin is a 52 amino acid peptide that is generated by proteolytic cleavage of CgA to comprise amino acid sequence positions 250−301 of the full originally translated CgA [27]. Specific and sensitive assays for pancreastatin have been developed in research centers. In a study of 92 patients with gastroenteropancreatic and lung NETs pancreastatin and CgA measurements were compared [41]. For pancreastatin and that particular CgA assay the respective sensitivities were 64% and 43% and the specificities were 100% and 64%. Notably, in 27 patients with NETs the CgA measurement was normal whereas pancreastatin was elevated.
- Pancreastatin has also shown promise as a biomarker correlating with survival and outcomes of treatment. Pancreastatin and CgA were compared as tumor markers in a large study including 98 patients with a SI-NET and 78 patients with a PNET [42]. Both preoperative and postoperative levels of pancreastatin were independent predictors of survival whereas only the postoperative CgA level was predictive of survival. Among 59 patients with gastroenteropancreatic NETs, a proportion showed a rapid rise in circulating pancreastatin level during somatostatin analog treatment [43]. This rise was associated with a poor survival outcome. In multivariate analysis pretreatment pancreastatin levels and the change in pancreastatin level were significant prognostic indicators. A large study was done of 122 patients with NETs from multiple sites having hepatic artery chemoembolization treatment [44]. The patients with a very high initial level of pancreastatin at >5000 pg/mL (normal <135) had no radiological evidence of tumor size reduction and had a decreased survival. Overall a greater than 20% decrease of pancreastatin level in response to treatment correlated with a radiological reduction of tumor burden [44].
- A substantial advantage of pancreastatin as a tumor marker is that its blood concentration is not affected by a patient's use of a PPI drug [39].

9.1.7.1.6 Neuron-specific enolase

- NSE is considered a cytoplasmic complex enzyme that is not thought to be part of the hormone secretory process. Its blood level is elevated in association with poorly differentiated NETs to a greater extent than CgA. This characteristic may be due to NSE being nonspecifically discharged from poorly differentiated tumor cells as a consequence of cell death.
- Circulating NSE has been studied in a variety of tumors including GI and PNETs, small-cell neuroendocrine carcinoma of the lung, medullary thyroid

cancer and pheochromocytoma [29]. Three studies of 128, 106, and 211 patients with NETs of multiple sites demonstrated low sensitivities of NSE (38%, 33%, and 46%, respectively) compared to CgA [28–30].

- NSE was predominately elevated in patients with poorly differentiated tumors [29] and small-cell lung carcinomas (SCLC) in particular [30].
- Consistently with these results, Korse et al. [31] found in patients with poorly differentiated neuroendocrine carcinomas of various sites that NSE was an independent predictor of survival. In a study of 113 patients with low to intermediate grade advanced PNETs, an elevated level of circulating NSE was associated with a shorter progression-free survival [45]. In addition the response of a decrease of circulating NSE level to treatment with everolimus predicted a decrease in tumor size, suggesting potential utility of NSE as a prognostic marker for response to treatment [45].
- In summary, NSE has demonstrated inferior characteristics compared to CgA for longitudinal surveillance of patients with NETs. However in a patient with a NET of poor differentiation and with no specific hormone or other general tumor marker elevation, NSE may be applicable as a general tumor biomarker for surveillance, particularly if there is clinical or radiological suspicion of progression.
- The NSE level may also serve as a prognostic marker for a clinical or radiological response to treatment.

9.1.7.1.7 Pancreatic polypeptide

- PP is a 36 amino acid peptide normally localized to the PP cells of the pancreas and intestine.
- Physiologically PP is released into the circulation due to vagus nerve stimulation in response to a meal and the peptide may play a role in the regulation of nutrition.
- It is a member of the "PP family" of peptides that include peptide YY (PYY) found in the endocrine cells of the small intestine and colon and neuropeptide Y (NPY) which is localized in the nervous system.
- Specific PNETs are considered as arising from the normal pancreatic PP cell and in the case of these tumors circulating PP may serve as a useful organ-specific tumor marker.
- NETs of the pancreas and intestine may also secrete PP as a general tumor marker. Panzuto et al. [46] compared circulating PP and CgA levels in 68 patients with NETs and 24 patients with non-NET tumors. Measuring PP, the sensitivity for PNETs was 63% and for GI NETs was 53%, both results being less sensitive than using measurement of CgA. The specificity of PP for NET patients using comparison with patients with non-NETs was 67%, similar to the 63% specificity of CgA.
- Walter et al. [47] studied 66 patients with PNETs and 49 patients with GI NETs and found an elevated circulating PP value in 31% of cases and increased CgA concentration in 69% of cases. In only seven patients PP was

elevated when the CgA level was normal. Combined use of both PP and CgA increased the sensitivity of positivity in the NET patients. Both tumor markers showed a significant elevation in patients with metastases.

- These studies indicate that PP is an inferior tumor marker compared to CgA for surveillance of patients with GI and PNETs.
- The reference normal range for PP increases substantially with advanced age.
- The circulating PP level is subject to elevation by renal failure, inflammatory conditions, liver disease, and bowel resection [48].

9.1.7.2 *Organ-specific neuroendocrine tumor tumor markers: hormones normally synthesized and secreted from a specific endocrine organ or cell such as serotonin from the small intestinal enterochromaffin cell or insulin from the pancreatic B cell*

- Specific hormones are secreted from syndromic PNETs such as
  - insulinomas
  - glucagonomas
  - gastrinomas
  - somatostatinomas and more rare PNETs
- If they are inappropriately elevated, they are crucial to making a correct diagnosis of a hormone-producing NET. Subsequently they can serve as tumor markers to assist with surveillance of the clinical course of the NET and the effect of treatment.
- Other peptides such as neurokinin A and substance P with less well understood physiological functions may serve as organ-specific tumor markers of SI-NETs [49,50].
- Ectopic hormone production refers to secretion of a hormone from a NET originating from a cell that normally does not synthesize that hormone. The most familiar example is the ectopic adrenocorticotropic hormone (ACTH) syndrome in which ACTH is produced in excess from a lung or PNET.
- Eutopic hormone production as applied to a NET refers to excess secretion of a hormone from a NET that originates from a cell that normally synthesizes the hormone, such as insulin from the pancreatic B cell.

KEY POINTS

CgA has been the most extensively studied and utilized NET tumor biomarker. It is most applicable to well-differentiated GI NETs and PNETs. The level of CgA correlates with survival, disease progression, and tumor burden. Pancreastatin is a promising alternative to CgA and has the advantage of not being elevated by PPI drugs. NSE is most applicable to poorly differentiated neuroendocrine carcinomas. PP, when combined with a CgA level, yields a higher sensitivity than that of either tumor marker alone. In clinical settings in which residual NET tissue is suspected and blood CgA is normal or is unreliable due to factors causing false positivity, PP may be helpful as an alternative general tumor marker to CgA.

9.1.8 LABORATORY CONSIDERATIONS FOR MEASUREMENT OF NEUROENDOCRINE TUMOR MARKERS

Most NET markers are relatively small proteins that are found in low concentrations. This predisposes these markers that are measured in the laboratory to preanalytical and analytical processes that can affect the reliability of laboratory measurements in biological fluids. The general preanalytical effects such as patient preparation, specimen type, and collection have been discussed in Chapter 1, Endocrine laboratory testing: Excellence, errors, and the need for collaboration between clinicians and chemists. Here and elsewhere in this chapter are briefly mentioned some of the variables that can affect specific test results.

9.1.8.1 Preanalytical considerations

In the preanalytical phase of the testing process the most important issue is collecting correct specimens from the correct person under proper conditions and labeling them correctly. For most NET markers the best tubes are red top tubes with no preservatives. For urgent cases requiring rapid results, it is often useful to collect specimens in tubes with clot activators (gold top), heparin (green top), or EDTA (lavender top) in the situation in which plasma is acceptable. These tubes will enable the lab to separate the serum or plasma faster; however, they may introduce problems in analysis.

For many analyses, medications and foods may interfere and, in those cases, abstinence for at least 48 h prior to collection is often necessary. In some cases, clinical conditions will also affect the interpretation of the results.

9.1.8.1.1 Specific test concerns

5-Hydroxyindoleacetic acid
- Measurement of *5-HIAA* has the most potentially interfering drugs and foods:
- Foods containing serotonin, also termed 5-HT, elevate 5-HIAA: eggplant, avocado, tomato, fruit including banana, pineapple, plum, kiwi, plantain, cantaloupe, and melon, and nuts including walnut, hickory nut, pecan, and butternut.
- Medications falsely elevating 5-HIAA measurement include: guaifenesin, antihistamines, phenothiazines such as chlorpromazine (Thorazine) and prochlorperazine (Compazine), aspirin, acetaminophen (Tylenol), methocarbamol (Robaxin), cyclobenzaprine (Flexeril), and diazepam (valium).
- Medications causing a physiological decrease include: chlorophenylalanine, ACTH, guanfacine, tricyclic antidepressants such as imipramine, isocarboxazid, isoniazid, levodopa, monoamine oxidase (MAO) inhibitors (MAO is necessary to convert serotonin to 5-HIAA), methyl dihydroxyphenylalanine (DOPA), moclobemide, and octreotide.
- Medications causing a physiological increase include: cisplatin, fluorouracil, melphalan, reserpine, and rauwolfia.

Chromogranin A

- A physiological increase in gastrin and, thus in *CgA* is caused by PPIs such as omeprazole. In this case, discontinuation for 1−2 weeks prior to collection may be necessary and is discussed later.
- Atrophic gastritis and pernicious anemia have a similar effect.

PP, pancreastatin, somatostatin, insulin, and *glucagon* require fasting of at least 8−10 h and abstention of drugs affecting insulin for 48 h. Atropine blocks secretion of PP.

Neuron-specific enolase (NSE) may be falsely elevated by PPIs, hemolytic anemia, hepatic failure and end stage renal failure, seizures, encephalitis, and dementia.

The most important point for all laboratory assays to remember is that if the result does not match the clinical findings, then one must consider the possibility of a technical error.

9.1.8.2 Analytical considerations

Almost all of the neuroendocrine markers are routinely measured by immunoassays in clinical laboratories. There is a detailed section on immunoassays and issues related to immunoassays in Chapter 1, Endocrine Laboratory Testing: Excellence, Errors, and the Need for Collaboration Between Clinicians and Chemists. Recently, some laboratories have implemented liquid chromatography tandem mass spectrometry (LC−MS/MS) technology to measure some of the neuroendocrine markers. Unlike immunoassay that suffers from cross reactivity and compromised specificity, LC−MS/MS has superior specificity and sensitivity. The only issue at this time with LC−MS/MS technology is lack of standardization among laboratories that use this system. Therefore we recommend LC−MS/MS technology as a method of choice for measuring all endocrine markers, especially when the results from immunoassays do not match the clinical findings.

KEY POINTS

The preanalytical, analytical, and postanalytical phases in the measurement of NET biomarkers are susceptible to errors. 5-HIAA is particularly sensitive to exposure to foods and medications although the others are to a lesser extent. Most are measured by one of several types of immunoassays. These are susceptible to interferences including cross reactivity, heterogeneous antigens, high-dose hook effect, signal interference, matrix effects, and endogenous or exogenous antibodies. Most of these analytical problems are avoided by use of the current gold standard, LC−MS/MS.

9.2 GASTROINTESTINAL NEUROENDOCRINE TUMORS

9.2.1 CLINICAL PRESENTATION DEPENDS UPON SITE OF TUMOR ORIGIN

- NETs may originate throughout the whole extent of the tubular gastrointestinal (GI) tract from the esophagus to the rectum [1−3]. Clinical presentations of GI NETs due to effects of tumor mass, local invasion of other organs or tissues, and metastases are more common than symptoms due to syndromes of hormone secretion.

- SI-NETs and rectal NETs are the most frequently encountered but with different clinical presentations.
- The SI-NETs arising from the EC cell distinctively produce excess serotonin that causes regional fibrosis near the site of the primary tumor and often the systemic carcinoid syndrome.
- In contrast, rectal NETs almost never produce a hormonal syndrome and may present with obstruction, rectal bleeding, pain, or are discovered incidentally on endoscopy or diagnostic imaging.
- Gastric NETs may produce excessive circulating histamine from the ECL cell and may exhibit a variant form of the carcinoid syndrome.
- Duodenal NETs can present as syndromic gastrinomas, somatostatinomas, and rarer ectopic hormone-producing syndromic tumors.
- NETs of the esophagus, appendix, colon, and rectum almost never cause syndromes of hormone hypersecretion. However, circulating CgA is a helpful tumor biomarker in almost all GI NETs. Nonhormonal presentations may include esophageal or abdominal pain, GI tract obstruction, GI bleeding, and anemia. Increasingly NETs are found incidentally on endoscopy and various diagnostic imaging modalities. The incidental discovery of NETs is almost certainly due to advancements in GI endoscopy and diagnostic imaging but also due to increasing awareness of NETs by the medical community.

Syndromic NETs are of great interest to clinical chemists and endocrinologists. Representative tissue or organ-specific hormones and biomarkers that are produced in GI NETs include circulating insulin, C-peptide, gastrin, ACTH, and urinary 5-HIAA which are readily available in most tertiary medical centers. A large number of more rarely required specific hormones or biomarkers are usually available in specialized independent laboratories and academic centers with a special interest in NETs. Examples are somatostatin, glucagon, vasoactive intestinal peptide (VIP), histamine, serotonin, parathyroid hormone—related peptide, neurokinin A, substance P, and other rare biomarkers.

Small intestinal and rectal NETs are the most common whereas gastric, colonic, and duodenal NETs have a lower incidence and prevalence, and esophageal NETs are rare. Small intestinal, gastric, and duodenal NETs will be emphasized because of the important role in patients with these neoplasms of measurement of hormones and biomarkers in diagnosis, surveillance of the clinical course for recurrence and progression, and assessment of the tumor response to a variety of treatments available such as surgery, biological agents such as somatostatin analogs, peptide receptor radiotherapy, chemotherapy, and therapies directed to liver metastases. Because of the similarity of hormonal syndromes between duodenal and PNETs, the duodenal NETs will be discussed in the section on PNETs.

9.2.2 SMALL INTESTINAL NEUROENDOCRINE TUMORS

9.2.2.1 Clinical presentation and course

- SI-NETs originate in the serotonin-producing EC cells of the mucosa with the most frequent site of the primary tumor being the ileum.
- Serotonin derives from the essential amino acid tryptophan which is converted to 5-hydroxytryptophan by the rate-limiting enzyme tryptophan hydroxylase. The conversion of 5-hydroxytryptophan to 5-HT (serotonin) is carried out by aromatic-L-amino acid decarboxylase.
- In addition SI-NETs synthesize members of the tachykinin family of small peptides such as neurokinin A and substance P [49,50].
- The primary tumor mass often causes local fibrosis in adjacent tissues believed to be mediated by serotonin and termed a desmoplastic reaction to a neoplasm. The local fibrotic process results in retraction or indrawing of adjacent small intestinal segments and their vasculature which may cause local bowel obstruction and vascular obstruction of the superior mesenteric artery and consequently intestinal ischemia.
- The primary tumor mass and its desmoplastic response may result in a classical "starburst" or "spokes of a wheel" appearance due to local tissue retraction evident on CT or MR imaging.
- Less often, SI-NETs present as a result of incidentally discovered hepatic metastases and biopsy findings of histologically diagnostic characteristics of a SI-NET such as immunohistochemical positivity for CgA, synaptophysin, and serotonin.

9.2.2.2 Carcinoid syndrome

- The carcinoid syndrome most frequently results from excessive secretion of serotonin and vasoactive peptides from SI-NETs.
- About 30% of patients with a SI-NET will have the carcinoid syndrome.
- Prominent symptoms are diarrhea, transient vasodilation described as flushing of the skin, wheezing due to bronchospasm and consequences of fibrotic changes to the heart, retroperitoneal organs, and skin [51].
- Serotonin is likely the causative agent of diarrhea and fibrosis whereas secretion of vasoactive tachykinins probably produces the flushing.
- Serotonin released from the small intestine into the portal vein is metabolically inactivated by the liver.
- In contrast, liver metastases from SI-NETs secrete the biologically active agents directly into the systemic circulation resulting in the symptoms and a fibrotic effect on the right side of the heart.
- The tricuspid and pulmonary valves and the endocardium are thickened and when dysfunctional produce carcinoid heart disease with consequent right-sided heart failure and pulmonary hypertension. The presence of carcinoid heart disease confers a poorer prognosis and survival than if absent [10].

- SI-NETs in the majority of patients metastasize to the liver and regional abdominal lymph nodes although less common metastases can occur to the lung, bone, and other distant sites.
- Approximately 90% of patients with the carcinoid syndrome have hepatic metastases. About half of patients with the carcinoid syndrome have carcinoid heart disease [1,3].
- Rarely NETs of the lung and ovary can secrete serotonin and other bioactive molecules directly into the general circulation producing the carcinoid syndrome in the absence of liver metastases.

9.2.2.3 Laboratory diagnosis and surveillance

The principal biomarker for diagnosis and surveillance of serotonin-producing EC cell SI-NETs is urinary 5-HIAA. This metabolite is the product of the action of MAO and aldehyde dehydrogenase on serotonin.

Measurement of 5-HIAA is subject to numerous factors that may result in false positive results.

- Patients may be ingesting tryptophan, 5-hydroxytryptophan or serotonin as "natural health" products. A clue to the possibility that a patient may be receiving a precursor to 5-HIAA is the situation where the measured level of 5-HIAA is much higher than expected for the clinical situation, such as a patient in which the carcinoid syndrome is suspected based on symptoms but without any diagnostic imaging evidence of anatomical disease. Very high levels of 5-HIAA may be present with an extensive tumor burden and correlate with a poorer prognosis [32].
- Several medications and foods may interfere with the assay of 5-HIAA, particularly fruits of the plantain and banana family of plants and also pineapples, kiwi, eggplant, pecans, walnuts, and avocadoes.
- Interfering medications include MAO inhibitors, selective serotonin reuptake inhibitors, nicotine, L-DOPA, salicylates, acetaminophen, 5-fluorouracil, streptozotocin, melphalan, phenothiazines, and phenobarbital.

For surveillance in a patient with a serotonin-producing SI-NET and, if present, the carcinoid syndrome, the general tumor marker, CgA, and the specific marker, urine 5-HIAA, are the most utilized and have the most evidence supporting their applicability. A 24 h urine 5-HIAA collection is most reliable due to the diurnal variability of urinary 5-HIAA excretion. SI-NETs may also secrete tachykinins, a family of small vasoactive peptides that includes neurokinin A and substance P as the most studied examples [49,50]. Serotonin is very likely the pathogenic factor responsible for the fibrogenic processes and diarrhea characteristic of the carcinoid syndrome. The tachykinins very probably cause the flushing of the carcinoid syndrome. A study using a RIA directed to the common carboxyl terminal region of the tachykinin peptides suggested that the tachykinins may also be associated with diarrhea in the carcinoid syndrome [52]. The specific circulating tachykinins, neurokinin A and substance P, have been proposed as organ-specific tumor markers for SI-NETs [49,50]. However, their availability and research evidence of their applicability as tumor markers are limited.

9.2.2.4 Case example

A 50-year-old woman presented to the emergency department after 2 days of central abdominal cramping pain, nausea, anorexia, and vomiting. She had a history of diarrhea for 8 weeks, and skin flushing and wheezing occurring daily for the past year. On examination she had reddish skin pigmentation of the upper half of her body, generalized abdominal tenderness, and a systolic cardiac murmur. "Carcinoid syndrome" was suspected.

> *Question: What biochemical tests should be ordered at this stage of diagnostic investigation?*

> *Answer:* Urinary 5-HIAA.

> *Question: At what stage of investigation should the doctor order one of the "general" NET biomarkers?*

> *Answer:* CgA prior to initiating any treatment as a baseline.

The endocrinologist notices that the CgA was measured while the patient was taking omeprazole, a PPI.

> *Question: To what extent can a PPI cause an elevation in CgA and would it be sufficient to give a false positive in the event that the diagnosis was not related to a NET?*

> *Answer:* A PPI drug may elevate CgA to very high levels. Raines and colleagues found that CgA levels in persons taking PPI drugs were 9 ± 13 (mean $\pm$ SD) times the average CgA level in people not on a PPI [39]. Expressed differently, there is about a 2.5% chance that use of a PPI drug will increase the CgA level by about 35 times the level seen without a PPI.

Abdominal X-ray and CT revealed partial obstruction of the small intestine, a central abdominal mass with a spiculated desmoplastic appearance, and liver metastases. An echocardiogram demonstrated carcinoid heart disease with thickening of the pulmonary and tricuspid valves, the endocardium and pulmonary hypertension. Urinary 5-HIAA was increased five times above the upper limit of normal (ULN), and drug-free blood CgA was elevated three times above the ULN. A laparotomy was undertaken to relieve the bowel obstruction by bypassing the small intestine tethered to the primary tumor mass, partial debulking of the primary tumor, and excision of two accessible liver metastases. Histology of the primary tumor and the liver metastases was diagnostic of a primary SI-NET metastatic to liver: positive immunohistochemistry to CgA, synaptophysin, and serotonin, and a Ki67 score of 1%. The tumor, using modern nomenclature, was a well-differentiated SI-NET, anatomical stage IV (distant metastases), histological grade 1 and syndromic (classical clinical carcinoid syndrome). Throughout her clinical course she was treated with a long acting somatostatin analog resulting in considerable relief of the diarrhea, flushing, and wheezing.

Question: After initiation of surgical and medical therapies, what biomarker tests would be useful for follow-up?

Answer: Urinary 5-HIAA and blood CgA in most cases are sufficient. If the clinical or diagnostic imaging surveillance suggest progression of disease in the absence of an increase in CgA, other general tumor markers may be helpful. A recommended priority ranking of the alternative general tumor markers would be PP, pancreastatin, and NSE.

Question: How much change must be seen in a measured biomarker in order to determine that there has likely been a clinical change (either positive or negative)?

Answer: An increase of the CgA level above the assay ULN would be concerning for disease progression. In the case of patients with a baseline CgA level that is above the ULN, a change by ± 1 SD of the mean of the assay baseline value would be clinically significant, based exclusively on the author's opinion. Understanding of the CgA assay properties, such as the variation (reproducibility) at different levels of CgA is important.

KEY POINTS

SI-NETs usually arise from the serotonin-producing EC cells. Their clinical presentation is most commonly abdominal pain, bowel obstruction, GI bleeding and incidental discovery on diagnostic imaging. The great majority of patients with the carcinoid syndrome (diarrhea, cutaneous flushing, wheezing, and systemic fibrosis) harbor liver metastases. The general tumor marker, CgA, and the organ-specific marker, 5-HIAA, are widely available and have correlated with disease status, progression, prognosis, and response to therapy. Practitioners should be cognizant of several factors that can falsely elevate or lower measurements of these markers.

9.2.3 GASTRIC NEUROENDOCRINE TUMORS

9.2.3.1 Clinical features and subtypes

Gastric NETs, frequently referred to as gastric carcinoid tumors, are less common than SI-NETs or rectal NETs. However, they present considerable challenges for diagnosis and management [3,33]. Gastric NETs arise from the stomach ECL cell which normally secretes histamine. In normal physiology, histamine subsequently stimulates the gastric parietal cell to secrete hydrochloric acid. The hormone gastrin normally arises from the G cells of the pyloric antrum of the stomach, the duodenum, and the pancreas and functions to regulate ECL secretion of histamine. Normally, gastric acidity regulates gastrin secretion in an inhibitory fashion.

In the clinical setting of an abnormally elevated gastric pH (achlorhydria) gastrin secretion is stimulated, in some individuals to levels several fold the ULN.

9.2.3.1.1 Three subtypes

Gastric NETs (carcinoids) are classified into three subtypes [3,33].

- Type I gastric NETs develop in the setting of atrophic gastritis and pernicious anemia and may present clinically with epigastric pain, nausea, and vomiting. These tumors are usually detected as gastric mucosal masses on upper GI endoscopy. Their biochemical profile usually features an abnormally elevated gastric pH and elevated blood levels of gastrin, histamine, and CgA. Type I gastric NETs comprise about 80% of all gastric NETs and are malignant or metastatic in only about 5% of individuals.
- Type II gastric NETs are rare neoplasms associated with gastrinomas arising in patients with the MEN1 syndrome. The primary tumor site of the MEN1 gastrinoma is most often the duodenum with a pancreatic primary being uncommon. The duodenal gastrinomas associated with the MEN1 syndrome are frequently multifocal and metastatic to regional lymph nodes. The Type II gastric NETs may present with upper abdominal pain, nausea, diarrhea, and upper GI bleeding from acid-mediated gastric ulcers. Laboratory features include elevation of circulating levels of gastrin, histamine, CgA, and gastric hyperacidity. Type II gastric NETs are malignant in only about 10% of individuals.
- Type III gastric NETs represent sporadic neoplasms of the ECL cell and in about half of patients are associated with metastases. Presenting symptoms are often upper abdominal pain, nausea, vomiting, and upper GI bleeding. On CT or MR imaging extension of the tumor mass outside the serosal confines of the stomach may be present. Typical biochemical results are elevation of circulating histamine and CgA but a normal level of gastrin.
- A variant type of the carcinoid syndrome may be caused by histamine secretion from gastric NETs. Flushing and diarrhea are present and the histamine excess also results in cutaneous edema, wheal formation, lacrimation, or wheezing.

9.2.3.2 Laboratory features

- Elevated blood gastrin occurs in type I gastric NETs associated with atrophic gastritis and achlorhydria and in type II gastric NETs caused by a gastrinoma and MEN1 syndrome.
- Excess histamine production is associated with all three types of gastric NETs and may be quantitated by measurement in urine of histamine, methyl histamine, or the subsequent metabolic product, methyl imidazole acetic acid, depending on availability [33]. We favor urinary histamine as the preferred test.
- CgA also serves as a circulating general tumor marker in all three types of gastric NET. If a patient is receiving a PPI drug, or to a lesser extent a histamine receptor antagonist, a false positive elevation of circulating CgA and gastrin likely will result.

> **KEY POINTS**
>
> Gastric NETs arise from the ECL cell of the stomach and secrete histamine as the major organ-specific tumor biomarker. They are of three types with distinct profiles of gastric acidity and levels of histamine, gastrin, and CgA.

9.3 PANCREATIC NEUROENDOCRINE TUMORS

9.3.1 OVERVIEW OF ENDOCRINE SYNDROMES

PNETs originate from the normal endocrine cells of the islets of Langerhans [1−3,48]. These cells and their peptide products are:

- A cell (glucagon)
- B cell (insulin)
- D cell (somatostatin)
- PP cell

Three of the cells are represented by distinct tumor syndromes due to hypersecretion of their hormones, referred to as eutopic hormone tumor syndromes, such as insulinoma, glucagonoma, and somatostatinoma. PP is secreted in excess from PNETs arising from the PP cell but these tumors are not associated with a recognized distinct clinical syndrome.

PNETs may secrete numerous hormones not usually produced by the islet cells, referred to as ectopic hormone tumor syndromes, for example gastrin, VIP, ACTH, corticotropin-releasing hormone (CRH), growth hormone−releasing hormone (GHRH), parathyroid hormone−related peptide, serotonin, calcitonin, renin, ghrelin, luteinizing hormone, erythropoietin, and insulin-like growth factor 2 [1].

Gastrinomas and insulinomas have the highest incidence among the syndromic PNETs with the other hormone-secreting tumors being rare. Among syndromic PNETs, the specific hormone produced in excess serves as a valuable tumor-specific biomarker for surveillance.

9.3.2 CLINICAL PRESENTATIONS

- Despite the multitude of hormonal syndromes that PNETs can cause, about 75% of PNETs are not associated with a clinical hormonal syndrome [25,26]. Their clinical presentation may feature local or metastatic tumor mass effects such as abdominal pain, obstruction of the hepatobiliary tract that may cause jaundice, or obstruction of the duodenum. Alternatively the primary PNET tumor may be found incidentally on diagnostic imaging for nonspecific abdominal symptoms. Detection of liver metastases during investigation of abdominal pain, hepatobiliary obstruction, or impaired liver function may also be the first clinical indication of a PNETs' existence. In this scenario the

PNET primary may be discovered on diagnostic imaging or by histological features characteristic of a PNET on biopsy of a liver metastasis.

- Syndromic PNETs can have their first clinical presentation being tumor mass effects with the hormone excess syndrome being recognized subsequently.
- The general NET biomarkers, CgA and PP, are appropriate for surveillance of both syndromic and nonsyndromic PNETs.

The incidence of PNETs is low, about two per million population per year, but has doubled between 1973 and 2004 likely due to increased physician awareness and advances in diagnostic imaging [53]. PNETs comprise only 1.3% of pancreatic neoplasms [53]. However, their prevalence is about 10% of pancreatic neoplasms because of the poor prognosis of the more common adenocarcinomas compared to PNETs.

Among hereditary PNETs about 80% are associated with MEN1 and approximately 15% with von Hippel Lindau syndrome [1,48]. Gastrinomas are the most frequent syndromic PNET in patients with MEN1 with insulinomas being second. PNETs in the von Hippel Lindau syndrome are nonsyndromic.

9.3.3 INSULINOMAS

Insulinomas are characterized by symptomatic hypoglycemia in the context of an inappropriately elevated blood level of insulin.

- In the extensive Mayo Clinic experience of 237 patients with insulinomas, hypoglycemia occurred only during the fasting state in 73%, in both fasting and postprandial states in 21% and only in the postprandial period in 6% [54].

9.3.3.1 Biochemical investigation for suspected insulinoma

- *The key to diagnosis of an insulinoma is demonstration of an elevated insulin level simultaneously with hypoglycemia*
  - The hypoglycemia may occur at various times such as during a morning fasting blood collection or at the time of visiting an emergency department with symptoms of hypoglycemia.
  - If hypoglycemia is documented in a specific blood collection, the clinician should request the laboratory to perform at least insulin and C-peptide assays on the sample.
  - Measurement of cortisol in such a blood collection on suspicion of adrenal insufficiency may be appropriate depending on the situation.
  - Assays for blood insulin are variable in their reference ranges among medical centers, which makes calculation of insulin to glucose ratios unreliable for diagnosis of an insulinoma [55].
  - Familiarity with one's local insulin assay is essential for investigation of hypoglycemia. A single insulin measurement done in isolation or obtained without a matched normal glucose level is unreliable and potentially misleading.

- As an example, obesity-related insulin resistance is highly prevalent and can result in a considerable elevation of circulating insulin during normoglycemia.
- Documentation of hyperinsulinemic hypoglycemia may require a period of controlled fasting, possibly extending up to 72 h. Owing to the risks of arrhythmias and seizures, controlled fasting studies should be done in hospital following a standardized protocol.

9.3.3.2 Localization of biochemically confirmed insulinoma

After establishing the probability of an insulinoma biochemically, localization of the tumor mass is undertaken by a variety of available imaging techniques including CT, MR, endoscopic ultrasound, and radiolabeled octreotide nuclear imaging. Selective angiographically guided injection of calcium into the arteries perfusing the pancreas combined with time-matched measurement of hepatic venous insulin levels may help localize an insulinoma to a specific region of the pancreas. Using these modern modalities all insulinomas were successfully localized preoperatively during the recent period of study between 1998 and 2007 at the Mayo Clinic [54].

9.3.3.3 Special situations of hyperinsulinemic hypoglycemia

- In the clinical situation of hyperinsulinemic hypoglycemia occurring in the postprandial period consideration should be given to the possible diagnosis of noninsulinoma pancreatogenous hypoglycemia syndrome (NIPHS) [56]. The pathogenesis of this condition involves islet cell hyperplasia and nesidioblastosis, a histological description of islet cells budding off and extending into the exocrine ductal system of the pancreas. NIPHS may arise sporadically but recently has been associated with the Roux-en-Y procedure of gastric bypass bariatric surgery [57]. Biochemical diagnosis of NIPHS may require a controlled meal or caloric intake test followed by monitoring of symptoms, blood glucose, and insulin serially for several hours. This test should be performed in an endocrine diagnostic testing center experienced with this protocol. Diagnostic imaging and treatment of NIPHS are distinct from those for an insulinoma and are reviewed in other sources.
- Measuring C-peptide (concomitantly with glucose and insulin) helps to eliminate the possible diagnosis of exogenous insulin administration since modern bioengineered insulins do not contain C-peptide.
- Sulfonylurea drug ingestion should be ruled out before diagnosing an insulinoma. Antiinsulin antibodies should also be measured to detect the rare hypoglycemic syndrome due to these antibodies.

Insulinomas are unique among PNETs in being metastatic or malignant in only 6% of cases, whereas all other functional PNETs are metastatic in the majority of patients [1–4,48]. Primary insulinomas are almost exclusively localized to the pancreas. About 5% of patients with insulinomas have MEN1. Insulinomas arising in patients with MEN1 may feature multicentric primary sites and a higher recurrence rate after surgery than sporadic insulinomas. For biochemical follow-up of a patient treated for an insulinoma, matched fasting blood glucose and insulin levels are the most helpful, in addition to obvious attention to possible symptomatic recurrence. In the rare patient with a recurrent or metastatic insulinoma, CgA and PP could be added to the surveillance protocol.

KEY POINTS

Insulinomas classically present with symptomatic hypoglycemia and hyperinsulinemia. Blood insulin levels must be interpreted in the context of a simultaneous blood glucose measurement. Hyperinsulinemic hypoglycemia may occur in either or both of the fasting and postprandial states. The NIPHS should be considered as a possible diagnosis of postprandial hypoglycemia. Insulinomas are unique among PNETs in only rarely being malignant. Measurement of simultaneous fasting blood glucose and insulin is indicated for follow-up.

9.3.4 GASTRINOMAS

9.3.4.1 Clinical presentation

The elevated blood gastrin caused by a gastrinoma results in the Zollinger–Ellison syndrome of excessive gastric acid production and its clinical consequences [1–4,48]. Severe acid peptic ulcer disease of the stomach and duodenum is manifested as abdominal pain, heartburn, nausea, vomiting, upper GI bleeding, and infrequently, perforation of the viscus. The hyperacidic gastric fluid flowing into the duodenum and intestine is also likely responsible for diarrhea in the gastrinoma syndrome.

9.3.4.2 Laboratory biomarker diagnosis

- The biochemical diagnosis of a gastrinoma requires documentation of an elevated blood gastrin level that is inappropriately high for the degree of gastric hyperacidity. In this setting an elevation of blood gastrin to over 10 times the ULN is diagnostic of a gastrinoma.
- However, gastrinomas may be associated with a lower degree of hypergastrinemia. At lower levels of hypergastrinemia the specificity of diagnosis is proportionally lower also. In this situation a secretin stimulation of gastrin test may be helpful in resolving the diagnosis.
- PPI drugs will raise the blood gastrin level and present the risk of a false positive result and an incorrect diagnosis of a gastrinoma. Discontinuing the PPI drug for at least a week will result in a reliable gastrin measurement but presents the risk of exacerbation of acid peptic disease, GI bleeding, or perforation [58]. Stopping a PPI should be done in collaboration with a gastroenterologist or surgeon with measures to ameliorate possible worsening of acid peptic disease and anticipation to intervene in the worst case scenario of acid peptic complications.

Gastrinomas are localized to the duodenum in 60%−80% of cases and to the pancreas in most other patients. Gastrinomas are sporadic in 70%−80% of patients and associated with the MEN1 syndrome in 20%−30% of patients. The latter are classically located in the duodenum and may have multiple primary sites. Among patients with PNETs in the setting of the MEN1 syndrome, about half will be gastrinomas. Importantly, 60%−80% of gastrinomas are malignant or metastatic. Recommended biochemical surveillance is by the tumor-specific hormone gastrin and the general NET biomarkers CgA and PP.

KEY POINTS

Gastrinomas cause severe acid peptic disease. Biochemical diagnosis may be difficult since there is not a standard diagnostic cut-off level of blood gastrin. PPI drugs will falsely increase blood gastrin, but their discontinuation should be done with caution. The possibility of underlying MEN1 syndrome should be considered in patients with a gastrinoma.

9.3.5 GLUCAGONOMAS

The classical glucagonoma syndrome has components of glucose intolerance, a characteristic rash termed necrolytic migratory erythema, oral and peri-oral inflammation (stomatitis, glossitis, angular cheilitis), thromboembolism, diarrhea, and anemia [1−3,48]. The blood glucagon level is increased. Potential false positive elevations can be caused by renal or hepatic failure, diabetes mellitus, fasting, and sepsis. Glucagonomas almost always arise in the pancreas and are rarely associated with the MEN1 syndrome. Metastases are present in 60%−80% of patients.

9.3.6 SOMATOSTATINOMAS

The somatostatinoma syndrome is a triad of glucose intolerance, cholelithiasis, and steatorrhea [1−3,48]. The measured blood somatostatin level is increased and can serve as a tumor-specific biomarker for follow-up. However, nonsyndromic PNETs with immunohistochemically positive staining for somatostatin are more common than syndromic somatostatinomas. The pancreas is the origin of 50%−60% of somatostatinomas with the remainder sited in the duodenum. The association of somatostatinomas with hereditary syndromes is unique. About half of duodenal somatostatinomas are associated with neurofibromatosis type 1 and are localized in proximity to the ampulla of Vater. Somatostatinomas are rarely associated with MEN1. Overall, somatostatinomas are metastatic in 70%−90% of cases with the pancreatic variety having a higher incidence of metastases than the duodenal counterpart.

9.3.7 VASOACTIVE INTESTINAL PEPTIDEOMAS

Normally, VIP is a peptide neurotransmitter in the digestive tract enteric nervous system. Physiologically, it promotes fluid secretion from the intestinal cell into the lumen and regulates intestinal blood flow. VIPomas arise from the pancreas as ectopic hormone-secreting tumors in over 90% of cases. The clinical presentation is large volume secretory (watery) diarrhea which may result in dehydration and hypokalemia [1–3,48]. The diarrhea is not ameliorated by fasting and may be difficult to control. The pathogenesis of diarrhea is VIP-receptor mediated activation of intestinal cell adenylate cyclase that has parallels with diarrhea caused by cholera toxin, which irreversibly activates adenylate cyclase by binding to the Gs subunit. The historical alternative name for the VIPoma syndrome was pancreatic cholera, which highlights the similarity in pathogenesis. Blood levels of VIP are elevated in the VIPoma syndrome. Only about 5% of VIPoma patients have the MEN1 syndrome. Similarly as for other PNETs, with the exception of insulinomas, VIPomas have a high incidence of metastases, in about 80% of patients.

9.3.8 CASE EXAMPLE

A 28-year-old man presented with nephrolithiasis, hypercalcemia, and an elevated blood level of parathyroid hormone. A diagnosis of primary hyperparathyroidism was made and treated by surgical excision of a parathyroid adenoma. Hyperparathyroidism recurred 2 years later and he had a second successful operation for a parathyroid neoplasm. At age 31 he developed symptoms of hypoglycemia (tachycardia, sweating, anxiety, and tremor) in the morning fasting state.

> *Question*: How should the next steps in investigation of suspected hypoglycemia proceed?

> *Answer*: If a patient has symptoms of hypoglycemia at any time, such as during a visit to the emergency department, matched blood glucose, insulin, and C-peptide levels should be obtained. For patients in whom the baseline levels of glucose and insulin are nondiagnostic, a controlled fast in hospital may be required, infrequently needing extension to 72 h.

On investigation, fasting blood glucose was 2.5 mmol/L (45 mg/dL) and matched blood insulin was 150 pmol/L (20.9 mU/L). The blood insulin level was inappropriately elevated for the degree of hypoglycemia.

> *Question*: What is the differential diagnosis of the inappropriate insulin elevation with the hypoglycemia and what tests may help clarify the diagnosis?

> *Answer*:
> - Exogenous insulin administration: Blood C-peptide would be inappropriately low.

- Exogenous sulfonylurea drug administration: Toxicology studies for sulfonylurea drugs would be positive.
- Immune insulin antibody syndrome: Assay of insulin antibodies would be positive.
- Insulinoma: C-peptide and proinsulin would be elevated. Diagnostic imaging studies or selective arterial injection of calcium testing would be indicated.
- NIPHS: Insulin and C-peptide would be elevated during a controlled meal or caloric intake test. Localizing diagnostic imaging studies would not demonstrate a mass consistent with an insulinoma. Selective arterial calcium injection may localize to a region of the pancreas or may fail to show regional differences. Surgery and pathological examination of the pancreatic tissue may be necessary for a diagnosis.
- Gastric bypass bariatric surgery: History of this surgery would be positive. Biochemical and localization studies as for NIPHS may be required to make a diagnosis.

Blood C-peptide was elevated. Magnetic resonance imaging detected a 2 cm diameter mass in the head of the pancreas. Surgical excision of an insulinoma was performed. A genetic test for MEN1 was positive. Biochemical follow-up surveillance included matched fasting blood glucose and insulin and CgA. Three years later recurrence of symptomatic and biochemical hyperinsulinemic hypoglycemia occurred and on diagnostic imaging a mass in the pancreatic area suspicious for recurrence and two liver metastases were detected. Subsequent treatments included chemotherapy and peptide receptor radionuclide therapy.

9.4 INTRATHORACIC (LUNG AND THYMUS) NEUROENDOCRINE TUMORS

9.4.1 OVERVIEW AND CLASSIFICATION

NETs of the lung and thymus are highlighted herein because lung NETs are among the most common of NETs and thymic NETs are especially aggressive and carry a poor prognosis [1–3,59]. Both lung and thymic NETs are associated with MEN1. The classification of lung and thymic NETs is different than that for digestive system NETs and utilizes carcinoid terminology and the mitotic count as a major classification criterion:

1. Well-differentiated tumors including:
   a. typical carcinoid
   b. atypical carcinoid
2. Poorly differentiated tumors including:
   a. SCLC
   b. large-cell neuroendocrine carcinoma (LCNEC) [11]

The guidelines for staging of lung NETs follow the 2010 AJCC Cancer Staging Manual (7th ed.) TNM rules for lung cancers [13]. Blood CgA may be utilized as a general tumor marker for surveillance.

9.4.2 LUNG NEUROENDOCRINE TUMORS

- Lung NETs usually present clinically with symptoms of local tumor mass effects including cough, hemoptysis, pneumonia due to bronchial obstruction, atelectasis, and pleuritic pain [59,60].
- They are often discovered incidentally on diagnostic imaging or bronchoscopy for nonspecific symptoms.
- Epidemiologically the incidence of well-differentiated (carcinoid) lung NETs has increased from about 0.3 per 100,000 in 1973 to 1.35 per 100,000 in 2004 likely due to an increased detection rate [12].
- Only a minority (4%−18%) of well-differentiated lung NETs (typical carcinoids or atypical carcinoids) cause hormonal syndromes although these are complex and represent several ectopic hormone tumor syndromes [60].
- Serotonin may be secreted causing the classical carcinoid syndrome with the majority of these having liver metastases. However, less often serotonin secretion from lung carcinoid tumors is directly into the systemic circulation with liver metastases being absent or not prominent. This is in distinction to the classical carcinoid syndrome caused by SI-NETs in which liver metastases are present in about 90% of symptomatic patients.
- Some lung NETs may not express the enzyme aromatic-L-amino acid decarboxylase that converts 5-hydroxytryptophan to serotonin. In this situation, production of urine 5-HIAA may be reduced and unreliable [59,60]. However, peripheral tissues such as kidney express this enzyme and generate serotonin.
- Assays for measurement of serotonin in blood and urine are available and may be helpful in the setting of clinically suspected carcinoid syndrome with a normal urinary 5-HIAA.
- Lung carcinoid tumors may also secrete histamine, producing a similar variant carcinoid syndrome as is seen with gastric NETs.
- Notable other ectopic hormone tumor syndromes associated with lung carcinoids include ACTH or CRH causing Cushing's syndrome and GHRH causing acromegaly.
- Association with MEN1 is present in about 5% of patients with lung carcinoid tumors and over 80% of these occur in females [59,60].

The poorly differentiated lung NETs, SCLC, and LCNEC are malignant neoplasms with a probability of 5-year survival of 27% for LCNEC and 9% for SCLC [61]. SCLC and LCNETs are more common than the lung carcinoid tumors. SCLC represents 14%−20% and LCNETs account for 3% of all lung carcinomas, respectively. They are associated with smoking but not MEN1. Notable SCLC ectopic hormonal conditions include the syndrome of inappropriate antidiuretic hormone (SIADH) with vasopressin secretion and the ectopic ACTH syndrome [60].

9.4.3 THYMIC NEUROENDOCRINE TUMORS

- Thymic NETs are classified using similar terminology as lung NETs but carry a higher malignant potential.
- Their incidence is very rare (2 per 10 million population per year) [12]. About one-third of thymic NETs are first detected incidentally on diagnostic imaging which includes the anterior mediastinum [3].
- Regional tumor mass effects include obstruction of the superior vena cava with a characteristic syndrome and hoarseness from invasion of the recurrent laryngeal nerve.
- Ectopic hormone syndromes occur in 40%−50% of sporadic thymic NET patients, in contrast to the low incidence in patients with lung NETs [1,3].
- Ectopic ACTH is the most frequently encountered hormonal syndrome with others being SIADH, atrial natriuretic peptide, and GHRH acromegaly.
- Carcinoid syndrome is very rare with thymic NETs.
- There is a moderate association between thymic NETs and MEN1. About 20% of thymic NET patients will have MEN1 and about 5% of MEN1 patients will have a thymic NET [2,59]. Over 90% of patients with a thymic NET and MEN1 are male. Ectopic hormone production may be less common in MEN1 positive thymic NET patients compared to their sporadic counterparts [2].

KEY POINTS

Lung NETs are the most frequently occurring of all NETs. The large majority are nonsyndromic. In the remaining minority the prominent hormone syndromes are classical and variant carcinoid syndromes due to serotonin or histamine secretion, respectively, and the ectopic ACTH syndrome. Thymic NETs are biologically aggressive and patients have reduced survival. Hormone syndromes are present in up to half of patients and include ectopic ACTH, SIADH, and GHRH. About 5% of patients with lung and 20% with thymic NETs have MEN1. Patients with MEN1 and lung NETs are predominately female whereas patients with MEN1 and thymic NETs are predominately male.

9.4.4 CASE EXAMPLE

A 60-year-old man developed symptoms and signs of Cushing's syndrome. Screening tests including urinary free cortisol, overnight 1 mg dexamethasone suppression test, and salivary cortisol indicated nonsuppressible hypercortisolism. Baseline blood ACTH was three times the ULN. An overnight high-dose dexamethasone (8 mg) test suppressed morning blood cortisol by 80%. MR of the sella turcica was normal. Inferior petrosal sinus sampling suggested a peripheral (nonpituitary) localization of the ACTH source. Computed tomography of the chest demonstrated a 2 cm diameter homogeneous tumor mass in the left lower lobe. The tumor mass displayed avid uptake on radiolabeled octreotide nuclear imaging. Thoracic surgery with intention to cure was performed. Histology showed an atypical carcinoid tumor of lung with 5 mitoses per 2 mm^2 and immunohistochemical positivity for CgA, synaptophysin, and ACTH.

9.5 RARE NEUROENDOCRINE TUMORS OF OTHER SITES

Consistent with the widespread distribution of neuroendocrine cells in the body, NETs may rarely arise in a multitude of organs and tissues. Usually a NET developing in these rare locations presents with local tumor mass effects [62]. Examples include NETs of the following sites:

- digestive tract: liver, gall bladder, biliary tree
- reproductive organs: ovary, uterine cervix, testis
- breast
- skin: Merkel cell carcinoma
- presacral NETs that are hypothesized to arise from an remnant of the embryological tailgut [63].

NETs of the ovary may rarely cause the carcinoid syndrome due to direct secretion of serotonin into the systemic circulation.

9.6 OVERUSE AND PITFALLS OF BIOMARKER MEASUREMENTS FOR NEUROENDOCRINE TUMORS

9.6.1 MEASUREMENT OF CHROMOGRANIN A AS A SCREENING OR DIAGNOSTIC TEST FOR NEUROENDOCRINE TUMORS

CgA has an established role as a general tumor biomarker to monitor the clinical course, and response to treatment of patients known to harbor a documented NET. Frequently, physicians obtain a CgA measurement as a screening diagnostic test for a NET in patients with nonspecific symptoms such as flushing or diarrhea. However, CgA is a poor quality screening or diagnostic test for NETs based on published studies of Italian [38] and Dutch [40] populations and consistent with our personal local experience. A low measured CgA level may falsely reassure that a NET has been excluded or, conversely, an elevated level may be falsely positive leading to further expensive and misdirected biochemical tests and diagnostic imaging. These comments also apply to other general tumor biomarkers such as pancreastatin, NSE, and PP.

9.6.2 FAILURE TO RECOGNIZE THAT USE OF A PROTON PUMP INHIBITOR DRUG WILL YIELD FALSE POSITIVE ELEVATIONS OF CHROMOGRANIN A AND GASTRIN

Raines et al. [39] demonstrated that patients receiving a PPI drug had much higher blood CgA and gastrin levels compared to control patients who had never received these medications [39]. Stopping a PPI may be required to obtain a reliable blood CgA and gastrin measurements. However, discontinuation of a member of this class of drugs may cause an exacerbation and complications of acid peptic disease. Consultation with a gastroenterologist or surgeon for prospective planning of management should be obtained before stopping a PPI drug.

9.6.3 FAILURE TO RECOGNIZE THAT SEVERAL FACTORS MAY FALSELY RAISE THE LEVEL OF URINARY 5-HYDROXYINDOLE ACETIC ACID

Numerous foods, natural product supplements, and prescription drugs will falsely elevate urinary 5-HIAA levels. These abundant sources of interference and false positivity represent commonly occurring pitfalls for measurement and are discussed in Section 9.1.8. Preparation for urine collection to measure 5-HIAA should certainly include avoidance of interfering foods and natural product supplements. If a potentially interfering prescription drug cannot be discontinued or substituted, then its effect in raising the 5-HIAA measurement should be recognized and taken into consideration in interpreting the result.

9.6.4 ROUTINE MEASUREMENT OF A "PANCREATIC PANEL" OF HORMONES AND BIOACTIVE MOLECULES IN A PATIENT WITH A PANCREATIC NEUROENDOCRINE TUMOR AND IN THE ABSENCE OF A CLINICAL SYNDROME OF HORMONE EXCESS

A routine "pancreatic panel" may include insulin, C-peptide, gastrin, glucagon, VIP, somatostatin, VIP, 5-HIAA, and potentially other hormones and biomarkers. Approximately 75% of PNETs are nonsyndromic and not associated with an elevated level of a hormone or bioactive molecule [25,26]. For the minority of patients with a syndromic PNET the selection of hormones and other biomarkers for investigation should be guided by the presence of clinical symptoms and signs of a disorder of hormone hypersecretion. These may include syndromic insulinomas, gastrinomas, glucagonomas, somatostatinomas, VIPomas, and syndromes of ectopic secretion of several hormones. We are not aware of evidence that routine measurement of a "pancreatic panel" in nonsyndromic patients with a PNET will result in improved patient care outcomes. In addition, obtaining a "pancreatic panel" is expensive and probably does not represent a cost effective utilization of health care resources.

REFERENCES

[1] Yalcin S, Oberg K, editors. Neuroendocrine tumours: diagnosis and management. Berlin, Heidelberg: Springer; 2015.
[2] Hay ID, Wass JAH, editors. Clinical endocrine oncology. 2nd ed Malden, MA: Blackwell; 2008.

[3] Modlin IM, Oberg K, editors. A century of advances in neuroendocrine tumor biology and treatment. C.C.C.P; Hannover, Germany: Felsenstein; 2008.

[4] Vinik AI, Silva MP, Woltering G, Go VLW, Warner R, Caplin M. Biochemical testing for neuroendocrine tumors. Pancreas 2009;38:876−89.

[5] Feyrter F. Uber diffuse endokrine epitheliale Organe. Leipzig: Barth JA; 1938.

[6] Pearse AGE. The cytochemical and ultrastructure of polypeptide hormone-producing cells of the APUD series and the embryologic, physiologic and pathologic implications of the concept. J Histochem Cytochem 1969;17:303−13.

[7] Williams ED, Sandler M. The classification of carcinoid tumours. Lancet 1963;1:238−9.

[8] Bosman FT, Carneiro F, Hruban RH, Theise ND, editors. WHO classification of tumours of the digestive system. Lyon: International Agency for Research on Cancer; 2010.

[9] Sasano H, Stefaneanu L, Kovaks K, editors. Molecular and cellular endocrine pathology. London: Arnold; 2000.

[10] Rorstad O. Prognostic indicators for gastrointestinal neuroendocrine tumors of the gastrointestinal tract. J Surg Oncol 2005;89:151−60.

[11] Travis WD, Brambilla E, Burke AP, Marx A, Nicholson AG. WHO classification of tumours of the lung, pleura, thymus and heart. Lyon: International Agency for Research on Cancer; 2015.

[12] Yao JC, Hassan M, Phan A, et al. One hundred years after "carcinoid": epidemiology of and prognostic factors for neuroendocrine tumors in 35,825 cases in the United States. J Clin Oncol 2008;26:3063−72.

[13] Edge SB, Byrd DR, Compton CC, Fritz AG, Greene F, Trotti III A, editors. AJCC cancer staging manual. 7th ed. New York: Springer; 2010.

[14] Oberndorfer S. Karzinoide tumoren des dunndarms. Frankf Z Pathol 1907;1:425−32.

[15] Pearson CM, Fitzgerald PJ. Carcinoid tumors—a re-emphasis of their malignant nature. Review of 140 cases. Cancer 1949;2:1005−26.

[16] Hemminki K, Li X. Incidence trends and risk factors of carcinoid tumors. A nationwide epidemiological survey from Sweden. Cancer 2001;92:2204−10.

[17] Rinke A, Muller HH, Schade-Brittinger C, et al. Placebo-controlled, double-blind, prospective, randomized study on the effect of octreotide LAR in the control of tumor growth in patients with metastatic neuroendocrine midgut tumors: a report from the PROMID study group. J Clin Oncol 2009;27:4656−63.

[18] Caplin ME, Pavel M, Cwikla JB, et al. Lanreotide in metastatic enteropancreatic neuroendocrine tumors. N Engl J Med 2014;371:224−33.

[19] Zhao J, de Krijger RR, Meier D, et al. Genomic alterations in well-differentiated gastrointestinal and bronchial neuroendocrine tumors (carcinoids): marked differences indicating diversity in molecular pathogenesis. Am J Pathol 2000;157:1431−8.

[20] Cunningham JL, Diaz de Stahl T, Sjoblom T, Westin G, Dumanski JP, Janson ET. Common pathogenetic mechanism involving human chromosome 18 in familial and sporadic ileal carcinoid tumors. Genes Chromosomes Cancer 2011;50:82−94.

[21] Edfeldt K, Ahmad T, Akerstrom G, et al. TCEB3C a putative tumor suppressor gene of small intestinal neuroendocrine tumors. Endocr Relat Cancer 2014;21:275−84.

[22] Oberg K. Circulating biomarkers in neuroendocrine tumours. In: Yalcin S, Oberg K, editors. Neuroendocrine tumours: diagnosis and management. Berlin, Heidelberg: Springer; 2015. p. 77−95.

[23] Granberg D. Biochemical testing in patients with neuroendocrine tumors. Front Horm Res 2015;44:24—39.

[24] Modlin IM, Gustafsson BI, Moss SF, Pavel M, Tsolakis AV, Kidd M. Chromogranin A—biological function and clinical utility in neuroendocrine tumor disease. Ann Surg Oncol 2010;17:2427—43.

[25] Raymond E, Dahan L, Raoul JL, et al. Sunitinib malate for the treatment of pancreatic neuroendocrine tumors. N Engl J Med 2011;364:501—13.

[26] Yao JC, Shah MH, Ito T, et al. Everolimus for advanced pancreatic neuroendocrine tumors. N Engl J Med 2011;364:514—23.

[27] Taupenot L, Harper KL, O'Connor DT. The chromogranin-secretogranin family. N Engl J Med 2003;348:1134—49.

[28] Bajetta E, Ferrari L, Martinetti A, et al. Chromogranin A, neuron specific enolase, carcinoembryonic antigen, and hydroxyindole acetic acid evaluation in patients with neuroendocrine tumors. Cancer 1999;86:858—65.

[29] Baudin E, Gigliotti A, Ducreux M, et al. Neuron-specific enolase and chromogranin A as markers of neuroendocrine tumours. Br J Cancer 1998;78:1102—7.

[30] Nobels TRE, Kwekkeboom DJ, Coopmans W, et al. Chromogranin A as serum marker for neuroendocrine neoplasia: comparison with neuron-specific enolase and the alpha-subunit of glycoprotein hormones. J Clin Endocrinol Metab 1997;82:2622—8.

[31] Korse CM, Taal BG, Vincent A, et al. Choice of tumour markers in patients with neuroendocrine tumours is dependent on the histological grade. A marker study of chromogranin A, neuron specific enolase, progastrin-releasing peptide and cytokeratin fragments. Eur J Cancer 2012;48:662—71.

[32] Janson ET, Holmberg L, Stridsberg M, et al. Carcinoid tumors: analysis of prognostic factors and survival in 301 patients from a referral center. Ann Oncol 1997;8:685—90.

[33] Granberg D, Wilander E, Stridsberg M, Granerus G, Skogseid B, Oberg K. Clinical symptoms, hormone profiles, treatment and prognosis in patients with gastric carcinoids. Gut 1998;43:223—8.

[34] Korse CM, Bonfrer JM, Aaronson NK, Hart AA, Taal BG. Chromogranin A as an alternative to 5-hydroxyindoleacetic acid in the evaluation of symptoms during treatment of patients with neuroendocrine tumors. Neuroendocrinology 2009;89:296—301.

[35] Arnold R, Wilke A, Rinke A, et al. Plasma chromogranin A as marker for survival in patients with metastatic endocrine gastroenteropancreatic tumors. Clin Gastroenterol Hepatol 2008;6:820—7.

[36] Nikou GC, Marinou K, Thomakos P, et al. Chromogranin a levels in diagnosis, treatment and follow-up of 42 patients with non-functioning pancreatic endocrine tumours. Pancreatology 2008;8:510—19.

[37] Welin S, Stridsberg M, Cunningham J, et al. Elevated plasma chromogranin A is the first indication of recurrence in radically operated midgut carcinoid tumors. Neuroendocrinology 2009;89:302—7.

[38] Marotta V, Nuzzo V, Ferrara T, et al. Limitations of chromogranin A in clinical practice. Biomarkers 2012;17:186—91.

[39] Raines D, Chester M, Diebold AE, et al. A prospective evaluation of the effect of chronic proton pump inhibitor use on plasma biomarkers levels in humans. Pancreas 2012;41:508—11.

[40] De Laat JM, Pieterman CRC, Weijmans M, et al. Low accuracy of tumor markers for diagnosing pancreatic neuroendocrine tumors in multiple endocrine neoplasia type 1 patients. J Clin Endocrinol Metab 2013;98:4143−51.

[41] Rustagi S, Warner RR, Divino CM. Serum pancreastatin: the next predictive neuroendocrine tumor marker. J Surg Oncol 2013;108:126−8.

[42] Sherman SK, Maxwell JE, O'Dorisio MS, O'Dorisio TM, Howe JR. Pancreastatin predicts survival in neuroendocrine tumors. Ann Surg Oncol 2014;21:2971−80.

[43] Stronge RL, Turner GB, Johnston BT, et al. A rapid rise in circulating pancreastatin in response to somatostatin analogue therapy is associated with poor survival in patients with neuroendocrine tumours. Ann Clin Biochem 2008;45:560−6.

[44] Bloomston M, Al-Saif O, Klemanski D, et al. Hepatic artery chemoembolization in 122 patients with metastatic carcinoid tumor: lessons learned. J Gastrointest Surg 2007;11:264−71.

[45] Yao JC, Pavel M, Phan AT, et al. Chromogranin A and neuron-specific enolase as prognostic markers in patients with advanced pNET treated with everolimus. J Clin Endocrinol Metab 2011;96:3741−9.

[46] Panzuto F, Severi C, Cannizzaro R, et al. Utility of combined use of plasma levels of chromogranin A and pancreatic polypeptide in the diagnosis of gastrointestinal and pancreatic endocrine tumors. J Endocrinol Invest 2004;27:6−11.

[47] Walter T, Chardon L, Chopin-Ialy X, et al. Is the combination of chromogranin A and pancreatic polypeptide serum determinations of interest in the diagnosis and follow-up of gastro-entero-pancreatic neuroendocrine tumors? Eur J Cancer 2012;48:1766−73.

[48] Metz DC, Jensen RT. Gastrointestinal neuroendocrine tumors: pancreatic endocrine tumors. Gastroenterology 2008;135:1469−92.

[49] Turner GB, Johnston BT, McCance DR, et al. Circulating markers of prognosis and response to treatment in patients with midgut carcinoid tumours. Gut 2006;55:1586−91.

[50] Vinik AI, Gonin J, England BG, Jackson T, McLeod MK, Cho K. Plasma substance-P in neuroendocrine tumors and idiopathic flushing: the value of pentagastrin stimulation tests and the effects of somatostatin analog. J Clin Endocrinol Metab 1990;70:1702−9.

[51] Thorson A, Biorck G, Bjorkman G, Waldenstrom J. Malignant carcinoid of the small intestine with metastases to the liver, valvular disease of the right side of the heart (pulmonary stenosis and tricuspid regurgitation without septal defects), peripheral vasomotor symptoms, bronchoconstriction, and an unusual type of cyanosis: a clinical and pathologic syndrome. Am Heart J 1954;47:795−817.

[52] Cunningham JL, Janson ET, Agarwal S, Grimelius L, Stridsberg M. Tachykinins in endocrine tumors and the carcinoid syndrome. Eur J Endocrinol 2008;159:275−82.

[53] Yao JC, Eisner MP, Leary C, et al. Population-based study of islet cell carcinoma. Ann Surg Oncol 2007;14:3492−500.

[54] Placzkowski KA, Vella A, Thompson GB, et al. Secular trends in the presentation and management of functioning insulinomas at the Mayo Clinic, 1987−2007. J Clin Endocrinol Metab 2009;94:1069−73.

[55] Cryer PE, Axelrod L, Grossman AB, et al. Evaluation and management of adult hypoglycemic disorders: an endocrine society clinical practice guideline. J Clin Endocrinol Metab 2009;94:709−28.

[56] Service FJ, Natt N, Thompson GB, et al. Noninsulinoma pancreatogenous hypoglyce-mia: a novel syndrome of hyperinsulinemic hypoglycemia in adults independent of mutations in Kir 6·2 and SUR1 genes. J Clin Endocrinol Metab 1999;84:1582—9.

[57] Service GJ, Thompson GB, Service FJ, Andrews JC, Collazo-Clavell ML, Lloyd RV. Hyperinsulinemic hypoglycemia with nesidioblastosis after gastric-bypass surgery. N Engl J Med 2005;353:249—54.

[58] Metz DC. Diagnosis of the Zollinger-Ellison syndrome. Clin Gastroenterol Hepatol 2012;10:126—30.

[59] Phan AT, Oberg K, Choi J, et al. NANETS consensus guideline for the diagnosis and management of neuroendocrine tumors: well-differentiated neuroendocrine tumors of the thorax (includes lung and thymus). Pancreas 2010;39:784—98.

[60] Granberg D. Neuroendocrine tumors of the lung. In: Yalcin S, Oberg K, editors. Neuroendocrine tumours: diagnosis and management. Berlin, Heidelberg: Springer; 2015. p. 143—64.

[61] Travis WD, Rush W, Flieder DB, et al. Survival analysis of 200 pulmonary neuroen-docrine tumors with clarification of criteria for atypical carcinoid and its separation from typical carcinoid. Am J Surg Pathol 1998;22:934—44.

[62] Modlin IM, Shapiro MD, Kidd M. An analysis of rare carcinoid tumors: clarifying these clinical conundrums. World J Surg 2005;29:92—101.

[63] La Rosa S, Boni L, Finzi G, et al. Ghrelin-producing well-differentiated neuroendo-crine tumor (carcinoid) of tailgut cyst. Morphological, immunohistochemical, ultra-structural, and RT-PCR study of a case and review of the literature. Endocr Pathol 2010;21:190—8.

Index